Tenth International Heavy Minerals Conference (HMC 2016)

Expanding the Horizon

Sun City, South Africa
16 – 17 August 2016

ISBN: 978-1-5108-3199-5

Printed from e-media with permission by:

Curran Associates, Inc.
57 Morehouse Lane
Red Hook, NY 12571

Some format issues inherent in the e-media version may also appear in this print version.

Printed by Curran Associates, Inc. (2016)

For permission requests, please contact Southern African Institute of Mining & Metallurgy
at the address below.

Southern African Institute of Mining & Metallurgy
P.O. Box 61127
Marshalltown 2107
South Africa

Phone: +27 (011) 834-1273/7
Fax: +27 (011) 838-5923
+27 (011) 833-8156

www.saimm.co.za

Additional copies of this publication are available from:

Curran Associates, Inc.
57 Morehouse Lane
Red Hook, NY 12571 USA
Phone: 845-758-0400
Fax: 845-758-2633
Email: curran@proceedings.com
Web: www.proceedings.com

THE SOUTHERN AFRICAN INSTITUTE OF MINING AND METALLURGY
JOHANNESBURG 2016

SYMPOSIUM SERIES S90

Heavy Minerals Conference

Expanding the Horizon

16–17 August 2016

Sun City, South Africa

Contents

FOREWORD

This Conference, hosted by the South African Institute of Mining and Metallurgy, is the tenth in this series on the heavy minerals industry. Previous conferences have been held in the areas of the globe where the more traditional heavy minerals are abundant and the industry is operational, namely southern Africa, Australia, India, and the USA.

This year we have departed somewhat from our previous focus in that we have included the chrome industry. The chromite deposits in South Africa are a result of gravity settling within a molten magma chamber, so the mineral may also be classed as a heavy mineral although the environment of concentration was at high temperatures as opposed to the atmospheric conditions of traditional heavy mineral deposits. As a result of the inclusion of chromite in our conference focus, Sun City was chosen as the venue. This location, in the crater of an extinct volcano, is ideally situated with respect to the chrome industry as the operating chrome mines are in the same region. In addition to its proximity to the mines, Sun City also provides the opportunity for both work and play. We hope you will take advantage of the facilities offered by the venue

Our conference is, as usual, strictly technical in nature and the efforts by the Organising Committee are aimed at preserving this focus. The benefit is that it allows the operators of businesses within this sector to discuss topics not normally covered in such forums. Our conference theme is 'Expanding the Horizon', which relates to the inclusion of a new member of the heavy minerals industry.

We do realize that 2016 has been a tough year for the mining and metallurgical industries, but as you all know we have been there before and the industry will again regain its previous profitability.

We look forward to welcoming you to our conference once again, to meeting old friends and making new ones.

J.H. Selby
Chairman,
Organising Committee

Published by The Southern African Institute of Mining and Metallurgy
Fifth Floor, Chamber of Mines Building, 5 Hollard Street, Johannesburg, 2107
Republic of South Africa

The abstracts in this volume have been for the most part prepared from Word documents supplied by the authors, with additional typesetting and formatting by The Southern African Institute of Mining and Metallurgy.

The Organising Committee expresses its deep appreciation for the voluntary contributions of the technical review panel in assisting to maintain high technical standards for the papers presented at this conference.

N. Barkhuysen	V. Hugo
S. Badehorst	H. Hutcheson
N. Bernard	R.T. Jones
V. Bruisnma	H. Kotze
H. Burger	B. Malan
P. den Hoed	P. Ramsaywok
G. Diener	C. Pistorius
W. Erasmus	J. Selby
D. Ferguson	E. Spicer
S. Gilman	G. Williams
J. Grobler	J. Walklate
R. Hattingh	C. Ware

Printed by Camera Press, Johannesburg

SAIMM PUBLICATIONS

THE MONOGRAPH SERIES

M1 **Lognormal-De Wijsian Geostatistics for Ore Evaluation**
(2nd ed 1981) D.G. Krige

M2 **An Introduction to Geostatistical Methods of Mineral Evaluation**
(2nd ed 1981) J.-M.M. Rendu

M3 **Principles of Flotation**
(1982) (3rd imp. 1986) Edited by R.P. King

M4 **Increased Underground Extraction of Coal**
(1982) Edited by C.J. Fauconnier and R.W.O. Kersten

M5 **Rock Mechanics in Mining Practice**
(1983) (3rd imp. 1986) Edited by S. Budavari

M6 **Assay and Analytical Practice in the South African Mining Industry**
(1986) W.C. Lenahan and R. de L. Murray-Smith

M7 **The Extractive Metallurgy of Gold in South Africa,**
2 volumes (1987) Edited by G.G Stanley

M8 **Mineral and Metal Extraction — An Overview**
(1994) L.C. Woollacott and R.H. Eric

M9 **Rock Fracture and Rockbursts—an illustrative study**
(1997) Edited by W.D. Ortlepp

THE SPECIAL PUBLICATIONS SERIES

SP1 **Proceedings, Underground Transport Symposium**
(1986) Edited by R.C.R. Edgar

SP2 **Backfill in South African Mines** (1988)

SP3 **Treatment and Re-use of Water in the Minerals Industry** (1989)

SP4 **COREX Symposium 1990**
(1990) Edited by H.M.W. Delport and P.J. Holaschke

SP5 **Measurement, Control, and Optimization in Mineral Processing**
(1994) Edited by H.W. Glen

SP6 **Handbook on Hard-rock Strata Control**
(1994) A.J.S. Spearing

SP7 **Rock Engineering for underground coal mining**
(2002) J. Nielen van der Merwe and Bernard J. Madden

SP8 **Second Edition: Rock Engineering for underground coal mining**
(2010) J. Nielen van der Merwe and Bernard J. Madden

SP9 **Theoretical Rock Mechanics for Professional Practice**
(2016) M. Handley

SUPPLEMENT TO THE SAIMM *JOURNAL*

The Metals and Minerals Industry in South Africa - Part 1 (1989)
Edited by H.W. Glen

THE SYMPOSIUM SERIES

S1 **Mathematical Statistics and Computer Applications in Ore Valuation** (1966)
S2 **Planning Open Pit Mines** (1970) (4th imp.) Edited by P.W.J. van Rensburg
S3 **Application of Computer Methods in the Mineral Industry** (APCOM 1973) Edited by M.D.G. Salamon
S4 **Infacon 1974** Edited by H.W. Glen
S5 **Proceedings of the 12th CMMI Congress** 2 volumes (1982) Edited by H.W. Glen
S6 **Rockbursts and Seismicity in Mines** (1984) Edited by N.C. Gay and E.H. Wainwright
S7 **The Planning and Operation of Open Pit and Strip Mines** (1986) Edited by J.P. Deetlefs
S8 **GOLD 100: Proceedings of the International Conference on Gold (1986)**
Volume 1: Gold Mining Technology Edited by H. Wagner and R.P. King
Volume 2: Extractive Metallurgy of Gold Edited by C.E. Fivaz and R.P. King
Volume 3: Industrial Uses of Gold Edited by G. Gafner and R.P. King
S9 **APCOM 87: Proceedings of the Twentieth International Symposium on the Application of Computers and Mathematics in the Mineral Industries (1987)**
Volume 1: Mining Edited by L. Wade, R.W.O. Kersten and J.R. Cutland
Volume 2: Metallurgy Edited by R.P. King and I.J. Barker
Volume 3: Geostatistics Edited by I.C. Lemmer, H. Schaum and F.A.G.M. Camisani-Calzolari
S10 **International Deep Mining Conference (1990)**
Volume 1: Innovations in Metallurgical Plant Edited by G.A. Brown and P. Smith and
Application of Materials Engineering in the Mining Industry Edited by B. Metcalfe
Volume 2: Technical Challenges in Deep Level Mining Edited by D.A.J. Ross-Watt and P.D.K. Robinson
S11 **Infacon 6 (Incorporating Incsac) (1992)** Edited by H.W. Glen
S12 **Massmin 92** Edited by H.W. Glen (out of print)
S13 **Minefill 93** Edited by H.W. Glen
S14 **XVth CMMI Congress Publications (1994)**
Volume 1: Mining Edited by H.W. Glen– **Volume 2: Metals Technology and Extractive Metallurgy (1994)** Edited by H.W. Glen
S15 **Surface Mining 1996** Edited by H.W. Glen
S16 **Hidden Wealth** (1996) Edited by H.W. Glen
S17 **Heavy Minerals 1997** Edited by R.E. Robinson
S18 **The 8th International Platinum Symposium (1998)**
S19 **Mining in Africa '98**
S20 **Extraction Metallurgy Africa '98**
S21 **Sixth International Symposium for Rock Fragmentation by Blasting (1999)**
S22 **Metallurgy Africa '99**
S23 **Heavy Minerals 1999** Edited by R.G. Stimson
S24 **Tunnels under Pressure** Technical Editor T.R. Stacey
S25 **Mine Hoisting 2000**
S26 **Coal–The Future (2000)**
S27 **The Fifth International Symposium on Rockburst and Seismicity in Mines (RaSim 5) (2001)**
S28 **6th International Symposium on Mine Mechanization and Automation (2001)**
S29 **XIV International Coal Preparation Congress and Exhibition**
S30 **Surface Mining 2002—Modern Developments for the New Millennium**
S31 **IFSA 2002, Industrial Fluidization South Africa**
S31a **APCOM 2003—Application of Computers and Operations Research in the Minerals Industries**
S32 **ISSA/Chamber of Mines Conference—Mines and Quarries: Prevention of Occupational Injury and Disease**
S33 **ISRM—Technology Roadmap for Rock Mechanics**
S34 **Heavy Minerals Conference 2003**
S35 **Safety in Mines Research Institutes (2003)**
S36 **VII International Conference on Molten Slags, Fluxes & Salts (2004)**
Tenth International Ferroalloys Congress INFACONX 2004
S37 **Deep and high stress mining 2004**
S38 **'Platinum adding value' 2004**
S39 **Base Metals—Southern Africa's response to changing global base metals market dynamics 2005**
S40 **Strategic versus Tactical approaches in mining 2005**
S41 **Best practices in rock engineering SARES 2005**
S42 **IFSA 2005, Industrial Fluidization South Africa**
S43 **Southern African Pyrometallurgy 2006 International Conference**
S44 **Stability of rock slopes in open pit mining and civil engineering situations**
S45 **'Platinum Surges Ahead' 2006**
S46 **Hydraulic Transport of Solids - Hydrotransport 17 (2007)**
S47 **Fourth Southern African Base Metals Conference 'Africa's base metals resurgence'**
S48 **Heavy Minerals Conference, 2007**
S49 **Cave Mining Conference, 2007**
S50 **International Symposium on Lead and Zinc processing – Lead & Zinc 2008**
S51 **Surface Mining 2008**
S52 **Platinum in transformation 2008**
S53 **IFSA 2008, Industrial Fluidization South Africa**
S54 **Hydrometallurgy Conference 2009**
S55 **Shotcrete for Africa 2009**
S56 **Base Metals Conference 2009**
S57 **Heavy Minerals Conference, 2009**
S58 **Hard Rock Safe—Safety Conference 2009**
S59 **Sampling and Blending 2009**
S60 **World Gold 2009 Conference**
S61 **Physical Beneficiation Conference 2010**
S62 **Second Hardrock Safe—Safety Conference 2010**
S63 **Platinum in Transition 'Boom or Bust' 2010**
S64 **South African Pyrometallurgy 2011 International Conference**
S65 **Minefill 2011 (10th International Symposium on Mining with Backfill)**
S66 **Base Metals Conference 2011**
S67 **MineSafe 2011 Conference**
S68 **Iron Ore and Manganese Ore Metallurgy 2011**
S69 **Percolation leaching: The status globally and in Southern Africa**
S70 **IFSA 2011, Industrial Fluidization South Africa**
S71 **Southern Hemisphere International Rock Michanics Symposium 2012**
S72 **Platinum Conference 'A catalyst for change' 2012**
S73 **Ferrous 2012—Ferrous and Base Metals Development Network Conference**
S74 **Diamonds - Source to use 2013**
S75 **Sampling and analysis: Best-practice in African mining 2013**
S76 **Base Metals Conference 2013**
S77 **Precious Metals 2013 Conference—The Precious Metals Development Network (PMDN)**
S78 **Physical Beneficiation 2013 Conference**
S79 **21st Century challenges to the southern African coal sector**
S80 **Furnace Tapping Conference 2014**
S81 **Platinum Conference 'Platinum—Metal for the Future' 2014**
S82 **IFSA 2014, Industrial Fluidization South Africa**
S83 **Copper Cobalt Africa 2015**
S84 **The Danie Krige Geostatistical Conference 2015 - Vol. 1**
The Danie Krige Geostatistical Conference 2015 - Vol. 2
S85 **World Gold Conference 2015**
S86 **Slope Stability 2015**
S87 **Nuclear Materials Development Network 2015**
S88 **Diamonds still sparkling Conference 2016**
S89 **Hydrometallurgy Conference 2016**

Heavy Minerals Conference 2016

Expanding the Horizon

Organising Committee

J. Selby (Conference Chairman)

H. Burger

N. Barkhuysen

W. Erasmus

C. Mienie

P. Hlabane

P. Ramsaywok

B. Malan

S. Mokubung

The organisers of the conference wish to thank the following sponsors for their support

-oOo-

ALICOCO MINERAL TECHNOLOGY

BLUECUBE measurement for control

DSI DERRICK SOLUTIONS INTERNATIONAL

DERRICK

ERIEZ

HATCH

Mineral Technologies

MULTOTEC

Polypipe

Saveway

tega INDUSTRIES SOUTH AFRICA PTY LTD

tharisa

WEIR

Naturally-occurring radioactive materials and the regulatory challenges to the zircon industry

K. Harlow

Zircon Industry Association (ZIA)

Zircon is a naturally-occurring radioactive material [NORM] and as such is subject to a variety of regulations around the world. The regulatory processes and their implementation are summarized and the key issues affecting zircon and zirconia highlighted. Having absorbed the activities of the former Zircon Minerals Committee, the strategy of the Zircon Industry Association (ZIA) is to be the 'go-to place' for information on these matters, and its activities in the NORM space is described. Chemical industry regulations such as REACH are spreading around the world, and the presentation provides an update of their status in the key jurisdictions. As mineral sands are transported to market in oceangoing vessels, they are subject to the regulations of the International Maritime Organization, which are summarized in the paper.

INTRODUCTION

From a global perspective, zircon is generally not found in high concentrations other than in placer and dune deposits where it has been deposited along with other heavy minerals such as ilmenite, rutile, and monazite. Such deposits have been sorted and concentrated over geological time by the action of tides, waves, and wind to form concentrated deposits of heavy minerals along old coastlines and in river beds and deltas. Consequently, zircon is mostly found in association with the principal titanium minerals; ilmenite and rutile. These secondary concentrations in placer deposits provide the commercial sources of zircon.

All geological materials contain natural radioactivity, although the term NORM (naturally occurring radioactive material) is mostly used for minerals and other materials with above-average uranium and thorium concentrations. Importantly, the radioactive nature of zircon has not changed, but the way in which radioactivity is viewed by regulators has, which can impact on the zircon industry through the value chain.

Based in the UK, the Zircon Industry Association (ZIA) currently has 23 members spanning a good part of the value chain; from mining to milling, to zirconia, and to chemicals and representing some 80% of the global traded zircon supply. Geographically, ZIA's members are in the USA, Latin America, Europe, Africa, and Asia. This paper describes some of the work that ZIA has commissioned over the last two years to support its members' NORM activities, as well as the impact of Europe's REACH (Registration, Evaluation and Authorisation of Chemicals) regulations and their derivatives around the world.

ZIRCON AND RADIATION EXPOSURE

Materials are classified based on the level of radioactivity they contain, and further classification i s based on the possible radiation exposure to workers or the public. The radioactive properties are an inherent feature of the material and cannot be changed without altering the material itself. The radiation exposure that can result from the material, however, can be changed by handling the material in a different manner.

The radioactivity of a material is measured in becquerels per gram (Bq/g), where one becquerel is equal to 1 count (disintegration) per second. NORMs contain radionuclides of the uranium and thorium decay series and a radioactive potassium isotope. The uranium decay series has 14 members, starting with U^{238} and ending with Pb^{206}, while the thorium series has 10 members, starting with Th^{232} and ending with Pb^{208} (Figure 1).

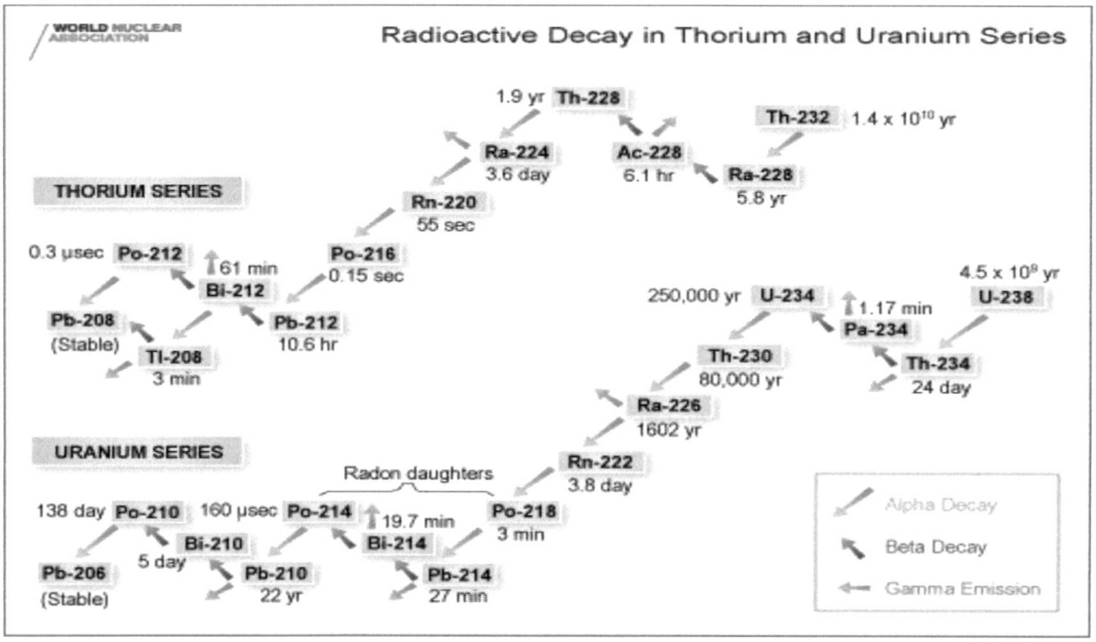

Figure 1. Radioactive decay in Th and U series.

If the decay series is not disturbed, each member will have the same activity level; known as secular equilibrium. Many NORMs show secular equilibrium due to their age, and in order to classify a material, the state of equilibrium needs to be known.

The activity levels in zircon vary depending on the source, but are generally in the ranges shown in Table I.

Table I. Typical activity concentrations of radionuclides in natural raw materials.

Nuclide	Activity range (Bq/g)
Uranium-238 in zircon	2 – 4
Thorium-232 in zircon	0.5 – 1.0
Potassium-40 in zircon	Approx. 0.002
Uranium-238 in natural zirconia	3 – 13
Thorium-232 in natural zirconia	0.1 – 26

Most zirconia (ZrO_2) is manufactured from zircon due to the scarcity of natural zirconia. The radioactivity content of manufactured zirconia depends on the activity of the source zircon and the production process used. Thermal zirconia contains higher levels of radionuclides than the parent zircon, whereas chemically-produced zirconia may contain very low levels of radionuclides.

REGULATION OF NORMs

In terms of legislation applied to materials, the levels where regulatory attention can be expected varies between countries, but a threshold value of 1 Bq/g is generally used internationally (International Atomic Energy Agency, 2004). Individual countries may apply stricter limits than these international standards, and if a material contains > 1 Bq/g it is classified as radioactive.

Zircon and zirconia containing above 1 Bq/g are subject to regulatory attention in all jurisdictions. Whether an individual country applies regulation to the handling of these materials will depend on the country in question; however, from the focus that NORMs have received from both the International Commission on Radiological Protection (ICRP) and the International Atomic Energy Agency (IAEA) it should be assumed that possession, handling, and use of zircon and zirconia will receive some level of regulatory attention in most countries. This attention may vary from simple 'notification' through 'registration' to the highest level – 'licensing'.

REGULATION OF RADIATION EXPOSURE

The radiation exposure of workers is closely regulated in all countries and the limits applied in most cases follow the ICRP and IAEA recommendations. The current internationally-agreed exposure limit is 20 millisieverts (mSv) per year for a worker. Exposure in a single year may exceed this value provided that the exposure over a 5-year period does not exceed 100 mSv, (*i.e.* an average of 20 mSv/a).

Regulation also requires that if a worker's exposure exceeds 1 mSv/a year the worker is considered as occupationally exposed and is classed as a radiation worker. If the operation has areas where exposure can exceed 1 mSv/a these areas are classified as *supervised* areas where a specific set of monitoring requirements is mandatory. If the exposure levels exceed 6 mSv/a, the areas where this occurs are classed as *controlled* areas and more severe forms of monitoring and control are required.

Radiation exposure of the public due to a zircon/zirconia handling operation is also regulated, with members of the public being limited to an exposure of 1 mSv/a from all sources, with constraints being applied for a single source. These single-source constraints are typically set at 0.3 mSv/a. For example, when zircon/zirconia is transported the exposure of the public from any spillages must be kept below 0.3 mSv/a. Corrective action would be required if this limit is exceeded.

Regulation of NORMs in the EU

EU Member States are required to comply with the latest version of the Basic Safety Standards – 'Council Directive, 2013/59/EURATOM' of 2013, by 6 February 2018. For the first time, zircon is specifically referenced in the annexes to this update; namely:

> *'Annex VI: List of industrial sectors involving naturally-occurring radioactive material as referred to in Article 23....When applying Article 23 the following list of industrial sectors involving naturally-occurring radioactive material, including research and relevant secondary processes, shall be taken into account: – Zircon and zirconium industry'.*

While the requirement of the 2013 version should be comfortably met by zircon operators in the EU, the fact that zircon is noted in the Annex could lead to greater examination from local authorities. In preparation for this, those operating in the EU should:

- Recognize that their operations will be subject to the scrutiny of the local regulator and prepare for an inspection
- Inform their regulators that they are processing 'radioactive materials' and conduct radiation risk assessments to ascertain the potential or actual exposure situation.
- Establish the levels in terms of U and Th activity of wastes and be prepared for negotiations with regulators
- Ensure that they have adequate dose data to support the claim that the impact on the public from their operation is less than 0.3 mSv/a.

The ZIA is maintaining a watching brief over the development of these regulations throughout the EU with the aim of forewarning members about any impending differences between Member States.

EU-REACH

The REACH (Registration, Evaluation and Authorisation of Chemicals) Regulation entered into force on 1st June 2007 to streamline and improve the former legislative framework for chemicals management in the European Union (EU). REACH also created the European Chemicals Agency (ECHA) in Helsinki, which has a central coordination and implementation role in the overall process by managing the registration, evaluation, and authorization steps to ensure consistency across the EU.

Two basis concepts are attributable to REACH: the shift of the burden-of-proof of safe use from authorities to industry, and the stated 'No data no market' (REACH Regulation, Article 5), whereby substances shall not be manufactured in or placed on the EU market unless they have been registered in accordance with the relevant provisions of the Regulation.

To date, some 45 000 registrations covering 9500 unique substances have been received by ECHA. Table II shows the top five registrations by country, and Table III, by substance.

Table II. Overview of top five countries to 13 May 2016 (ECHA).

Country	Number of registrations
Germany	11 677
United Kingdom	5 488
France	4 008
Netherlands	3 932
Italy	511

Table III. Overview of top five substances to 13 May 2016 (ECHA).

Substance	Number of registrations
Calcium dihydroxide	457
Ethanol	459
Ethylene oxide	398
Iron	391
Calcium sulphate	300

EU REACH AND ZIRCON

Although the term 'chemical' is not generally associated with metals, ores, and minerals, it is clear that reference to chemical substance, or substance, does include all organic, inorganic, and natural substances found in the periodic table along with all combinations, or mixtures, of these. In fact, substance is defined in REACH Article 3(1) as:

'... a chemical element and its compounds in the natural state or obtained by any manufacturing process, including any additive necessary to preserve its stability and any impurity deriving from the process used, but excluding any solvent which may be separated without affecting the stability of the substance or changing its composition'.

Importantly, there are some notable exemptions from registration obligations as detailed in Annexes IV and V of the Regulation. For example, substances for R&D purposes, substance used in the interests of defence, and those substances managed by other legislation such as food additives, radioactive substances, or medical applications are exempt for a period of time.

For the minerals industry, exempted substances in Annex V(7) include: *'The following substances which occur in nature, if they are not chemically modified. Minerals, ores, ore concentrates, cement clinker, natural gas, liquefied petroleum gas, natural gas condensate, process gases and components thereof, crude oil, coal, coke.'*

Zircon, along with other minerals, is therefore exempted from REACH registration obligations. Chemical derivatives of zircon are, however, subject to REACH registration at >1 metric ton per year per legal entity registrant. Analyses of some zirconia-based substances is given in Table IV

Table IV. The top five REACH-registered Zr-based substances

Name	Chemical Abstracts Service (CAS) number	No. of registrants	Total tonnage band (t/a)
Zirconium	7440-67-7	4	1000 -10 000
Zirconium dioxide	1314-23-4	37	10 000 -100 000
Zirconium dioxide Y-doped	64417-98-7	7	100 -1000
Calcium zirconium oxide	11129-15-0	3	1000 -10 000
Zirconium sulphate	14644-61-2	10	1000 -10 000

Parallel to the REACH registration process was the introduction of the Classification, Labelling, and Packaging (CLP) regulation – the EUs equivalent to the Globally Harmonised System of Classification and Labelling (or GHS) – where all substances, including those exempt from REACH registration obligations, are classified through a process of notification. It is here that we see zircon and entries for zirconium orthosilicate and zirconium silicate. These notifications include a number of different classifications for the same substance: *e.g.* zircon is classified variously by different producers as:

- Unclassified
- Skin irritation 2 [H315 - causes skin irritation]
- Eye irritation 2 [H319 - causes serious eye irritation]
- Carcinogen 1A [H350 - may cause cancer].

It is not unusual for there to be different classifications for the same substance: some may be default classifications; others may be precautionary. However, these classifications are in the public domain and must be addressed to reflect an industry-agreed, single classification.

REACH AROUND THE WORLD

While REACH refers to the EU, regulatory derivatives are developing in other jurisdictions as industrialized nations grapple with the complexities of chemicals management. A strong motivation for this is the UN Strategic Approach to International Chemicals Management (SAICM); a policy framework to foster the sound management of chemicals around the world. SAICM aims to achieve the sound management of chemicals throughout their life cycle so that, by 2020, chemicals are produced and used in ways that minimize significant adverse impacts on human health and the environment. This '2020 goal' was adopted by the World Summit on Sustainable Development in 2002 as part of the Johannesburg Plan of Implementation.

In order to meet this goal, many regions have been encouraged by the reported successes of EU-REACH to look at implementation of a similar legislative framework, some with an almost copy/paste approach, from China to South Korea and Turkey, each of which have implemented, or are in the stages of introducing, REACH-like laws. While many of the legislative terms may be familiar to those involved with EU-REACH (*e.g.* 'downstream user', 'only representative', 'exposure scenario'), care should be taken to understand the detail of these derivatives as there are some important differences, *e.g.* the requirement for annual tonnage reporting in Korea's K-REACH.

Other industrialized countries that have not fully implemented regulatory change are starting with the cataloguing of all existing chemicals on their market (*e.g.* Brazil, Taiwan, and India). Care has to be taken to monitor each list as it is published, as these often then lead to the classification and prioritization of substances for future registration.

OCEAN SHIPPING AND THE INTERNATIONAL MARITIME ORGANIZATION

While regulatory developments progress on land, the updating of various ocean shipping codes by the International Maritime Organization (IMO) also has an impact on international trade. Based in London, the IMO is a specialized agency of the UN that is responsible for measures to improve the safety and security of international shipping and to prevent pollution from ships. Its stated objective is *'safe, secure and efficient shipping on clean oceans'*.

IMO's governing body is the Assembly that is made up of all 171 Member States, and a Council of 40 Member States acts as its governing body between Assembly sessions. In practice, the main technical work is carried out by five committees - Maritime Safety, Marine Environment Protection, Legal, Technical Co-operation, and Facilitation – supported by a number of sub-committees. Outputs from these committees include several codes that lay out the legal framework for the shipping of goods in bulk and in containers.

Two codes can have an impact on the shipping of zircon and other minerals; namely the International Maritime Solid Bulk Cargoes (IMSBC) Code and the International Convention for the Prevention of Pollution from Ships (MARPOL).

The aim of the IMSBC Code is to enable the safe stowage and shipment of solid bulk cargoes by providing information on the dangers and precautions associated with the shipment of certain types of cargo in bulk. The IMSBC Code categorizes cargoes into three groups:

Group A:	cargoes that may liquefy
Group B:	cargoes that possess a chemical hazard that could give rise to a dangerous situation on a ship
Group C:	other cargoes that can still be hazardous (*e.g.* very dense cargoes that can damage a vessel's structure due to poor loading procedures).

Importantly, these characteristics refer to the bulk cargo properties rather than a hazard from the material itself. For example, liquefaction (where a bulk cargo becomes fluid) is a property displayed only in bulk form. Cargoes prone to liquefaction contain a certain quantity of moisture and small particles, although they may look relatively dry and granular when loaded. Liquefaction can lead to vessel instability and even to capsize and total loss of the ship, and can occur even when cargoes are cohesive and trimmed level.

Zircon sand is currently cargo Group C, whereas other minerals are assigned different codes – for example ilmenite sand is Group A or C.

Importantly, there is an ongoing review to classify ilmenite sand as Group A due to its liquefaction potential. In addition, the reference to 'sand' in the cargo description erroneously introduces an association with the health effects due to the inhalation of silica dust. Efforts are underway to inform and educate the relevant IMO technical committees that such reference to sand in mineral cargo descriptions relates to the physical form of the cargo rather than the presence of silica sand.

Transport of materials by ocean freight also has to comply with the requirements of the International Maritime Dangerous Goods Code (IMDG). This code is controlled by the IMO, but fortunately the IMO has aligned its standards with those of the IAEA so the same limits apply. By not exceeding the 10 Bq/g limit, zircon and zirconia may be shipped internationally without the requirement to comply with the IAEA transport regulations or the IMDG Code. Being not classified as a radioactive material in the IMDG Code means the concentrates avoid classification as 'Hazard Class 7 radioactive substances.' Shipping companies are reluctant to handle Class 7 materials, so this is a distinct advantage for the export of this material.

The status of radiation-based regulations in the destination port also needs to be confirmed. Although the country may subscribe to the IAEA standards, there have been situations where port officials apply different standards to those defined internationally. This inconsistency may also vary between ports in the same country. The radioactivity of zircon/zirconia does not require it to be governed by the IAEA transport regulations, and storage as part of transportation is also not mandated by the international standards. However, the detection of a radioactive signature from the material may trigger some reaction from port officials, which could delay the movement of cargo.

CONCLUSIONS

- Although NORM and radiation developments are at the end of a cycle, zircon operations should aim to proactively manage regulatory developments (*e.g.* meeting with relevant authorities and targeted advocacy)
- For those areas of potential higher worker exposure to NORM dust, such as dry milling of zircon, appropriate risk assessments should be undertaken and kept updated
- Regulations are harmonizing globally (land and sea), becoming more stringent and shifting the burden of proof from regulators to industry. It is vital to be aware of developments globally and plan/act accordingly.

The Zircon Industry Association (ZIA) is working to support its members' knowledge and compliance with major regulations impacting the full supply chain.

ACKNOWLEDGEMENTS

The author would like to express his thanks to colleagues at ZIA and to J.H. Selby for his various reports to the ZIA.

REFERENCE

International Atomic Energy Agency. (2004). Application of the Concepts of Exclusion, Exemption and Clearance. *IAEA Safety Standards Series*, Safety Guide No. RS-G-1.7. Vienna.

Keven Harlow

Executive Director
Zircon Industry Association

A metallurgist and materials scientist graduate, Dr Harlow spent over 20 years in the commercial area of the global metals industry where he was able to marry his technical knowledge with sales and marketing experience. He served as Managing Director of Höganäs (GB) Limited for five years. Not forgetting his technical roots, Dr Harlow became involved with the EU-REACH regulation in 2006, supporting various companies and trade associations with REACH compliance strategies and consortia management. In April 2016 Dr Harlow assumed the role of Executive Director of the Zircon Industry Association - the trade association those organisations involved in the zircon / zirconia / zirconium chemicals value chain.

Dr Harlow holds his PhD (metallurgy) at The University of Nottingham, UK.

The influence of trash minerals and conglomerate particles on spiral separation performance

J.D. Grobler[1] and J.H. Zietsman[2]

[1]Exxaro Metallurgical Services, Pretoria, South Africa
[2]Ex-Mente, Pretoria, South Africa

Understanding the impact of feed material mineral assemblage on spiral separation is paramount for the successful and consistent operation of a spiral plant. Plant stability is required to maintain high recovery of valuable heavy mineral (VHM). Standard material analysis methods used in the heavy mineral industry to evaluate spiral separation behaviour are unable to quantify performance differences with large variations in feed assemblage. This study illustrates a systematic approach to evaluate spiral separation performance. This technique was applied on a four-stage spiral plant. The performance information was used in a circuit simulation to clearly illustrate the importance of correct setting of the spiral splitter position and resulting spiral mass loading to ensure optimal circuit stability and maximum VHM recovery.

INTRODUCTION

Heavy mineral sands deposits generally consist of naturally fine-grained, free-flowing sand due to the way in which the deposit was formed. The size of the heavy minerals market is roughly 8 Mt/a of zircon and titanium combined (Iluka, 2013). If an average combined feed grade of 5% zircon and titanium, and an average combined overall product recovery of 80% are assumed, the resulting run-of-mine feed tons that would require primary spiral beneficiation is 200 Mt/a. The number of spirals required for the processing of this material, assuming 6500 operating hours per annum, an average of 2 t/h per start, and the same number of rougher spirals to other duty spirals, is over 30 000 spirals for the heavy mineral sands business alone. These numbers clearly illustrate the importance of the spiral in the heavy mineral industry.

The aim of a spiral plant in the primary processing value chain of a heavy mineral process is to reject non-valuable quartz (light mineral) from various valuable heavy minerals. The product from the spiral plant is a heavy mineral concentrate that has a reasonable grade (more than 90% heavy minerals) at a reasonable heavy mineral recovery (greater than 90%). This benchmark is fairly easy to achieve if all the mineral particles are well liberated (like in the case of free-flowing sand), the feed is properly deslimed (-45 um fraction removed to less than 5% of the solids mass), and the spirals are fed within design solids concentration (water to solids ratio), and design spiral load (dry tons per hour per start).

Consistently achieving a high-grade heavy mineral concentrate at high valuable mineral recovery is paramount to business success. Forecast valuable mineral recovery (especially zircon) is an important input into monthly and annual budgets. A 10% decline in zircon recovery over a month equates to significant revenue losses for a medium-sized operation, not considering other valuable mineral losses. The quick investigation and understanding of the reasons for decline in zircon recovery and appropriate response with corrective actions can save the business millions of dollars. The purpose of this article is to illustrate the systematic approach that was followed to identify recovery losses of a typical four-stage spiral circuit and the response with corrective actions to minimize the impact.

PROCEDURE

The spiral plant that was evaluated in this study has four separation stages, as indicated in Figure 1. The first three duties are all equipped with Mineral Technology MG6.3 spirals and the final cleaner stage utilizes Mineral Technology HG10S spirals. The primary performance indicator of this spiral plant is zircon recovery, determined through the two-product formula (using XRF assay of ZrO_2 in the feed, concentrate. and combined tail stream). Three Vezin samplers take composite samples of these three streams on a four-hourly basis. This study was initiated after a decline in 10% zircon recovery persisted for a considerable period. A zircon recovery decline was correlated with changes in plant feed material. However, since there were large volumes of this feed material, the plant had to process it at the highest achievable zircon recovery.

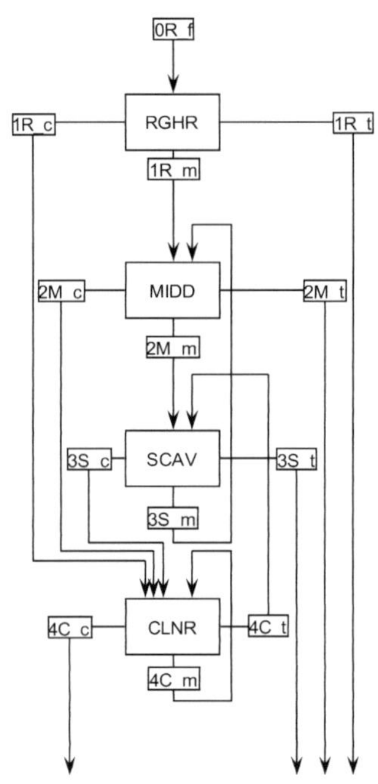

Figure 1. Flow sheet of four-stage spiral circuit. RGHR (1R) = rougher duty, MIDD (2M) = middling duty, SCAV (3S) = Scavenger duty, CLNR (4C) = cleaner duty. Spiral concentrate = _c, spiral middling = _m, spiral tail = _t.

The following procedure was applied to minimize the zircon losses while maintaining a reasonable zircon product grade for the downstream process.

1. Theoretical maximum recovery: determine how much of the zircon in the feed is recoverable by spiral separation

2. Feed mineral assemblage impact: determine the mineralogical composition of the feed material and assess what the impact is on the individual spiral separation
3. Spiral circuit performance assessment: evaluate the performance of the spiral circuit by sampling the four duties. Apply mass balance techniques using the spiral duty mass splits and stream analyses
4. Corrective measures: implement corrective measures based on the interpretation of the circuit performance results.

Each step is presented in the next four sections with a method description followed by illustration and discussion of the results.

STEP 1: DETERMINE THEORETICAL MAXIMUM RECOVERY

A spiral circuit feed sample was fractionated on a laboratory shaking table under controlled feed conditions. The sample was divided into four mass fractions based on the band colours. The first band concentrated the clear black band, the second band concentrated the 'other' coloured heavy minerals, the third band concentrated the fine quartz, and the last band concentrated the coarse tail particles. Figure 2 illustrates the coloured bands on the shaking table during fractionation for a condition of low trash, while Figure 3 illustrates the coloured bands for high trash content. Each mass fraction was assayed by XRF and a recovery-yield curve was drawn (Figure 4). The shaking table result presents the best case, or theoretical maximum recovery that can be achieved for a specific feed material, since the shaking table provides sufficient separation space and separation time for each particle to reach its equilibrium position from a particle density, size, and shape perspective.

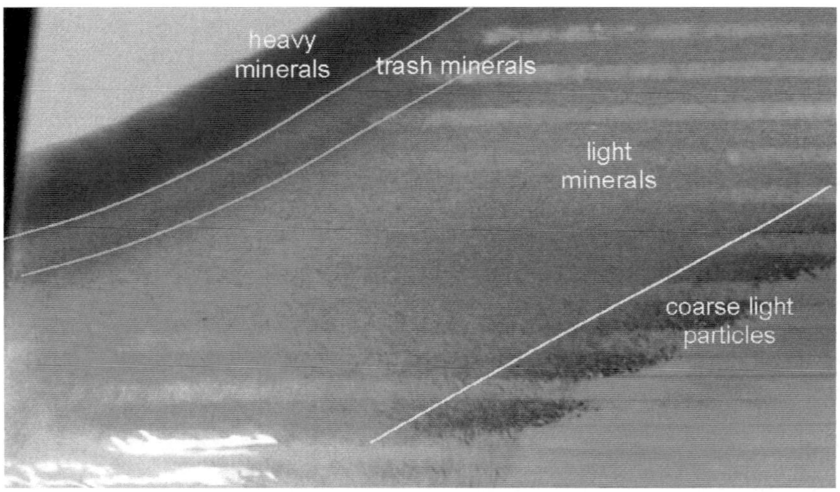

Figure 2. Shaking table fractionation with low trash content.

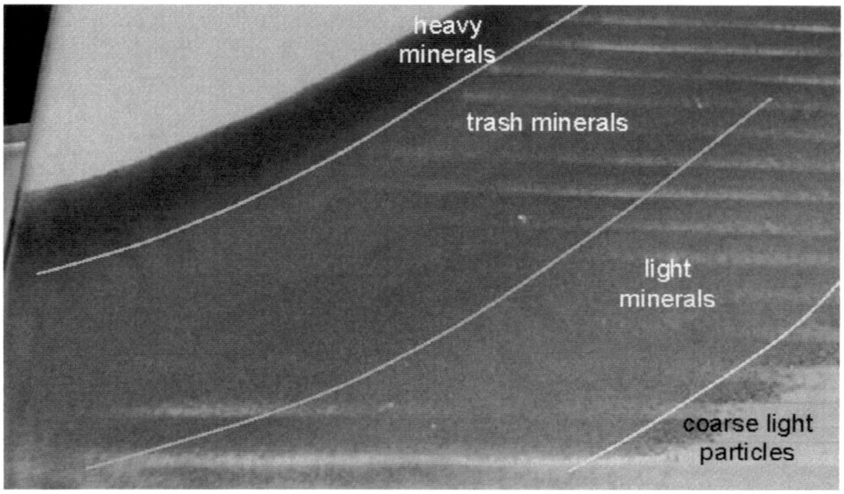

Figure 3. Shaking table fractionation with high trash content.

Figure 4 illustrates a 91% zircon recovery at 21% mass yield, which implies that 9% of the zircon is trapped in the remaining 79% of the mass. This zircon recovery is quite low, and further investigation indicated that fine conglomerates (less than 500 µm) were present in the feed. These conglomerates trapped valuable zircon particles (and other valuable minerals) in a low-density clay and quartz matrix, causing them to be misplaced to the tail fraction. The same feed sample was attrition-scrubbed in a laboratory unit and processed on the shaking table after desliming. The result from the scrubbed feed fractionation is presented in Figure 5. A clear step change (5–10%) in recovery is apparent for all three of the heavy elements (Zr, Fe, and Ti), which shows that the valuable heavy mineral particles were liberated by the scrubbing process. The detailed QEMSCAN analyses of some of these conglomerates are illustrated in Figure 6. The spiral circuit is unable to recover the zircon contained in these conglomerates due to their combined low density, larger size, and irregular shape. This material needs to be liberated before it reaches the spiral circuit, and the spiral separation performance should be measured against a new 'baseline'. That is, a theoretical maximum recovery as achieved by the shaking table; 91% in the case of Figure 4. The results indicated in Figure 5 can be achieved only with improved feed preparation in the plant (milling and/or scrubbing) to liberate valuable minerals from conglomerates prior spiral separation with significant power input.

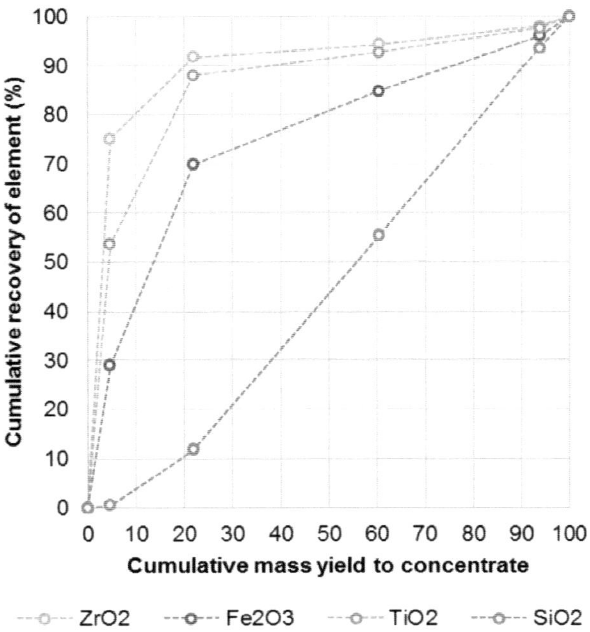

Figure 4. Recovery-yield relationship of the four main elements in spiral feed fractionated on a shaking table.

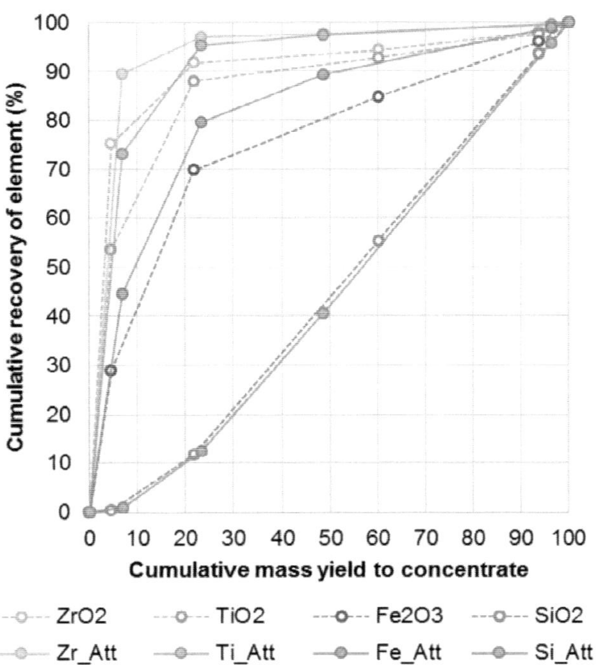

Figure 5. Recovery-yield relationship of the four main elements in spiral feed fractionated on a shaking table – before and after attrition scrubbing.

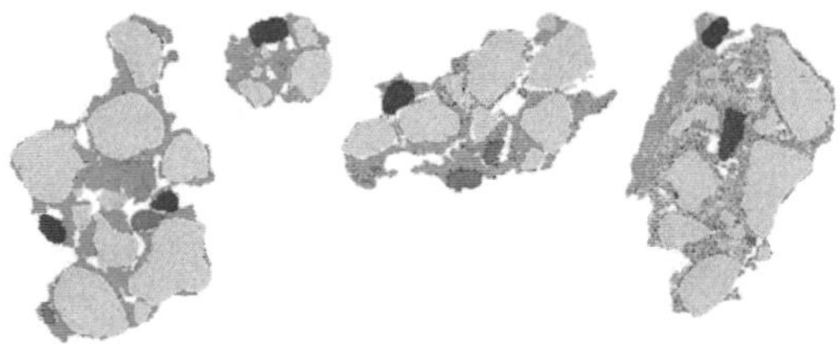

Figure 6. Conglomerate cross-section: blue particles are zircon, green is clay matrix, light pink particles are quartz. The round particle, second from the left, is 200 μm in diameter.

STEP 2: DETERMINE IMPACT OF FEED MINERAL ASSEMBLAGE

The feed mineral assemblage can have a significant impact on spiral recovery, especially the concentration of near-density heavy mineral particles. The heavy mineral industry gets its name from the particle class 'heavy minerals'. Traditionally this class contains mineral particles that sink in an organic medium (usually tetrabromoethane or bromoform) with a liquid density of 2.96 g/cm^3, and is called total heavy minerals (THM). This implies that the THM have a density greater than 2.96 g/cm^3. These heavy mineral particles are further classified. Valuable heavy minerals (VHM, for example zircon, rutile, and ilmenite) usually have a density greater than 4 g/cm^3. The term 'trash' is used to describe a particle class with a specific density range. In this article that range is greater than 3 g/cm^3 and less than 4 g/cm^3, and therefore this material has an intermediate density between 'light' particles and 'heavy' particles. It is called trash because it contains no valuable or saleable minerals, but is considered as 'heavy' minerals since their density is greater than 2.96 g/cm^3. This paper makes reference to three distinct particle classes, heavy minerals (>4 g/cm^3), trash minerals (3–4 g/cm^3), and light minerals (<3 g/cm^3). To illustrate the difference in separation response between these different mineral groupings, a shaking table fractionation was carried out on the cleaner spiral feed. The fractions were analysed with QEMSCAN to quantify the different minerals present in the assemblage. Figure 7 illustrates the three density response bands, and Table I lists the mineral types and percentages in each band.

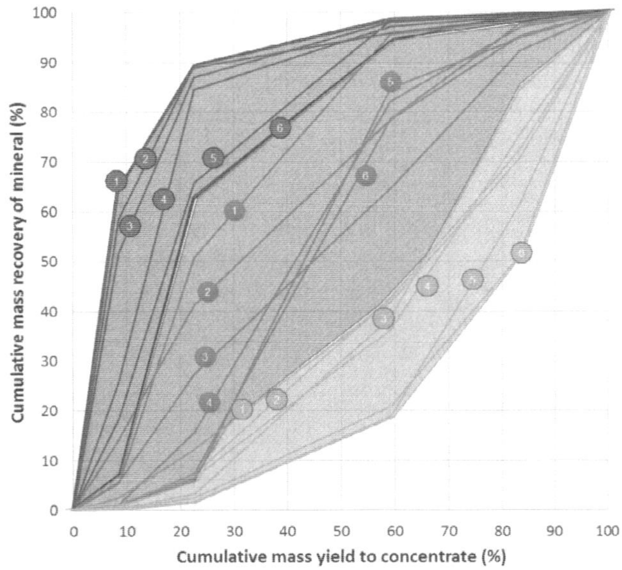

Figure 7. Shaking table density fractionation of cleaner spiral feed. Numbers correspond with ranking in Table I. Actual data points (not shown) were connected with straight lines – no curve smoothing was applied.

Table I. Mineralogical composition and classification of the sample used for density fractionation in Figure 7.

No.	Heavy minerals	%	No.	'Trash' minerals	%	No.	Light minerals	%
1	Monazite	0.1	1	Almandine	8.7	1	Mica	2.3
2	Zircon	2.2	2	Fe oxides	3.0	2	Kyanite	0.9
3	Ilmenite	10.8	3	Grossular	1.3	3	Dolomite	0.2
4	Fe (HiTi) oxides	.5	4	Apatite	3.0	4	Plagioclase	2.4
5	Rutile	1.8	5	Pyroxene	23.0	5	Quartz	38.0
6	Spessartine (garnet)	0.4	6	Chlorite	0.9	6	K-feldspar	0.5
	Total (blue)	15.8		Total (green)	39.9		Total (yellow)	44.3

Since there is sufficient separation space on a shaking table the trash minerals can easily achieve their respective equilibrium positions. However, on a spiral the separation space is limited, and residence time for separation is short (less than 15 seconds). Trash minerals will concentrate towards the middle of the spiral, between the heavy minerals and light minerals (Figure 8). This implies that there is probably a threshold of trash mineral content that can be accommodated on the spiral, beyond which separation will be adversely affected.

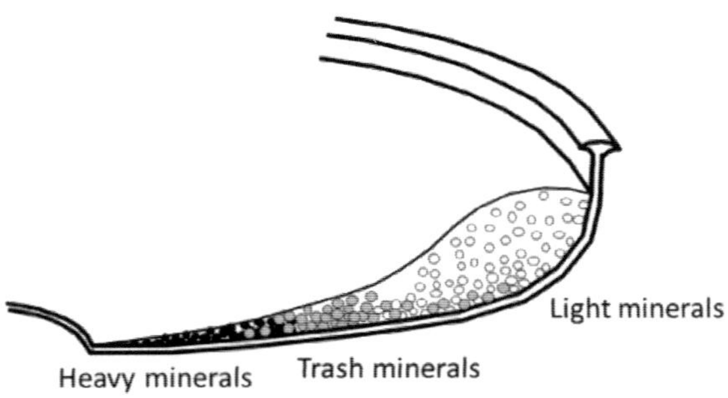

Figure 8. Positions of different mineral density classes on the spiral cross-section.

Literature reporting on the influence of trash minerals on spiral separation is limited, since most mineral sands deposits contain low trash concentrations in the feed, seldom exceeding 5% of the mineral assemblage. In this investigation the trash mineral content was up to 25%. To measure the impact of trash on spiral separation, shaking table separation response (ideal separation) was compared with spiral separation response (on the same spiral profile) for completely liberated particles with increasing trash content (three trash content scenarios). QEMSCAN particle mineral analysis (PMA) enables the classification of particles into different quality classes. This analytical technique has been effectively applied in many other density fractionation studies (Grobler and Bosman, 2011). One of the PMA mineral classes is Zircon I, a clean, completely liberated particle with ZrO_2 content greater than 66%. Zircon I was used as an indicator particle class to compare its recovery at different trash contents. The pyroxene particle class was selected as the most predominant trash mineral particle, and quartz was selected as the most predominant light mineral particle.

Figure 9 illustrates the three trash content scenarios. The enhanced Holland-Batt equation was used to fit the test work data points. This fitting technique for spiral data was described by Grobler *et al.* (2016. In all three feed scenarios the Zircon I recovery was above 95% for the shaking table. However, the Zircon I recovery for the spiral decreased significantly with increase in trash (from 95% to 83% to 78% at 20% mass yield to concentrate in Figure 9a, b, and c). These three graphs illustrate the adverse effect of trash content in feed.

The impact of unliberated zircon particles as well as trash content has been illustrated for the spiral in isolation; the next step is to determine its impact on the spiral circuit as a whole.

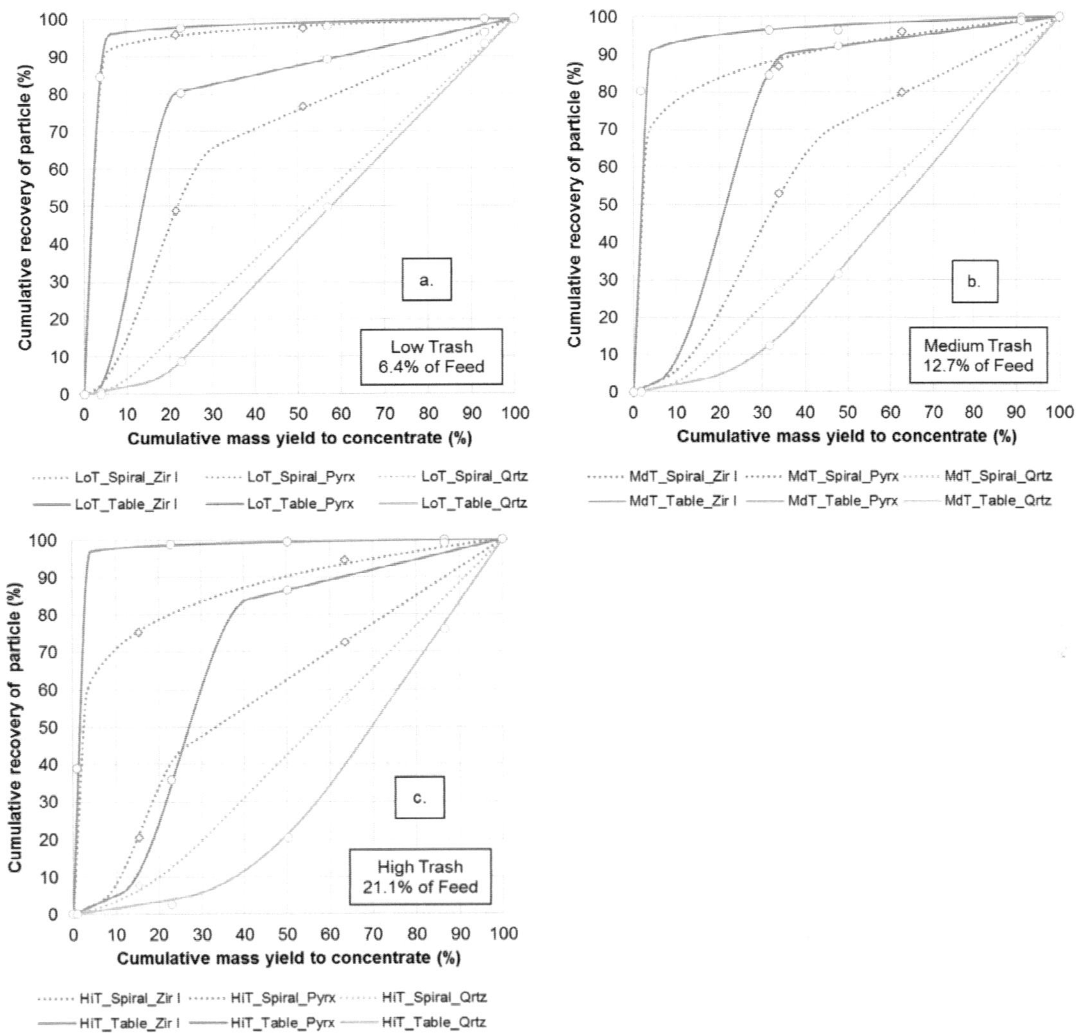

Figure 9. Yield-recovery relationship for three different trash content scenarios. (a) 6.4% trash, (b) 12.7% trash, (c) 21.1% trash).

STEP 3: SPIRAL CIRCUIT PERFORMANCE ASSESSMENT

The four-stage spiral circuit was sampled by selecting an individual spiral column in each spiral duty and taking a timed sample of the feed, concentrate, middling, and tail streams. Feed distribution checks were done beforehand to confirm that the selected spiral column was representative of the entire spiral duty. The feed to the rougher spiral bank was fractionated into four samples with the laboratory shaking table to determine the theoretical maximum recovery. The four spiral duty sample sets with the four shaking table fractions resulted in 20 samples, which were deslimed and assayed by XRF. No sink-float analyses were done. The XRF assays were used in a mass balance model using the Limn flow sheet processor (Excel add-in) to determine the mass balanced zircon recovery considering the main elements in the mass balance solver. The circuit mass balance calculated the total mass yield contributions of the rougher duty tail, middling duty tail, and scavenger duty tail in relation to the mass yield to concentrate. Refer to Figure 1 for the spiral circuit flow sheet.

The following three feed scenarios were used to illustrate the impact of trash and conglomerates on the spiral circuit performance.

<u>Case 1</u>: the circuit was sampled under normal feed conditions when the circuit zircon recovery was at acceptable levels (above 90%).

<u>Case 2</u>: the circuit was sampled after the feed had changed and the zircon recovery declined, with more than 15% creating an unacceptable performance condition.

<u>Case 3</u>: the circuit was sampled after corrective measures were employed on the new feed condition to minimize zircon losses.

The three feed conditions are depicted in Table II. The trash content in the feed increased in Case 2 and Case 3, while the ZrO_2 grade decreased. The most significant difference is shown in the zircon recovery for the three cases. Each of the three cases is discussed in the next three sections. Table III demonstrates the normalized mass yield to the different tail and concentrate streams. The design load for middling and scavenger spirals is 2 t/h per start, and for the cleaner 1.85 t/h per start. Figure 10 gives the zircon recovery to concentrate response, and Figure 11 the zircon grade to concentrate response, over the entire mass yield spectrum.

Table II. Feed and circuit response for the three cases.

Case no.	Feed grade %ZrO_2	Feed grade %trash	Recovery %ZrO_2	Conc. grade %ZrO_2
1	0.61	11.1	90.1	3.49
2	0.44	17.2	72.2	3.80
3	0.46	19.0	83.2	5.12

Table III. Normalized mass balance and circuit loading summary for the three cases.

Case no.	Rougher tail tass%	Midd tail mass%	Scav tail mass%	Clnr conc. mass%	Midd load t/h/start	Scav load t/h/start	Clnr .load t/h/ start
1	64.6	12.6	7.0	15.8	1.7	1.3	1.9
2	54.7	16.3	20.7	8.3	3.0	3.4	3.4
3	64.0	19.2	9.3	7.6	2.5	1.6	2.6

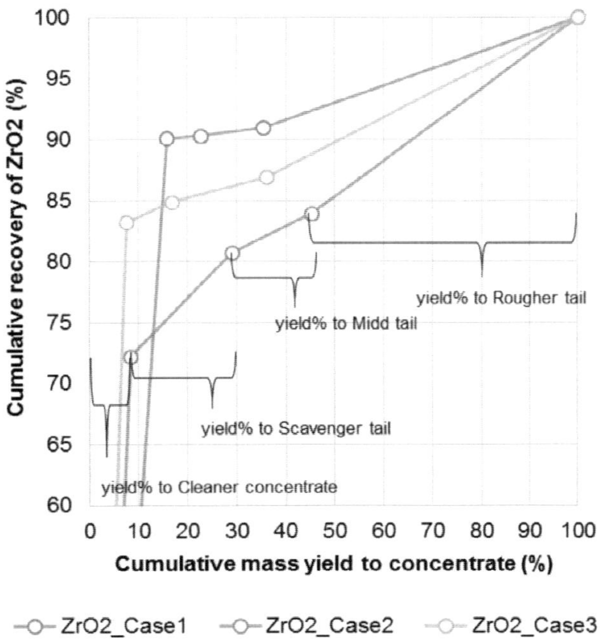

Figure 10. Recovery-yield relationship for the three spiral circuit separation cases. The respective yields to the different exiting process streams are indicated by the four horizontal lines (for Case 2 only).

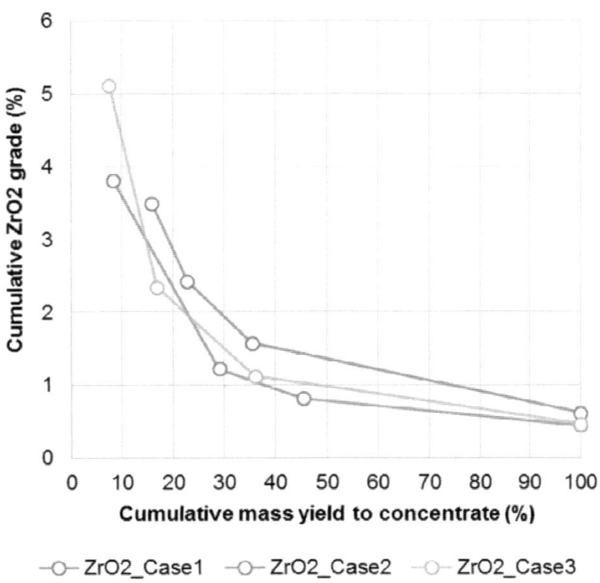

Figure 11. Grade-yield relationship for the three spiral circuit separation cases.

Case 1: Low Trash Feed - Normal Feed Conditions

Figure 12 compares the ideal zircon recovery with spiral circuit zircon recovery. The ideal recovery was determined to be 94%, indicating that 6% of the zircon is not available for spiral concentration due to lack of liberation (indicated by red arrow in Figure 12). The circuit zircon recovery response was 90%, which implies that 96% of the available zircon was successfully recovered by the four-stage spiral circuit. The trash content was on the low end at 11.1% (Table II). The downstream spiral loading (midd, scavenger, and cleaner) was acceptably low. Case 1 demonstrates an acceptable circuit separation response with high zircon recovery (96% with consideration of irrecoverable zircon) to concentrate, and the bulk of the zircon losses occurring in the rougher tail stream.

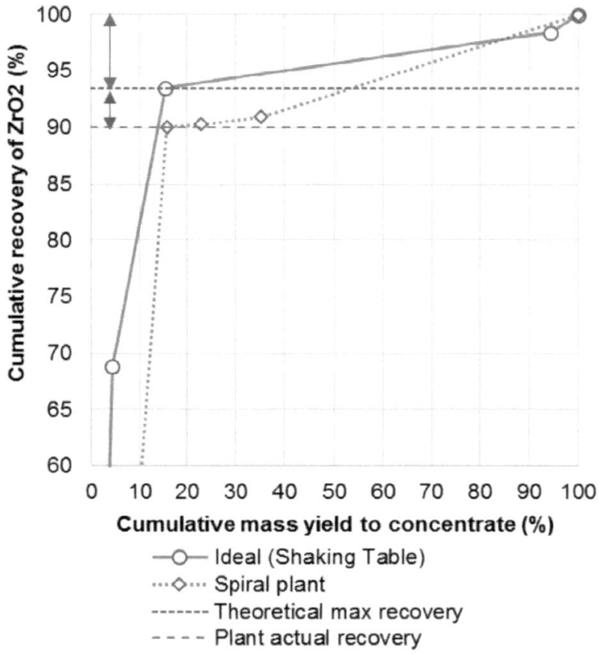

Figure 12. Case 1 ideal recovery versus spiral circuit recovery.

Case 2: High Trash Feed Conditions - Normal Operating Conditions
Figure 13 compares the ideal zircon recovery with spiral circuit zircon recovery. The ideal zircon recovery was determined to be 87%, indicating that 13% of the zircon is not available for spiral concentration due to lack of liberation (indicated by red arrow in Figure 13). The circuit zircon recovery response was 72%, which implies that 83% of the available zircon was successfully recovered by the four-stage spiral circuit. The trash content was on the high end at 17.2% (Table II). The downstream spiral loading (midd, scavenger, and cleaner) was unacceptably high, with every spiral duty exceeding the design loading. Case 2 illustrates the impact that high feed trash content has on the spiral circuit. Firstly, it decreases the separation space and reduces the recovery potential on the spiral. Secondly, it increases the loading (tons per hour per start) on the downstream spiral duty, causing a further decrease in zircon recovery. These are the two main reasons for the zircon recovery drop of 13% from 96% in Case 1 to 83% in Case 2. Case 2 demonstrates an unacceptable circuit separation response with low zircon recovery (only 83% with consideration of irrecoverable zircon) to concentrate, and zircon losses occurring in all three tail streams. A high zircon loss in the scavenger tail stream is a clear indication of overload in the spiral circuit.

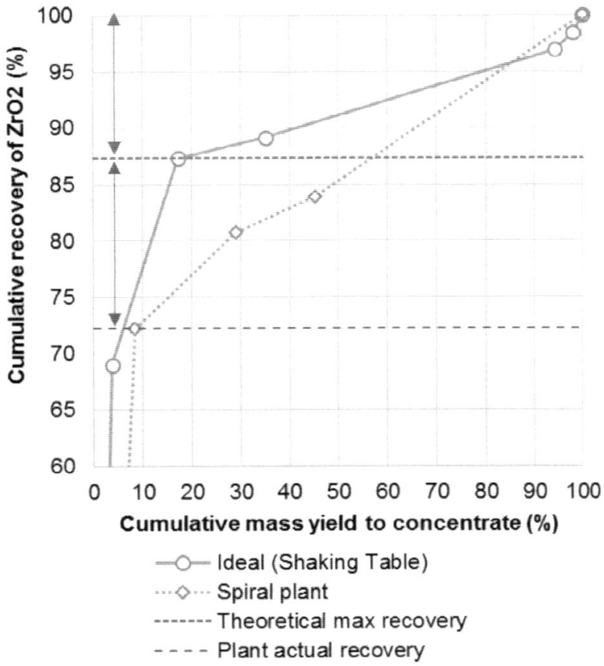

Figure 13. Case 2 ideal recovery versus spiral circuit recovery.

STEP 4: IMPLEMENT CORRECTIVE MEASURES

Since Case 2 resulted in an unacceptable zircon recovery, corrective measures were implemented in Case 3. The feed composition cannot be changed, therefore the only remaining control parameters are the operating conditions, total feed to the plant, but more specifically the spiral splitter positions.

Case 3: High Trash Feed Conditions - Changed Operating Conditions

Figure 14 compares the ideal zircon recovery with spiral circuit zircon recovery. The ideal zircon recovery was determined to be 90%, indicating that 10% of the zircon is not available for spiral concentration due to lack of liberation (indicated by red arrow in Figure 14). No corrective measures could be applied to the ideal zircon recovery, since this depends on the nature of the feed material that has entered the spiral circuit and can only be addressed by upgrades in the feed preparation circuits, which was not part of this investigation. The trash content was on the high end (19%) as indicated in Table II. The feed trash content can be diluted by blending with other low-trash mining zones, but this was not part of this investigation since this feed condition persisted for a considerable period due to large volumes of this type of material being mined. The negative impact of trash on individual spiral performance has to be tolerated (Figure 9). However, the increased loading on the downstream spiral duties can be managed by the operational personnel.

In Case 3 the tail splitters on the rougher and midd duties were opened 'abnormally' wide to reject mass to the rougher tail and midd tail streams. The rougher and midd duty outer splitter position was changed by more than 60 mm (set inward) due to the mass flow shift that occurred with the high percentage of trash in the feed. A splitter template was designed and implemented to set every spiral splitter in each duty at the exact same spiral trough position. This counter-intuitive operational change rejected 85% of the feed mass to the rougher tail and midd tail steams (Case 3) instead of only 71% in Case 2, as shown in Table III. Table III further shows that the mass loading per spiral duty reduced to more acceptable levels for Case 3, with a significant decrease in recirculating loads.

After this intervention the overall circuit zircon recovery response was 83%, which implies that 93% of the recoverable zircon was successfully recovered to concentrate by the four-stage spiral circuit (indicated by blue arrow in Figure 14). The change in spiral control approach resulted in a recovery increase of 10% (from Case 2 to Case 3 with consideration of irrecoverable zircon) while at the same time increasing the concentrate grade by more than 1% ZrO_2 as indicated in Figure 14.

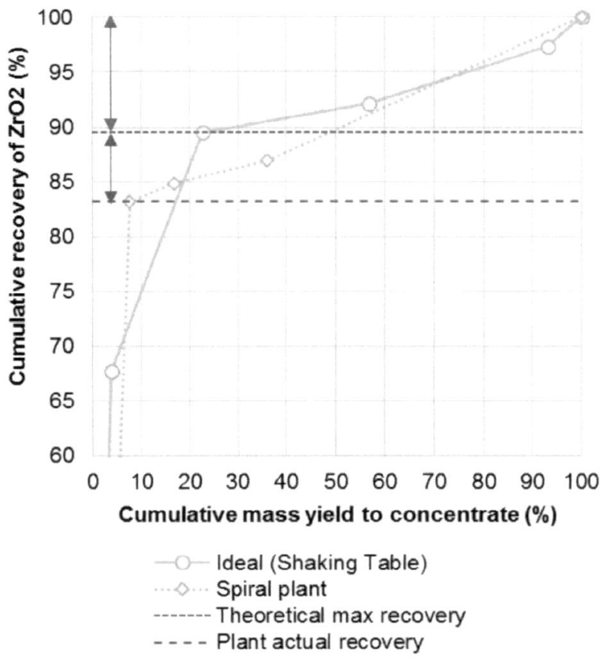

Figure 14. Case 3 ideal recovery versus spiral circuit recovery.

CONCLUSIONS

Trash mineral content is increasing as more 'difficult' mineral deposits are being considered for beneficiation. It is therefore important to understand the impact of increased trash mineral content and unliberated mineral on an individual spiral profile, as well as on the spiral circuit as a whole.

QEMSCAN analyses provided important insight for differentiating between trash minerals and valuable heavy minerals, which can easily be misinterpreted when considering heavy mineral content (typical sink-float result) in new deposits or new mining areas of existing deposits. The characterization of ideal feed material separation response using a laboratory shaking table assisted considerably in identifying the portion of the recovery problem that is within the reach of the spiral operator, and the portion that is not.

Using different techniques simultaneously to measure the same initial observation can speed up understanding of the problem. The application of attrition scrubbing to liberate conglomerates and QEMSCAN to define the content of the conglomerates is an excellent example of this, resulting in a clear explanation of the 'irrecoverable' zircon.

An increase in feed trash content was shown to negatively impact zircon recovery, decreasing the separation space on the spiral trough. An increase of trash from 6% to 21% resulted in a decrease in liberated zircon particle recovery from 95% to 80%. Limited operational control can be applied to reduce this impact, other than reducing the spiral loading to below design feed capacity.

An increase in feed trash content was shown to negatively impact the zircon recovery of the total circuit by increasing the load to the downstream circuits and increasing recirculating load, since a large portion of the mass has moved to the middle of the spiral trough. This zircon recovery loss can be minimized through the application of spiral splitter control to reject as much mass as possible (more than 85%) in the first two separation stages, resulting in a recovery improvement for recoverable zircon of more than 10%.

The mass balance of the spiral circuit enabled the measurement of the total circuit response. Although some effort is required to configure the calculations initially, it provides invaluable insight to understand the recirculating loads and interaction between different spiral separation duties. It is important to verify the mass balanced information with all available measurement data.

REFERENCES

Grobler, J.D. and Bosman, J.B. (2011). Gravity separator performance evaluation using Qemscan® particle mineral analysis. *Journal of the South African Institute of Mining and Metallurgy,* 111, 401–408.

Grobler, J.D, Naude, N., and Zietsman, J.H. (2016). Enhanced Holland-Batt spline for describing spiral concentrator performance. *Minerals Engineering*, 92, 189-195.

Iluka (2013). Mineral Sands Industry. Fact Book. Iluka. http://www.zirconassociation.org/Websites/zircon/images/The_Mineral_Sands_Industry_Factbook.pdf (accessed 4 January 2015).

Jac De Clercq Grobler

Principal Process Engineer, Heavy Minerals
Exxaro Resources

Graduated from University of Pretoria in 2001 with distinction and appointed as metallurgical engineer at Exxaro Resources. Completed honours degree in metallurgy and was appointed as chief scientist in 2005. Completed the SAIM programme in Business and Strategic management at UNISA in 2008. Appointed as principal process engineer heavy minerals in 2008. Registered as professional engineer in 2009. Completed Masters degree in Metallurgy at University of Pretoria in 2015. Fourteen years of experience in heavy mineral sands processing.

Johan Zietsman

Managing Director
Ex Mente Technologies

Johan Zietsman is the owner of, and consulting engineer at Ex Mente Technologies, a metallurgical engineering systems and consulting firm. He is also involved as extraordinary senior lecturer in the Department of Materials Science and Metallurgical Engineering of the University of Pretoria, and supervises a number of postgraduate students. His interests include mathematical modelling of material properties and metallurgical processes, thermochemical analysis, multiphysics modelling and techno-economic modelling of metallurgical operations, as well as performance improvement of metallurgical processes.

Characterization of thin coatings on heavy minerals, with application to surface cleaning technologies

A. Walliser and J. Grobler

Exxaro Resources, Pretoria, South Africa

Thin coatings (2 to 15 µm) on economic heavy mineral (EHM) particles can adversely affect the processing behaviour, causing misplacement, resulting in lower recovery and affecting final product grade by increasing detrimental elements such as silica and phosphorus in an already tight product specification environment. In response to these two adverse effects, the mineral sands industry commonly uses surface cleaning techniques such as attrition scrubbing to reduce these coatings.

The measurement of the thin coating reduction, or surface cleaning efficiency, remains a significant challenge in the mineral sands industry since the quantification of these thin coatings is an analytical challenge. This paper demonstrates that QEMSCAN, in combination with a customized measurement program, can characterize thin coatings at high resolution over an acceptable duration. This in turn allows for the accurate measurement of surface cleaning technology efficiency as well as provides reasons for why the surface cleaning efficiencies can vary.

INTRODUCTION

Particle coatings in the mineral sands industry

Various types of coating, coating thickness and different degrees of coating level can exist on heavy mineral particles, depending on the origin and maturity of the heavy mineral placer deposit. Processed heavy mineral particles often consist of several mineral phases. In this paper, any gangue mineral attached to an economic heavy mineral is classified as a coating - examples are illustrated in Figure 1. These coatings often affect the behaviour of particles during processing. The result is the misplacement of minerals, which reduces the final grade and yield of a product.

Coatings can include cryptocrystalline silica, various clays (mainly sepiolite and kaolinite), Fe oxide, and calcium-containing coatings. The extent of coating depends on the occurrence of the mineral in the deposit and the durability of that mineral in remaining attached to an economic heavy mineral particle.

Application of QEMSCAN for determination of coatings.

QEMSCAN (Quantitative Evaluation of Minerals by Scanning Electron Microscopy) is a valuable analytical application used in quantitative mineralogy (Gottlieb *et al.*, 2000; Adams, 2007). Heavy mineral particles commonly consist of more than one mineral phase. For this reason, QEMSCAN complex particle categorizers are used to classify these particles according to mineral content and/or element-oxide content in the different particle classes. This classification also makes it possible to evaluate the economic heavy mineral (EHM) particles according to grade, on a particle-by-particle basis. Coatings with a thickness of >15 μm can be successfully quantified by QEMSCAN under routine heavy mineral measurement conditions, while higher resolution settings are required to quantify the under-reported thinner coatings. Quantification of thin coatings plays a critical role in the evaluation of off-specification products and surface cleaning technologies. Furthermore, QEMSCAN is able to distinguish between impurities on the particle surface (coatings) and inside the particle (inclusions). This has significant benefit in understanding the performance and limitations of surface cleaning technologies.

Higher resolution settings increase the measuring time exponentially, which is usually time-consuming and costly. In specific cases, if there is a difference in backscattered electron (BSE) signals between the target particles and coatings, the adaptability of the QEMSCAN software can be used to create a measurement program to characterize thin coatings with measurements of acceptable duration.

METHOD

Sample preparation

The standard sample preparation technique was followed to prepare duplicate polished epoxy blocks for each sample; micro-riffler splitting to ensure representivity and to minimize segregation with a thick paste epoxy sample mixture.

High-resolution particle mapping

QEMSCAN produces energy-dispersive X-ray (EDX) spectra and a BSE signal at each measurement point. This reflects the elemental composition and the mean atomic number, used as contrast, at each measurement point. Decreased point spacing's increase the resolution and cause less mixed spectra to be resolved at boundary phases. Routine QEMSCAN measurements are usually taken at 5 μm intervals and at 1000 counts per second (cps) at each point. The resolution may be refined by operating at smaller intervals.

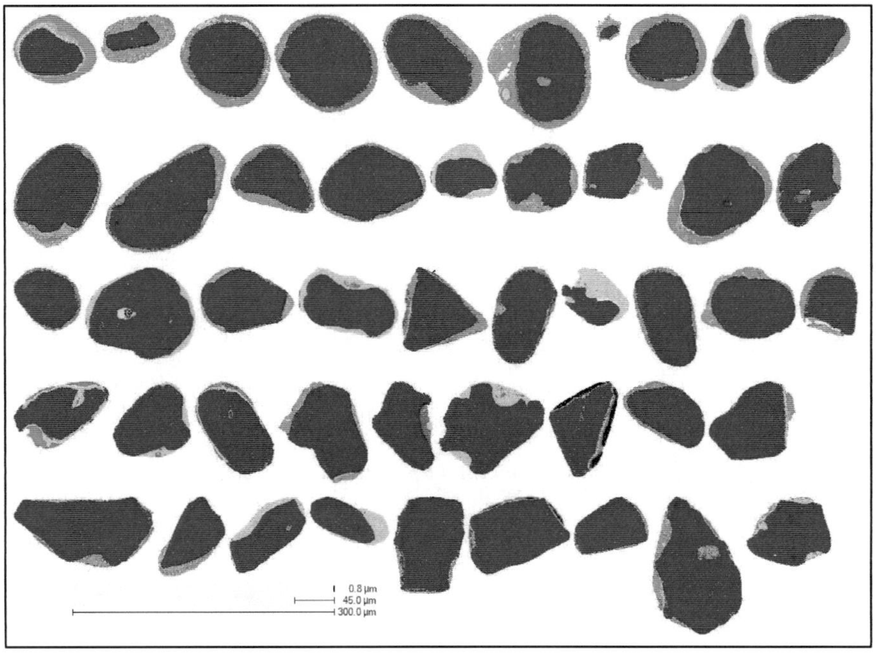

Figure 1. High-resolution particle map of zircon particles (blue) presenting various coatings; quartz (pink), sepiolite (green), kaolinite (orange), and Fe-oxide-containing coatings (black). Note the multi layered coatings in several combinations. Some particles also contain inclusions, which will be unaffected by surface cleaning technologies.

Coating characterization

A qualitative characterization of coatings was done with a PMA scan at 1 µm intervals. These high-resolution particle maps identified the thin under-reported coatings and the occurrence of multi layered coatings. These 1 µm, 3000 cps, high-resolution particle maps are costly to produce, and were used as a base to determine what the new measurement conditions should be to quantify and characterize these coatings.

High-resolution QEMSCAN analyses clearly detect the different mineral coating phases attached to the target particles, as illustrated in Figure 1. Analyses at a lower resolution detect combinations of the mineralogical compositions of these layered coatings, thus adversely influencing the identification of these coatings and therefore potentially influencing the measurement of surface cleaning efficiency.

Two case studies are presented with a PMA scan at 3 µm intervals, 1000 cps. Case study 1 evaluates the benefit of an additional stage of zircon attrition scrubbing; case study 2 compares the efficiencies of two very different surface cleaning technologies on ilmenite.

CASE STUDY 1

Background

In this case study a zircon-rich concentrate undergoes two-stage attrition scrubbing (referred to as stage 1 and stage 2). The primary purpose of the scrubbing process is to reduce surface coatings to improve the efficiency of the next process, which is the hot acid leach. The thin coatings on these economic particles constitute pacifying layers and limit effective penetration by the acid, which adversely affects the downstream circuit efficiencies. Previous plant trials indicated that longer scrubbing times would improve downstream recoveries (Maluleke *et al.*, 2011). A third stage of attrition scrubbing was implemented for a trial period.

As it is difficult to isolate the impact of a single process change in a plant environment, this case study provided an ideal opportunity to demonstrate the ability of QEMSCAN to directly measure thin coating reduction. The increase in attrition scrubbing energy could be directly correlated with decrease in particle surface impurities.

Basic feed characterization

From the simplified mineralogy (Table I), it can be seen that there are various minerals present in the feed material. It is therefore impossible to simply measure the reduction in problematic elements by XRF analysis, since these elements form part of the economic mineral structures (SiO_2 in zircon), as well as various gangue minerals. QEMSCAN can isolate specific mineral particles of interest, zircon in this case, and evaluate the reduction in their specific coatings. The zircon particles are therefore classified into different 'purity' classes, with zircon I being the purest and zircon II being less pure with more associated gangue coatings and/or inclusions, as illustrated in Figure 2. The classification criteria for the zircon particle categories can be seen in Table II.

Table I. Simplified mineralogical composition of the feed to the three-stage attrition scrubbing process.

Mineral grouping	Mass %
Zircon	58.9
Total rutile	17.1
Total ilmenite	1.8
Total gangue silicate	18.7
Total phosphate	2.8
Fe oxide	0.6
Others	0.1
Total	**100.0**

Table II. Zircon particle category criteria, used to evaluate zircon particle grades according to mineral mass percentage content and/or element-oxide content.

Particle category	Classification criteria
Zircon I	(≥98 mass% zircon) "Prime" ≥66% ZrO_2
Zircon II	(≥92 mass% zircon) ≥62% ZrO_2
Zircon III	(≥77 mass% zircon) ≥52% ZrO_2
Zircon IV	(≥17 mass% zircon) ≥25% ZrO_2

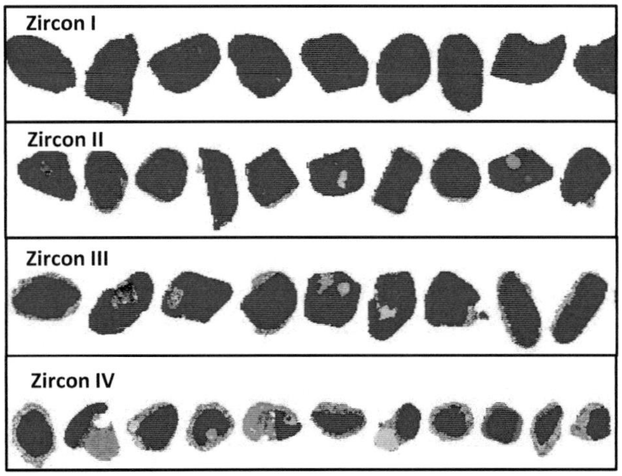

Figure 2. Particle maps of the different zircon particle categories according to the criteria set out in Table II. Zircon I represents the prime particles, with decreasing grades in zircon II – IV. Zircon particles (blue) present various coatings and inclusions; quartz (light pink), sepiolite (green), kaolinite (orange), apatite (cherry pink), and Fe-oxide-containing coatings (black).

Results

The normalized mass (%) zircon results reported in Table III indicate that the percentage of pure zircon particles (zircon I) increased while the percentages of the less pure particles (zircon II, III and IV) all decreased with a third stage of attrition scrubbing. Sufficient amounts of surface impurities were removed on these particles; in other words, the particles were upgraded and categorized according to the classification criteria in Table II. This demonstrates that the third stage of attrition scrubbing contributed to the improvement of zircon quality through further reduction of surface coatings. From previous studies, the recovery of the different zircon particle categories were tracked in all the downstream processing units and recovery to final zircon product could be closely correlated to the mass percentage in each particle category. The normalized mass (%) values in Table III could be easily converted to recoveries at each stage, but this information is not disclosed owing to the confidentiality of these values. This recovery improvement could be evaluated against the capital and operating costs (electricity and impeller wear) of an additional attrition scrubbing stage.

Table III. Normalized mass (%) zircon distribution across zircon particle categories.

Particle category	Feed	Stage 1	Stage 2	Stage 3
Zircon I	71.1	74.8	77.4	80.3
Zircon II	22.5	20.3	18.3	16.4
Zircon III	5.7	4.1	3.6	2.7
Zircon IV	0.7	0.9	0.7	0.6

Different types of coating behave differently during attrition scrubbing, hence the importance of coating characterization. Table IV quantifies the mass percentages of the main four coating types attached to zircon. These types are illustrated in Figure 1. The sepiolite and Fe oxide coatings were significantly reduced by 65% and 71% respectively during the three scrubbing stages, while the kaolinite and quartz coatings were only reduced slightly (by less than 10%). This coating removal response indicates possibly multi layered coatings, with Fe oxide and sepiolite coatings being removed first. Hence the different origins of coating minerals; clays being weathering products and quartz coatings precipitates from fluids (Deer, Howie, and Zussman, 1992), the durability's differ and cause different responds in attrition scrubbing. Attrition scrubbing had little impact on quartz coatings.

Table IV. Coating type and mass% on zircon particles during consecutive attrition scrubbing stages. Colours refer to the particle coatings in Figure 1.

Coating type	Feed	Stage 1	Stage 2	Stage 3
Sepiolite (*green*)	0.19	0.13	0.09	0.07
Kaolinite (*orange*)	0.05	0.04	0.04	0.04
Fe-oxide (*black*)	0.08	0.05	0.04	0.02
Quartz (*pink*)	0.11	0.10	0.09	0.09
Total coating	**0.43**	**0.32**	**0.26**	**0.22**

CASE STUDY 2

Background

Attrition scrubbing is applied to cleaning of ilmenite surfaces prior to electrostatic separation and smelting. A significant coating level is required for ilmenite particles, which are highly conductive, to be misplaced in electrostatic separation and this is therefore not a major concern. However, thin coatings on the conductive ilmenite particles increase the SiO_2 content in the final product and therefore reduce the mass yield to final product. This in turn increases SiO_2 levels in the titanium slag, the high-value product of the smelting process. These thin quartz-rich coatings on the ilmenite grains are commonly referred to as 'associated silica'. Several plant studies have indicated the correlation between associated SiO_2 and mass yield to final product.

This case study applied QEMSCAN to ilmenite particles to compare at a high level the coating removal efficiency between two surface cleaning methods; method 1 and method 2. The specific focus was on the quartz and other coatings contributing SiO_2.

Basic feed characterization

Table V provides the basic mineralogical composition of the feed sample for case study 2. The major mineral in the total silicate class is almandine, an Fe-rich garnet. From the mineralogical analyses reported in Table V, it can be seen that XRF, or total elemental analysis, will provide no information on the problematic Si-rich coatings on the ilmenite particles. 'Associated silica' in ilmenite is determined by performing a bench-top electrostatic separation test to isolate the silicates in a non-conducting fraction, after which the conducting fraction is analysed by XRF. The percentage SiO_2 in the conductor fraction is termed 'associated silica'. The challenge with this approach is that bench-top electrostatic separation has poor repeatability as the technique is operator-sensitive as well as sensitive to atmospheric conditions, as indicated by internal company reports.

The QEMSCAN technique is able to isolate the specific ilmenite particles and directly quantify and characterize the coatings on the total ilmenite population. As in case study 1, ilmenite particles are also classified according to 'purity' but only in two major classes, ilmenite I and ilmenite II, as illustrated in Figure 3. The classification criteria are shown in Table VI.

Table V. Simplified mineralogical composition of the feed to the two surface cleaning technologies.

Mineral grouping	Mass %
Total ilmenite	81.2
Total Fe oxide	10.5
Total silicate	8.3
Total	**100.0**

Table VI. Criteria, used to evaluate ilmenite particle grades.

Particle category	Classification criteria
Ilmenite I	($\leq$45 TiO_2 <55 mass%) alt-ilmenite and ilmenite $\geq$ 50 mass% and <5 mass% gangue
Ilmenite II	($\leq$45 TiO_2 <55 mass%) alt-ilmenite and ilmenite $\geq$ 50 mass% and <15 mass% gangue

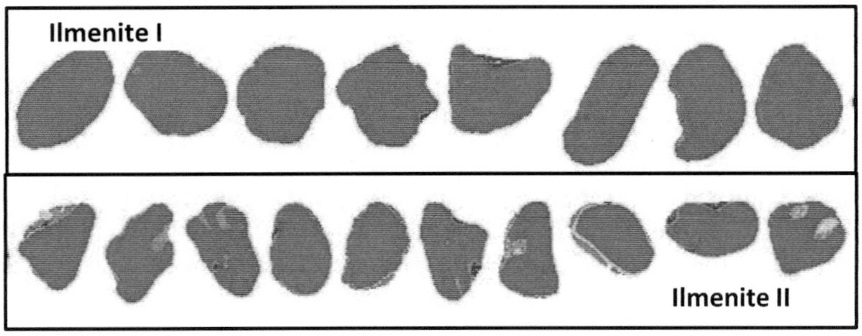

Figure 3. Particle map of the different ilmenite particle categories according to the criteria set out in Table VI. Ilmenite I represent the prime ilmenite particles, with decreasing grade for ilmenite II. Ilmenite particles (cherry pink) presenting various coatings and inclusions; quartz (light pink), sepiolite (green), kaolinite (orange), and Fe- oxide-containing coatings (black).

Results

The normalized mass (%) ilmenite results reported in Table VII illustrate that both methods increased the percentage of purer ilmenite particles (ilmenite I), while the less pure particles decreased (ilmenite II). Sufficient amounts of surface impurities were removed from these particles for them to be reclassified according to the criteria in Table VI. The purity increase with method 2 was more than double that of method 1, measured as difference between the normalized mass (%) ilmenite in the feed and that obtained by the two methods.

Table VII. Normalized mass (%) ilmenite distribution across ilmenite particle categories.

Particle category	Feed	Method 1	Method 2
Ilmenite I	91.9	92.3	92.8
Ilmenite II	8.1	7.7	7.2

Although the ilmenite particle quality increases in this case study was less than the zircon particle quality increases in case study 1, the coating levels (Table VIII) were three times greater than those on the zircon. Method 1 achieved only a 5% coating reduction, while method 2 achieved a 32% coating reduction. This result stand in contrast with the particle quality improvement result in Table VII, and it is therefore important to evaluate the particle impact from two different angles: (1) particle quality improvement and (2) coating reduction. The quartz coating, which is recognized as hard to remove, was significantly reduced with method 2, while method 1 was unable to reduce the quartz coating.

Table VIII. Coating type and mass% on ilmenite particles for different surface cleaning methods.

Coating type	Feed	Method 1	Method 2
Sepiolite (*green*)	0.26	0.20	0.08
Kaolinite (*orange*)	0.86	0.85	0.73
Fe-oxides (*black*)	0.12	0.11	0.09
Quartz (*pink*)	0.55	0.55	0.34
Total coating	**1.80**	**1.71**	**1.23**

CONCLUSIONS

Thin coatings on economic heavy mineral particles can cause losses in recovery by influencing the behaviour of these particles in the different separation environments, and thus decrease the grade or yield of the product. It is therefore important to characterize these coatings accurately at higher resolutions when evaluating surface cleaning technology efficiency.

The ability of QEMSCAN to distinguish the relevant particle classes and quantify the amount and type of coating on the particle surface provides valuable information for surface cleaning technology evaluation. Elemental analyses of total particle classes are unable to isolate the contribution of the problematic surface impurities. QEMSCAN reports quantitative mineralogy, which entails mineral compositions and discrete distributions that can be quantified on a particle-by-particle basis.

The impact of the surface cleaning technology on particle surfaces should be evaluated from two different angles - firstly, the increase in the amount of pure economic particles; and secondly, the decrease in the total coating percentage on the total economic particle class.

Different coating types have different origins and therefore significantly different behaviours during surface cleaning. The characterization and quantification of their content during different stages of cleaning provides valuable insight into the removal efficiency as well as the expectation from the cleaning technology (maximum achievable coating removal).

REFERENCES

Adams, M.D. (2007). Towards a virtual metallurgical plant 2: Application of mineralogical data. *Minerals Engineering*, 20, 472 - 479.

Deer , W.A., Howie, R.A., and Zussman, J. (1992). *An Introduction to the Rock-forming Minerals*, 2nd edn. Prentice Hall/Pearson Education UK.

Gottlieb, P., Wilkie, G., Sutherland, D., Ho-Tun, E., Suthers, S., Perera, K., Jenkins, B., Spencer, S., Butcher, A., and Rayner, J. (2000). Using quantitative electron microscopy for process mineralogy applications. *Journal of Materials*, 52 (4), 24 - 25.

Maluleke, O.S. and Grobler J.D. 2011. Attrition scrubber performance evaluation. *Proceedings of the Eighth International Heavy Minerals Conference*. Australasian Institute of Mining and Metallurgy, Melbourne. pp. 185–196.

Annabé Walliser (Johanna Elizabeth)

Mineralogist
Exxaro

Electron microprobe analyst for 12 years at the Council for Geosciences and then joined Exxaro Resources in 2008 as Mineralogist. Responsible for Heavy Mineral Sands Characterisation for the last 8 years.

Experience in Base-metal, Fe-ore characterisation and multiple SEM-investigations on Off-specification products and Refractory brick failures.

Also a Coal accredited petrographer in the Single coal program, through the International Committee for Coal and Organic Petrology (ICCP), since 2010.

Development of an enhanced heavy mineral spiral

C. Ramotsabi

Multotec Process Equipment (Pty) Ltd

Spiral concentrators are simple low-energy consuming devices that separate minerals mainly on the basis of density. In the mineral sands industry, spirals are mainly used in the primary concentration stage for low-grade beach sand orebodies. The heavy mineral (HM) content usually ranges between 0.5% and 15%. The product specification requirement is a selective recovery of between 85 and 98% HM concentrate. The limitation on spirals is that high selectivity is required to separate near-dense materials from one another. Selectivity between high density and medium density remains a challenge.

The study investigated the influence of various spiral design parameters and characteristics of slurry flow on recovery of economic heavy minerals (EHMs) and selectivity of separation on a spiral. The test work was conducted on the Multotec SC21/5, NHM, and the 12-turn prototype spirals. The 12-turn spiral is a combination of the NHM and the SC21/5. In this spiral the configuration is such that the SC21/5 is on top and the NHM at the bottom. The 12-turn was also fitted with detachable tails diverters and the effect on grade and recovery of EHMs evaluated. The effect of solid concentration and solid feed rate on grade and recovery of EHM was also evaluated.

The 12-turn spiral showed potential in providing a good compromise between grade and recovery of EHMs in the rougher application. Recoveries of EHMs as high as 99.40% were achieved at a mass yield of 10% and a grade of 31.98% EHM. These results were achieved at 2 t/h and 40% solid concentration. The preliminary results suggest that, individually, the features of the SC21/5 enhances grade while the NHM enhances recovery of the EHMs, consequently a combination of both spirals provides both enhanced grade and recovery.

INTRODUCTION

Spiral concentrators are simple low-energy consuming devices that separate minerals mainly on the basis of density. Spirals are widely used in mineral processing as a method for preconcentration and have proven to be metallurgically efficient and cost-efficient. Spirals have a wide application; they are used in coal, gold, chromite, mineral sands, glass sands, iron ore processing, and in soil cleaning.

In the mineral sands industry, spirals are used for preconcentration of low-grade heavy mineral (HM) material, typically 0.5% to 15% HMs. The requirement in wet concentration plants is to recover as much as possible of the HMs at a concentrate grade between 85 and 98% HMs while minimizing gangue in the concentrate (Jones, 2009.). The challenge in processing of HM sands is that the spiral is not selective; some gangue in heavy mineral concentrate (HMC) is dense, and since a spiral separates on density the economic heavy minerals (EHMs) and gangue heavy minerals (GHMs) are both recovered as concentrate. Selectivity between high density and medium density remains a challenge. Spirals are also limited in that they generally have higher Ecart probable (Ep) values than other gravity concentrators.

The Tenth International Heavy Minerals Conference
Sun City, 16–17 August 2016
Southern African Institute of Mining and Metallurgy

Separation on a spiral is achieved through a combination of forces that act on particles as they move down the trough of the spiral. The main forces known to act on the particle on a spiral are the gravitational forces, centrifugal force, hydrodynamic drag, and lift and friction forces (Kapur and Meloy, 1998). Apart from the forces acting on a spiral, the design parameters that govern the performance of the spiral can be identified as the diameter, height, number of turns, pitch and slope, and shape of the trough. The properties of the slurry flowing on a spiral, including ore texture, percentage solids, particle size distribution, particle shape, density distribution, and flow rate, also play an important role in the separation on the spiral.

The aim of this study is to investigate the influence of various spiral design parameters and characteristics of slurry flow on recovery of HMs and selectivity of separation on a spiral. The advantage of a variable pitch will also be investigated. The results are anticipated to provide the best spiral design to optimize recovery of the EHMs and increase selectivity of separation with near-dense material.

LITERATURE REVIEW

The variety of minerals treated on spirals has brought about the need for a detailed understanding of the separation mechanism and the choice of different components for a spiral. Spirals available in the market possesses different design parameters for a variety of applications aimed at addressing various needs of the industry, and a great deal of development work has been carried out to improve process efficiency. Improving the performance of spirals is done by assessing and optimizing the design parameters while ensuring maximum recovery of minerals and minimizing costs.

When separation occurs on the spiral, the heavy particles migrate toward the inner region of the trough where they become concentrated. As a result, the ease of flow of material in this region becomes reduced. This can hinder the further separation of particles on a spiral without wash water. This necessitates the use of splitters to remove the material that has already separated from the trough before they start to 'beach'. There are different types of splitters available, *viz.* small finger splitters, banana splitters, and slide splitters. Continuous removal of the concentrate along the length of the spiral provides more area for further separation. The reduced fluidity in the inner region of the trough can be restored by employing repulpers, which spray back the dilute material from the outside of the trough in a shallow fan over the inner part of the trough. This also provides the misplaced HMs on the outside of the trough with another chance of separation. A schematic of a repulper in action is depicted in Figure 1.

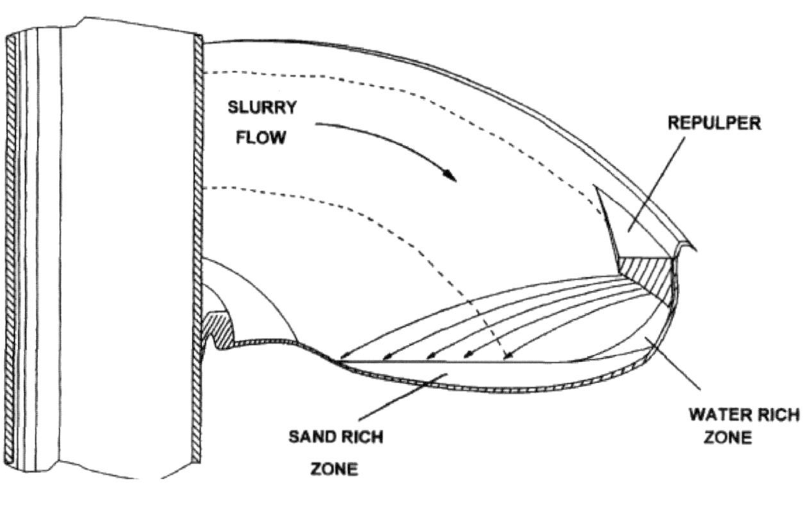

Figure 1. Schematic of a repulper in action (Holland-Batt, 1995).

Holland-Batt (1995) demonstrated that removal of product followed by employing a repulper was effective in restoring the initial higher rate of recovery. The relationship between separation efficiency and feed rate was shown to be linear, with a decrease in efficiency at higher flow rates. Previous work on flow profile by Holland-Batt (1990) indicated that any increase in the feed rate to a spiral results in a greater part of the additional flow volume reporting to the outside of the trough. The flow on the outside of the trough becomes energized and consequently the heavy particles are hindered from settling and escaping from this region to the inner region. This has a high impact on the recovery of heavy minerals, since any increase in volumetric flow on the spiral is not made available for recovery in the inner region of the spiral.

Figure 2 illustrates TiO_2 recovery as a function of feed rate on an FM1 spiral. The highest recovery is achieved at lower feed rates, and recovery decreases as feed rate increases. This figure also shows the impact of percentage solids in the feed on recovery of TiO_2, with higher recoveries in the 36% to 40% solid concentration range.

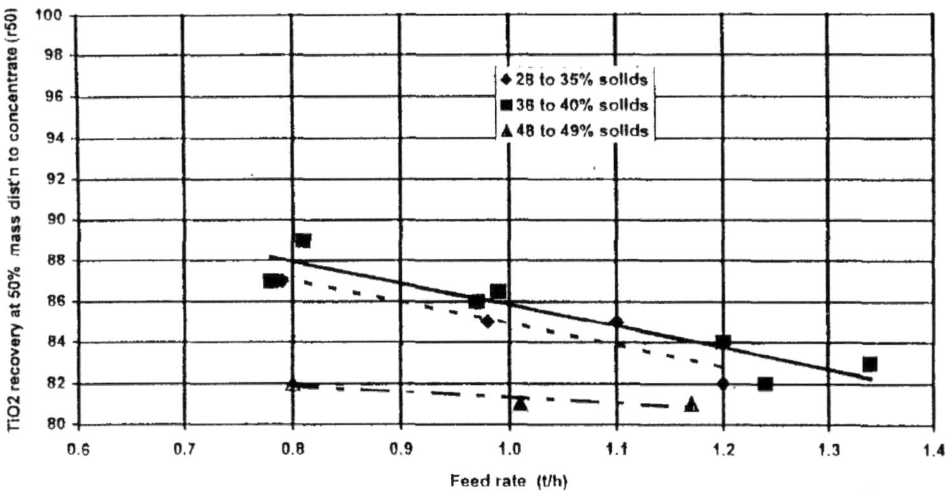

Figure 2. FM 1 spiral performance on fine-grained Australian mineral sand – effect of feed rate on TiO_2 recovery (Richards and Palmer, 1997).

Control of solid concentration on a spiral is very important as it influences separation and recovery. Recovery on a spiral has been shown to increase with increasing solid concentration up to a point where it is optimum, and then drops at relatively high solid concentrations.

EXPERIMENTAL

Material

The sample was classified as low grade with an average head grade of 7.19% THMs, with only 2.77% measured and calculated as EHMs (ilmenite, rutile, and zircon). The sample was blended and a representative sample removed for particle size distribution (PSD) determination and size by assay. The results of the size by assay of the sample are shown in Table I. The PSD shows less than 5% of the 38 µm fraction, which indicates that the slime content was not significant in the sample. Rhe sample was therefore processed as-is on the spirals without desliming upfront.

Table I. Feed sample assay by size fraction.

Size (μm)	Average size (μm)	% mass retained	Cum% retained	% THM	Cum % THM	% THM distribution	Cum % THM distribution	% floats	Cum % floats	% floats distribution	Cum % floats distribution
-850+600	725	0.59	0.59	2.63	2.63	0.21	0.21	97.37	97.37	0.62	0.62
-600+425	513	3.94	4.53	3.23	3.15	1.77	1.98	96.77	96.85	4.11	4.73
-425+300	363	31.51	36.04	2.44	2.53	10.70	12.69	97.56	97.47	33.12	37.85
-300+212	256	45.61	81.65	3.02	2.80	19.14	31.83	96.98	97.20	47.66	85.51
-212+150	181	13.06	94.71	11.58	4.01	21.01	52.84	88.42	95.99	12.44	97.96
-150+106	128	3.79	98.50	59.26	6.14	31.23	84.07	40.74	93.86	1.66	99.62
-106+75	91	1.31	99.81	84.84	7.17	15.44	99.51	15.16	92.83	0.21	99.83
-75+53	64	0.11	99.92	29.54	7.20	0.44	99.96	70.46	92.80	0.08	99.92
-53+38	46	0.05	99.97	6.22	7.20	0.04	100.00	93.78	92.80	0.05	99.96
-38	19	0.03	100.00	0.34	7.19	0.00	100.00	99.66	92.81	0.04	100.00
		100.00		**7.19**		**100.00**			**92.81**	**100.00**	

Methods

The test work was conducted on the Multotec SC21/5, NHM, and 12-turn prototype spirals fitted with a mouth organ product box as shown in Figure 3. Mouth organ (MO) fraction A represents the sample collected from the inside of the trough. It is the heaviest sample and should contain most of the HMs.

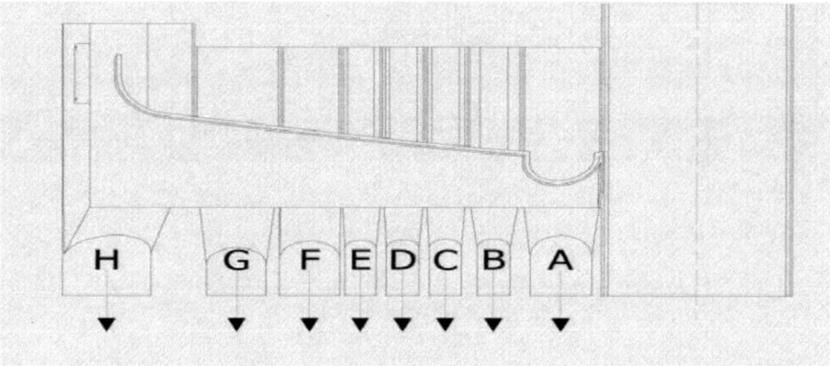

Figure 3. Mouth organ configuration.

The 12-turn spiral is a combination of the NHM and SC21/5 spirals. The SC21/5 is positioned on top of the NHM spiral.

Figure 4(a) shows a photograph of a tails diverter in action. Two tails diverters were designed for two positions on the NHM spiral - after the first 3.5 turns and 1.5 turns after the first diverter. The tails diverter design was such that the diverter could be clamped on and removed without changing the trough surface and to allow for the investigation of the effect of using one or two diverters. Figure 4(b) shows the distribution of the minerals on the mouth organ product box.

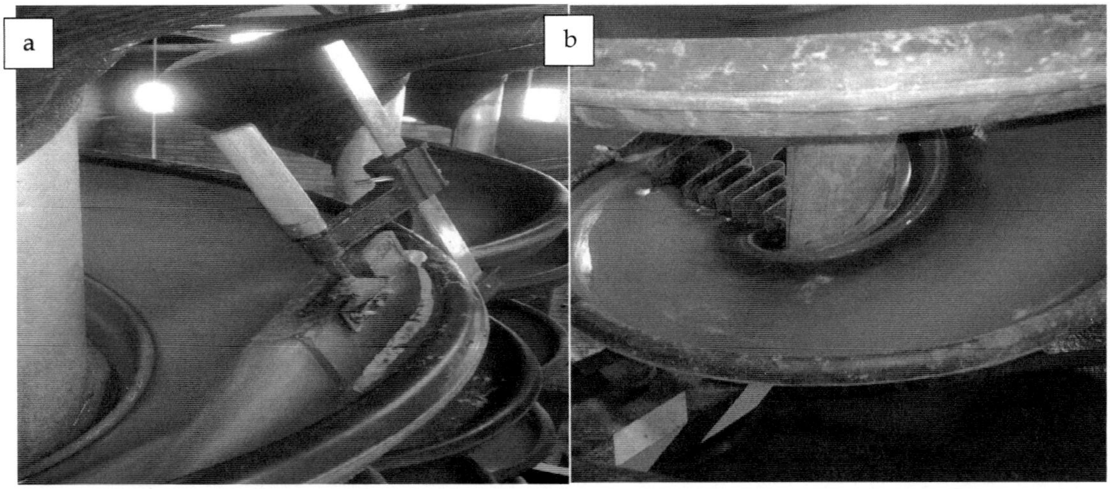

Figure 4. (a) Tails diverter in action; (b) HM band on last turn following a tails diverter.

The spiral was operated in closed circuit as shown in Figure 5. The slurry was fed to the spiral feed box via a distributor with an overflow to ensure constant feed conditions during the test.

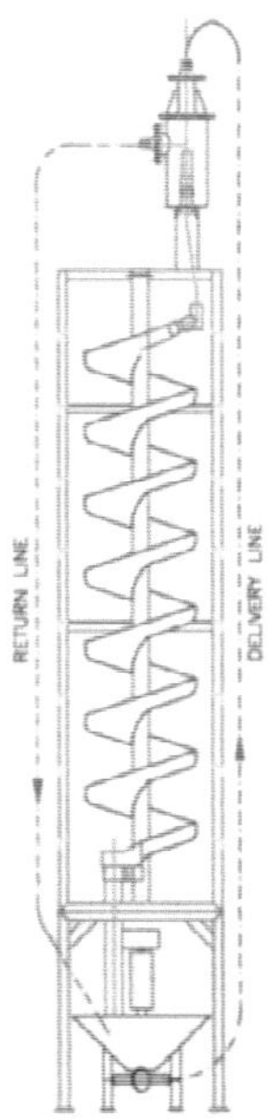

Figure 1. Spiral test rig set-up schematic.

Each spiral stage was run until steady-state conditions were reached before relative density (RD), solid concentration, and feed rate were measured at the top of the spiral rig. Three consistent RD measurements and three consistent flow rate measurements were recorded before taking simultaneous timed samples (fraction A to H) from the mouth organ, as shown in Figure 3.

Experimental Plan

The test plan is shown in Table II. A benchmark was established by testing the SC1/5, NHM, and the 12-turn spiral. The effect of the variable spiral profile on the recovery of HMs was evaluated by testing the 12-turn spiral. The 12-turn spiral was also fitted with removable tails diverters. Solid concentration was varied from 30% to 40% to establish the effect of varying concentration on recovery on the NHM spiral. The feed rate was also varied from 1.5 t/h to 2.5 t/h to establish the rate suitable to yield optimum recovery on the NHM.

Table II. Test matrix.

	Test parameters		
Spiral Types	**Solid concentration (wt%)**	**Tails diverter**	**Solid feed rate (t/h)**
SC21	35	0	2
NHM	35	0	2
12-turn	35	0	2
12turn	35	0	2
12-turn	40	0	2
12-turn	45	0	2
12-turn	40	0	1.5
12-turn	40	0	2
12-turn	40	0	2.5
12-turn	35	1 at pos. 1	2
12-turn	35	1 at pos. 2	2
12-turn	35	2 (pos. 1 and pos. 2)	2

Limitations

The sample available for this work was insufficient for completion of the test matrix as initially planned and to check repeatability of the results. As a result only two points were tested for solid concentration variation, 35% and 40%; solid feed rate, 2 and 2.5 t/h; one variation of the tails diverter was tested. A compromise had to be made to perform selected tests, and for each test performed only a single run was conducted.

Sampling and Analysis

Mouth organ product samples from each spiral test were collected and analysed by heavy liquid separation (HLS) for THM and float content. HLS was carried out at density of 2.85 using TBE and acetone. Approximately 800 ml of the heavy liquid was poured into a glass funnel and about 250 g of sample added to the heavy liquid, the mixture was stirred for 1 minute and left for 10 minutes to settle. Sinks were then tapped off into a glass beaker and the floats stirred again for 1 minute and allowed another 5 minutes for more sinks to settle. The sinks were then washed, dried, and weighed, then pulverized and submitted for major elements by borate fusion X-ray fluorescence (XRF). A split aliquot of the pulverized material was analysed by X-ray diffraction (XRD) to determine major mineral phases, with quantification by Rietveld refinement (guided by the assay data). The samples were analysed using a Panalytical X'pert Pro diffractometer, employing Co radiation. The data was processed using Highscore Plus analytical software. The limit of detection for phases with low crystallinity or concentrations was 3%.

To ensure a fair comparison, the results from various tests are compared at a constant mass yield of 10%. The concentrations of ilmenite, rutile, and zircon were combined to form the EHM.

RESULTS

Benchmark

Figure 6 shows a recovery (THMs and floats) *versus* yield relationship for the SC21/5, NHM, and the 12-turn spirals.

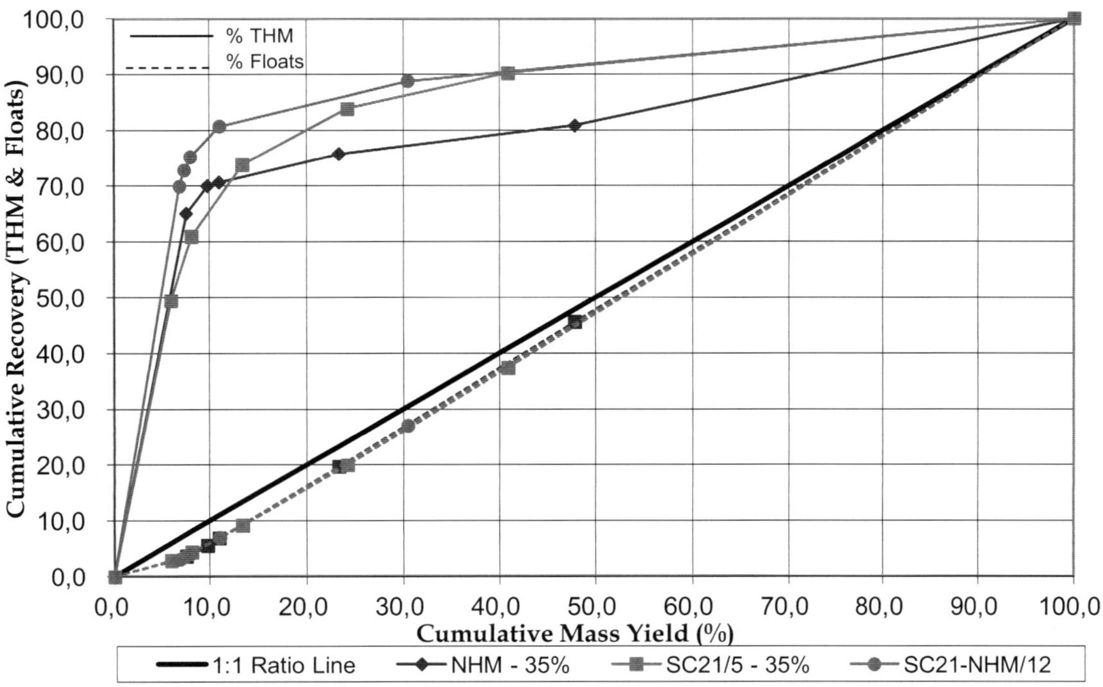

Figure 6. Comparison between the SC21/5, NHM, and the 12-turn spiral at 2 t/h and 35% solids and 2 t/h in the feed.

The results show a higher recovery of the THM on the 12-turn spiral than on the SC21/5 and the NHM spirals. At a mass yield of 10% to the concentrate, the delta in recovery between the three spirals is 14%.

Figure 7 shows a comparison of the SC21/5, NHM, and 12-turn spiral. For this interpretation the results were compared at a mass yield of 10%. The results indicate that the EHMs were as upgraded to 29.39%, 28.13%, and 26.14% with the SC21/5, NHM, and the 12-turn spirals respectively. The corresponding EHM recoveries were 90.08%, 97.95%, and 98.82%.

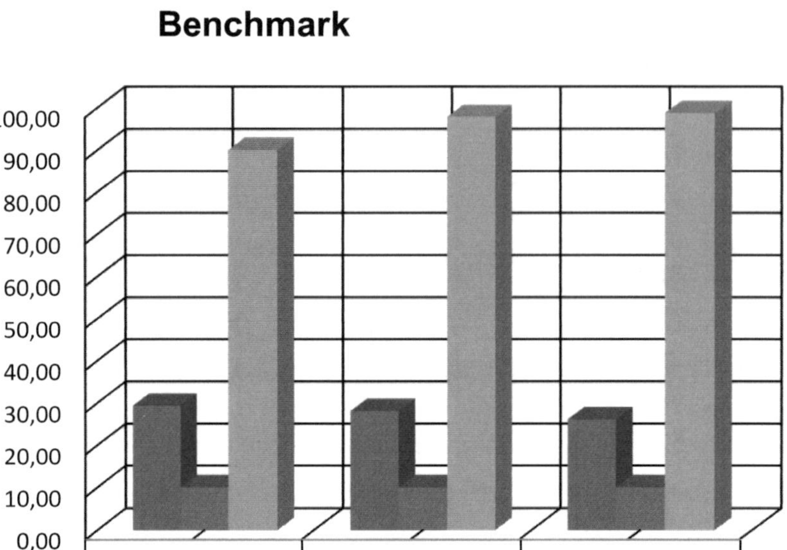

	SC21	NHM	12 turn
% EHM Grade in Concentrate	29,39	28,13	26,14
% Mass Yield	10,00	10,00	10,00
% EHM Recovery	90,08	97,95	98,82

Figure 7. Mass yield vs EHM grade and recovery – comparison between the SC21/5, NHM, and the 12-turn spiral at 35% solids and 2 t/h.

The results suggest that, individually, the features of the SC21/5 enhances grade while the NHM enhances recovery of the EHMs, consequently a combination of both spirals provides both enhanced grade and recovery.

Effect of Solid Concentration on Grade and Recovery on the 12-turn Spiral

Figure 8 illustrates the effect of solid concentration on EHM grade, yield, and recovery on the 12-turn spiral. EHM grades of 26.14% and 31.98% were achieved at 35% and 40% solid concentration respectively. The corresponding recoveries were 98.82% and 99.40%.

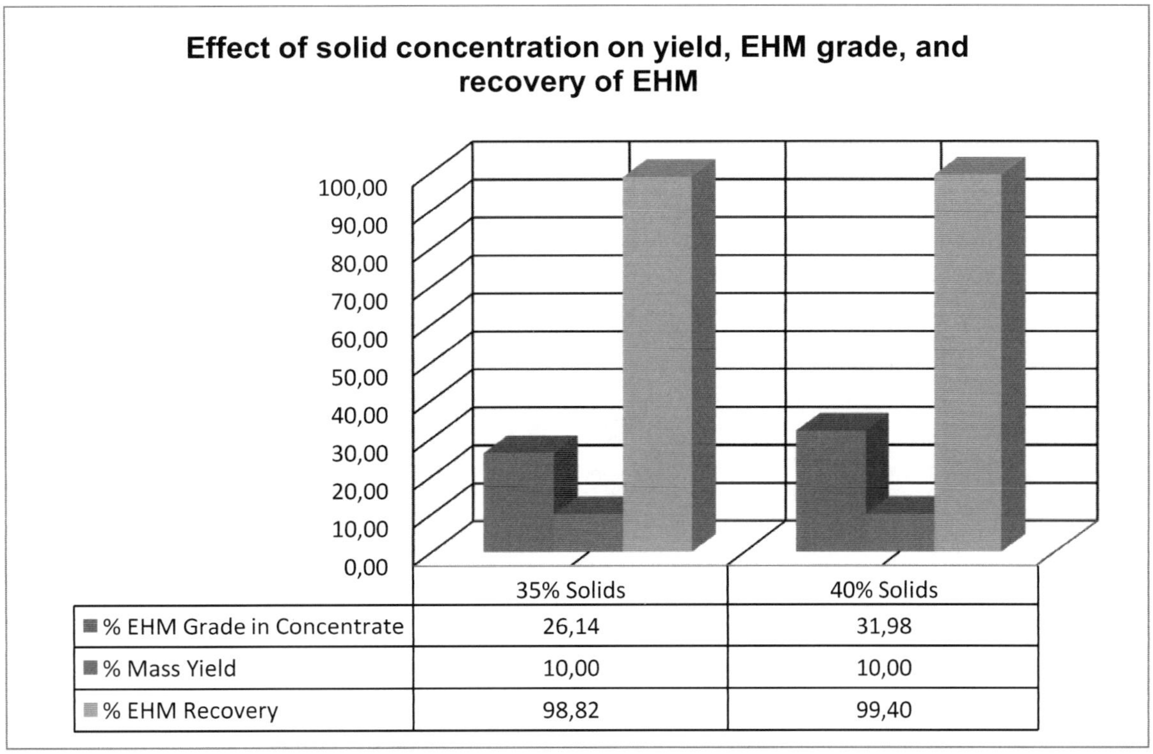

	35% Solids	40% Solids
% EHM Grade in Concentrate	26,14	31,98
% Mass Yield	10,00	10,00
% EHM Recovery	98,82	99,40

Figure 8. Effect of solid concentration on mass yield, EHM grade, and recovery on the 12-turn spiral at 2 t/h with no tails diverter.

The highest grade of EHMs is, surprisingly, achieved at higher solid concentration. An increase in solid concentration on the spiral is known to increase gangue entrapment and is usually expected to result in a decrease in grade of the product. An increase in solid concentration also resulted in an increase in EHM recovery.

Effect of Solid feed Rate on EHM Grade and Recovery on the 12-turn Spiral

Figure 9 demonstrates the effect of solid feed rate on EHM grade and recovery on the 12-turn spiral. The material was upgraded to 31.98% and 27.70% EHM in the concentrate; these results were achieved at 2 t/h and 2.5 t/h solid feed rate respectively. The corresponding EHM recoveries were 99.40% and 98.05%. The highest grade and recovery were achieved at 2 t/h.

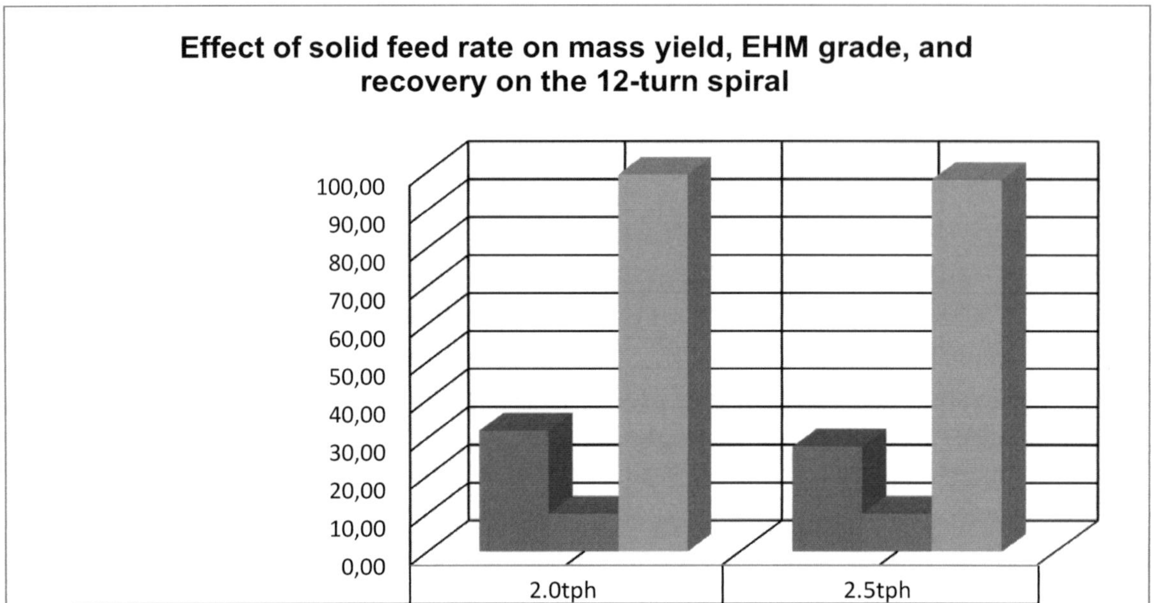

	2.0tph	2.5tph
% EHM Grade in Concentrate	31,98	27,70
% Mass Yield	10,00	10,00
% EHM Recovery	99,40	98,05

Figure 9. Effect of feed rate on mass yield, EHM grade, and recovery on the 12-turn spiral at 40% solids and no tails diverter

The findings are in line with previous work on flow profile by Holland-Batt (1990), showing that grade and recovery decrease with an increase in solid feed rate. This suggests that the increase in particles on the trough without any adjustment in sliding splitters does not result in additional recovery to the concentrate and that the additional solids on the spiral decreases the surface area available for particles to move freely and consequently results in gangue entrapment.

Effect of Tails Diverters on EHM Grade and Recovery on the 12-turn Spiral

Figure 10 shows the effect of using tails diverters on EHM grade and recovery on the 12-turn spiral.

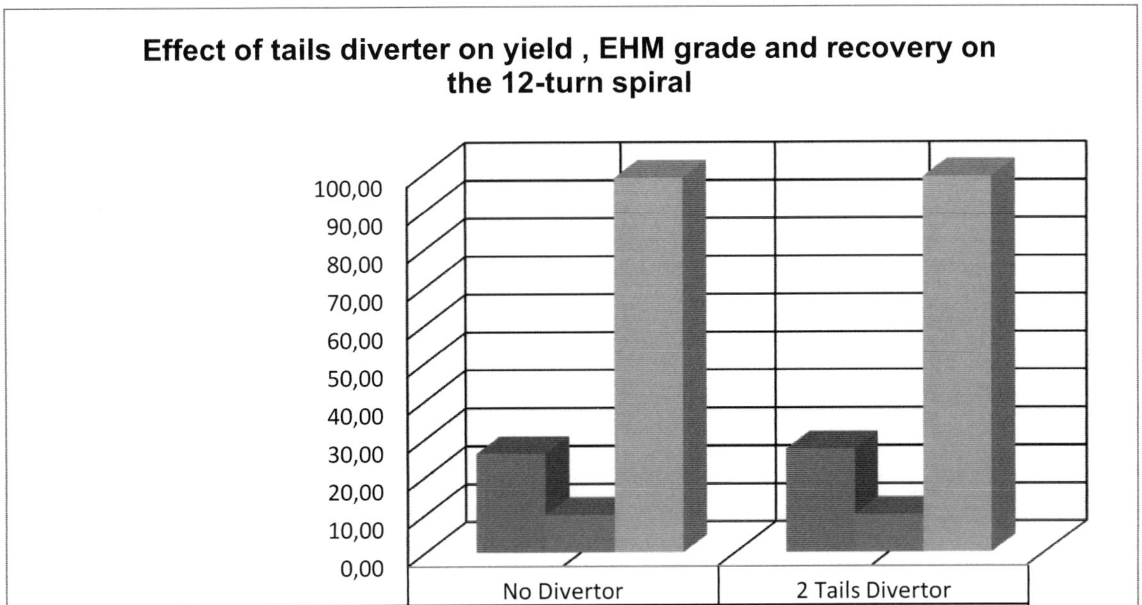

	No Divertor	2 Tails Divertor
% EHM Grade in Concentrate	26,14	27,29
% Mass Yield	10,00	10,00
% EHM Recovery	98,82	99,14

Figure 10. Effect of solid concentration on mass yield, EHM grade, and recovery on the 12-turn spiral at 35% solids and 2 t/h.

EHM grades of 26.14% and 27.29% in the concentrate were achieved without and with the use of the tails diverters respectively. The corresponding recoveries were 98.82%, 99.14% for EHMs. The use of the tails diverters had a positive effect on both grade and recovery of the EHMs.

CONCLUSIONS

The influence of various spiral design parameters and characteristics of slurry flow on recovery of economic heavy minerals (EHMs) and selectivity of separation on a spiral were investigated. The test work was conducted on the Multotec SC21/5, NHM, and the 12-turn prototype spirals. The 12-turn spiral is a combination of the NHM and the SC21/5, with the SC21/5 on top and the NHM at the bottom. The 12-turn was also fitted with detachable tails diverters and the effect on grade and recovery of EHMs evaluated. The effects of solid concentration and solid feed rate on grade and recovery of EHMs were also evaluated.

The combination of the NHM and SC21 spiral which make up the 12-turn spiral shows potential in providing a good compromise between grade and recovery of EHMs in the rougher application Recoveries of EHMs as high as 99.40% were achieved at mass yield of 10% and a grade of 31.98%. These results were achieved at 2 t/h and 40% solid concentration. The preliminary results suggest that, individually, the features of the SC21/5 enhance grade while the NHM enhances recovery of the EHM, consequently a combination of both spirals provides both enhanced grade and recovery.

Solid concentration was varied between 35% and 40% on the 12-turn spiral. The EHM grade was upgraded to 26.14% and 31.98% at 35% and 40% solid concentration respectively. The corresponding recoveries were 98.82% and 99.40%. The higher EHM grade was, surprisingly, achieved at the higher solid concentration. An increase in solid concentration on the spiral is known to increase gangue entrapment and is usually expected to result in a decrease in grade in the product.

The solid feed rate was varied between 2 and 2.5 t/h. Both EHM grade and recovery have been shown to decrease with an increase in solid feed rate. Holland-Batt (1990) and other authors in the literature have shown that grade and recovery decrease with an increase in solid feed rate. This suggests that the increase in particles on the trough without any adjustment of the sliding splitters does not result in additional recovery to the concentrate, and that the additional solids on the spiral decreases the surface area available for particles to move freely, and consequently results in gangue entrapment.

The use of two tails diverters at 35% solid concentration had a positive effect on both grade and recovery of the EHM on the 12-turn spiral. The highest grade and recovery were achieved with the two tails diverter.

The preliminary results suggests that with the 12-turn spiral, 2 t/h and 40% solid concentration result in a good compromise between grade and recovery of the EHMs. Further test work is recommended to establish the full effects of the solid concentration, feed rate, and tails diverters and sliding splitters on grade and recovery on the 12-turn spiral.

REFERENCES

Holland-Batt, A.B. (1990). Interpretation of spirals and sluice tests. *Transactions of the Institution of Mining and Metallurgy Section C: Mineral Processing and Extractive Metallurgy*, 99, C11-20.

Holland-Batt, A.B. (1995). Some design considerations for spiral separators. *Minerals Engineering,* 8 (1/2), 3-21.

Jones, G. (2009). Mineral sands: an overview of the industry. www.Illuka.com [accessed 24 May 2010].

Kapur, P.C. and Meloy, T. P. (1998). Spirals observed. *International Journal of Mineral Processing*, 53, 15-28.

Richards, R.G. and Palmer, M.K. (1997). High capacity gravity separators - a review of current status. *Minerals Engineering*, 10 (9), 973-982.

Christina Ramotsabi

Process Engineer
Multotec

Christina Ramotsabi is a Process Engineer at Gravity Concentration Department at Multotec. Christina has 9 years working experience in the mineral processing industry and is registered with the Engineering Council of South Africa (ECSA) as a Professional Technologist. Christina is also a part time student at the University of Pretoria studying towards a Master's Degree in Project Management. She holds an honours degree in Metallurgy from University of Pretoria, a B Tech Degree in Chemical Engineering and a National Diploma in Chemical Engineering from the Vaal University of Technology.

Fine chrome beneficiation and reverse PGM upgrade of existing slimes with vertical ring WHIMS technology

W.P. Slabbert

Multotec Process Equipment, Johannesburg, South Africa

The main differences between a vertical and a horizontal ring WHIMS machine are described. Four UG2 samples with P_{80}=100 µm were delivered from four different slimes dams on the western limb of the Bushveld Complex containing gravity separation tailings. A pilot-scale Longi-Multotec LGS 500 vertical ring WHIMS unit was successfully used to upgrade chromite by 6–14% with a recovery of 44–62% at 0.4 T, and by 4–6% with a recovery of 55–72% at 1.3 T. A horizontal ring WHIMS of comparable size was not available for testing. The broad range of grades and recoveries are believed to result from increased viscosity effects attributed to unreacted PGM flotation chemicals present in some samples. The viscosity effects were less pronounced at higher background magnetic field intensities, and this needs more investigation. The reverse upgrade of PGMs in the WHIMS tailings stream, through chromite removal to the magnetic concentrate stream, resulted in PGM grades of up to 6.5 ppm at a recovery 50%, which could make it economical to re-float this material, thereby increasing revenue for existing operations.

INTRODUCTION

Chromite $(Fe,Mg)(Cr,Al)_2O_4$ is a paramagnetic mineral that occurs as a gangue mineral associated with platinum group metal (PGM) orebodies in the western limb of the Bushveld Complex. Chromite grades vary between 3–5% in the Merensky Reef and the Platreef, and 5–15% in the UG2 reef (Engelbrecht, 2012).

The chromite could adversely influence flotation of the PGM minerals, and recent flow sheet configurations use gravity separation spiral circuits to remove chromite before flotation. This can be done safely without losing PGMs to the spiral heavy concentrate, since the PGM-bearing silicates are normally liberated from the chromite for UG2 material (Engelbrecht, 2012). Post-flotation spiral circuits are typically installed to separate the chromite - now the valuable product - from the PGM-stripped silicates which are no longer economic to float, even though PGMs might still be present in concentrations of less than 5 ppm.

A large number of slimes dams exist in the area with typical chromium and PGM grades of 10–25% and 1–5 ppm respectively (estimates from experience with the samples tested as presented in this paper).

Even with desliming and advancements in spiral diameter, pitch and slope, like in the HX5 spiral, these spirals are believed to be limited in effectively treating particles finer than 100 µm. The post-flotation particles are typically 5–300 µm (Engelbrecht, 2012).

There are vast amounts of fine chromite available in the slimes dams that currently cannot be effectively recovered. Beneficiation of this material holds potential economic benefits of cost savings, since mining, milling, and separation have already been done. A spin-off advantage could be the reverse upgrade of the far more valuable PGM minerals, where flotation could be considered feasible for streams with PGM concentrations of more than 6 ppm.

Wet high-intensity magnetic separation (WHIMS) is able to recover 5 µm haematite particles with 50% efficiency (Forssberg and Kostkevicius, 1982) and can also be applied to other paramagnetic minerals such as chromite.

LITERATURE SURVEY

Chromite Recovery with WHIMS

Guney *et al.*, as referenced by Murthy *et al.* (2011), used WHIMS and column flotation to beneficiate chromite tailings from a Turkish gravity separation circuit with a particle size 83% smaller than 100 µm. The tailings were successfully upgraded from 20–22% wt.% Cr_2O_3 to 45–48%, with an overall recovery of 60–65%.

Tripathy *et al.* (2013) reported chromite upgrades from 26.4% to 40.1% Cr_2O_3 with 37.6% recovery using WHIMS on an Indian chrome plant tailings stream.

Background Field Intensity *versus* Field Intensity in the Matrix

The magnetic field intensity generated by an electromagnetic coil wrapped around a core is carried to the area of separation (the matrix) by a yoke of suitable material; typically mild steel. The shape, geometry, and electromagnetic properties of the matrix material affect the way the background field intensity is concentrated in the matrix, which will result in different field intensities and gradients experienced by the particle - ultimately affecting the force applied to the particle.

In some vertical ring WHIMS machines the background intensity can be as high as 1.3 T. This intensity is sufficient to allow a concentration factor of 2 in the matrix (where the background intensity is 0.5 or 1.0 T, the intensity in the matrix would be 1.0 or 2.0 T respectively) - except for the highest intensity setting of 1.3 T where the concentration factor is only 1.9 and the intensity in the matrix 2.5 T. A magnetic intensity diagram of the field intensity distribution can be seen in Figure 1.

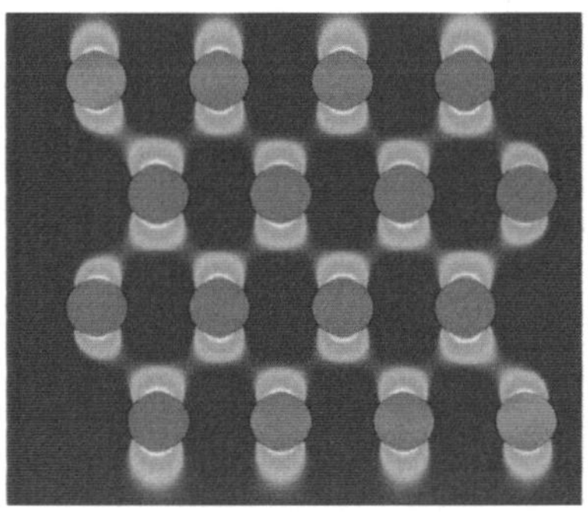

Figure 1. Magnetic field Intensity of a typical vertical ring WHIMS matrix showing a background intensity (dark blue) of 1.0 T and field intensity concentration of up to 2.0 T (red), with intermediate intensities of 1.7 T (yellow), 1.35 T (green), and 1.15 T (light blue).

Horizontal ring WHIMS models can generate up to 1.0 T (maybe more). The field intensities on the surface of the profiled plate matrix are expected to be of the same order of magnitude as in the vertical ring WHIMs at 1.7 T.

Comparison of Vertical and Horizontal Ring WHIMS

All WHIMS machines (or high-gradient separators) have the following major parts:

- A feed arrangement (with or without oversize protection) that presents the feed slurry to the separation matrix, and an associated level control system
- The rotor or ring that contains the separation matrix which governs the separation method, the rotor's drive system, and product collection
- The matrix, where separation occurs on a specific design surface comprising cylinders, rods, profiled plates, steel wool, or other surfaces
- The yoke, which is generally a sufficiently large piece of mild steel that 'conducts' the magnetic field from the coil to the rotor, where it is presented in a proprietary field distribution
- The magnetic coil(s) that generates the magnetic field, and an associated heat exchange system
- A product stream collection arrangement with a wash water system to wash the products from the matrix
- Control panel.

Figure 2 and Figure 3 depict the layout of the major components of vertical and horizontal ring WHIMS machines.

Feed Arrangement and Level Control

The vertical ring WHIMS is fed on both sides of the rotor via two feed inlets. Various oversize protection devices can be installed at the top of the inlets. The chambers are joined again deeper in the machine in the slots in the yoke and the slurry-containing tank (Figure 2), so reasonable imbalances between the two incoming feed rates are eliminated in the common tank. It is also possible to install level control on the visible overflow weir of the machine to control the feed and/or product discharge rate. The slurry bath provides some buffering against fluctuations in the feed, which promotes ease of operation. The slurry bath also offers a buffer against incoming flow directly impacting on the matrix, which could generate turbulence and shear, resulting not only in the removal of non-magnetic material entrapped in the magnetic concentrate layer, but also possibly the misplacement of magnetic concentrate to the nonmagnetic product stream.

The horizontal ring WHIMS has eight feed points. A level, well-designed and monitored feed distributor is required to feed all feed points at equal rates to prevent over- or under loading of individual points, which will result in underperformance of a feed point with regard to grade and recovery. The pulp distributor outlets feed into eight boil boxes, which may contain oversize protection trash screens. Manual level control between feed rate and rotor speed is required, especially when fluctuating feed conditions are experienced, against which the top-in bottom-out feed arrangement offers no buffer. The feed impacts directly on the matrix, where turbulence and shear rates are high and could result in effective washing of entrapped nonmagnetic particles, but in the extreme case may also result in loss of the desired magnetic material.

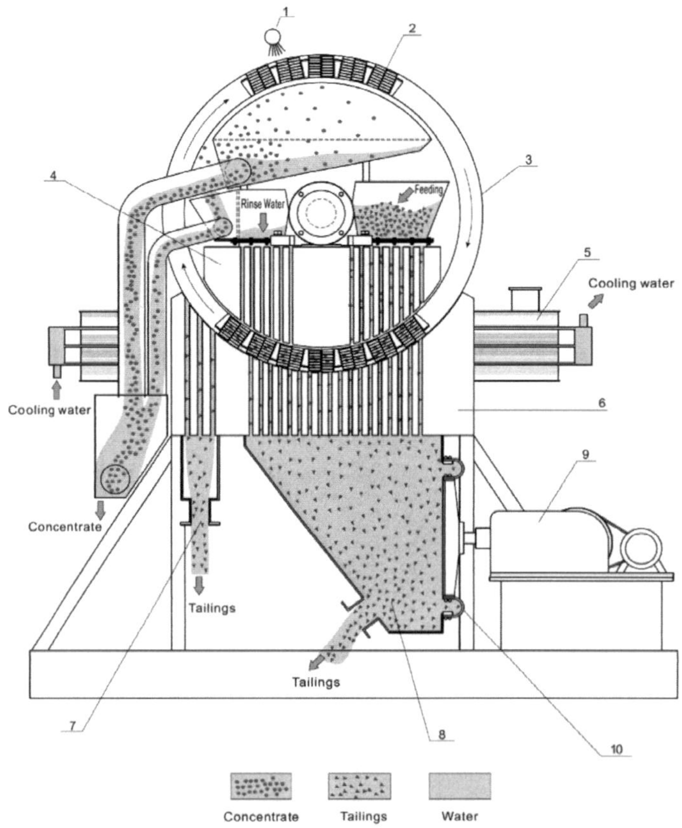

Figure 2. General cross-sectional layout of a vertical ring WHIMS showing (1) wash water system, (2) matrix boxes making up the matrix, (3) rotor rotating in the vertical plane, (4) magnetic yoke north pole, (5) magnetic coil, (6) magnetic yoke south pole, (7) middlings product, (8) tailings product, (9) pulsation system drive, and (10) pulsation membrane.

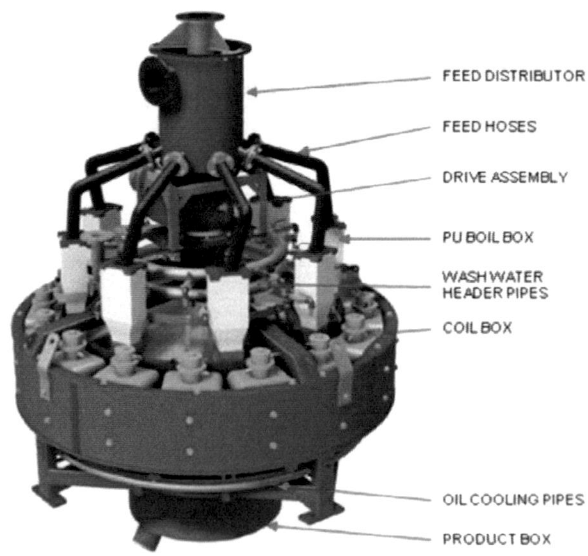

Figure 3. General layout of a horizontal ring WHIMS.

Ring Orientation, Drive System, Method of Separation, and Product Collection

The vertical ring rotates in the magnetic zone between the north and south poles of the yoke. The matrix is charged by the magnetic field and magnetic particles attach to the matrix. Magnetic particles are carried sideways with ring rotation and finally washed out at the top of the machine, while nonmagnetic particles pass through the matrix and out at the bottom. The slurry bath ensures a fully wetted matrix, which provides water as medium for movement and also protection against shear due to incoming slurry, which can contribute to misplacement of magnetic material from the matrix elements and adversely affect the recovery of magnetic minerals. A rubber membrane pulsator (Figure 2) acts on the slurry to keep the solids in suspension and impose a scrubbing mechanism to increase product grade and recovery, by allowing more time for interaction with the matrix. This pulsation is adjustable and independent of feed rate, allowing scope for optimization.

The horizontal ring is fed in a top-in bottom-out configuration into each of the eight magnetic zones (Figure 4). It is important that the ring speed is adjusted in accordance with the prescribed flow rate to ensure proper wetting of the matrix. Each feed point has three associated product collection boxes (thus 24 in total) with adjustable splitter blades for grade and/or recovery optimization. The lack of a slurry-containing bath makes the unit sensitive to changes in feed rate, but does allow high shear forces in fast flowing fluid that assist with cleaning of any entrapped nonmagnetic particles on the profiled plate surface.

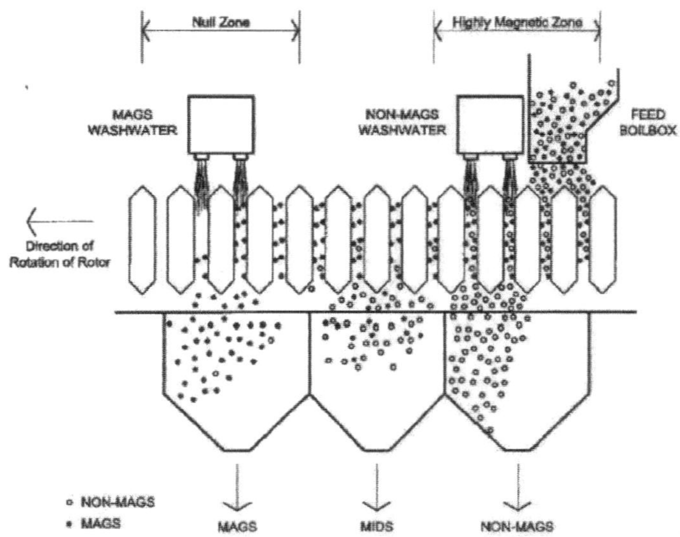

Figure 4. Separation principle of the horizontal ring WHIMS indicating one of the eight feed points with the associated nonmagnetic, middlings, and magnetic products.

Matrix Construction

The vertical ring matrix consists of individual matrix boxes fixed in place with a bolt-in wedge that can be replaced *in situ*. Each matrix box has tens or hundreds (depending on the machine model, size, and matrix specification) of matrix rods (or other innovative elements) providing the separation surface area.

The horizontal ring WHIMS rotor is installed in the middle of the machine as a complete unit with all of the profiled plates (typically 224) welded in place, hindering maintenance operations.

Magnetic Coil(s) and Cooling System

The vertical ring WHIMS uses one pancake of turned wire in a horizontal nonmagnetic stainless steel housing. Two, three, or four pancakes can be stacked and welded on top of each other to make up a coil that can generate background intensities in the matrix of 0.6, 1.0, and 1.3 T, respectively. Channels between the pancakes allow cooling water to be pumped through.

The horizontal ring WHIMS uses three smaller wire pancakes in a stainless steel housing to make up one coil. Oil is pumped through the coil for cooling purposes; in some models the coils are air-cooled. There are a total of 16 coils that generate up to 1.0 T background magnetic intensity in the matrix.

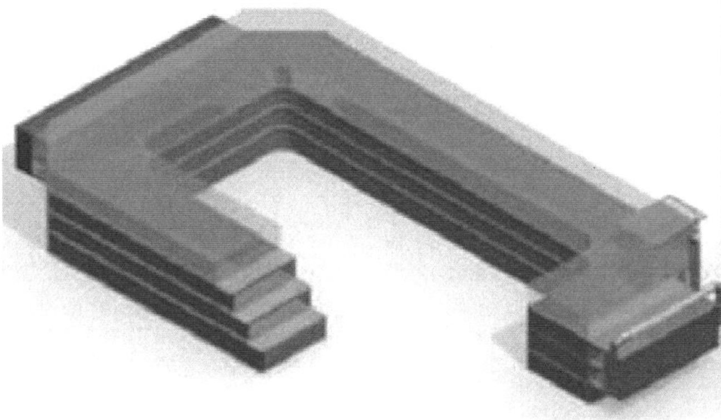

Figure 5. A vertical ring WHIMS, four-pancake coil capable of generating 1.3 T background intensity in the matrix.

Figure 6. Two of the 16 coils making up the electromagnet of a vertical ring WHIMS unit capable of generating 1.0 T background intensity; note the oil outlets at the top of the coil box.

EXPERIMENTAL

Test Equipment and Parameters

Test runs were conducted using a Longi-Multotec LGS 500 pilot-scale vertical ring WHIMS with associated feed sump and pump with valve-controlled recycle and agitator, header tank with two-way distributor, and control panel. A pilot-scale horizontal ring WHIMS of comparable size was not available for testing.

Magnetic field intensity was set at 0.4, 0.7, 1.0, and 1.3 T. Chromite feed grades were not controlled and the samples were tested as received. The solids feed rate was controlled around 80–120 kg/h, which is the mean of the machine capacity (50–250 kg/h). Solids in the feed were controlled at 20% by mass, which is the mean of the recommended 10–30%. The ring speed was set at the maximum of 4 r/min to prevent overloading of chromite onto the matrix. Pulsation was kept at 100 oscillations per minute to allow maximized cleaning and recovery of chromite interacting with the matrix.

Other parameters that could affect performance were not considered or controlled in this study. These parameters are particle size, particle shape, slurry viscosity, and the presence of magnetite, which will block the matrix due to its ferromagnetic property.

Experimental Method

Material from various sources was inspected for oversize and moisture, then screened at 1 mm and pan dried in an oven controlled at 40–60°C until dry and flowing. All samples were claimed to have P_{80} of less than 100 μm; this was not verified.

In cases where as-received samples were too large in volume the material was blended and then coned and quartered to remove around 50 kg of material for test work. Stickiness resulting from moisture prevented rotary splitting, and samples were deemed to be representative of the source.

A 20–30 kg portion (dry solids basis) of sample was weighed out and added to a predetermined volume of fresh potable water from Rand Water supply to make up a standard feed slurry of 20% solids by mass in the feed sump. The recycle and agitator were run at this time to ensure proper mixing. It was noted that cavitation of the feed pump, attributed to air in the feed slurry, resulted when the agitator was on while slurry was pumped to the header box; agitation was terminated moments before the feed valve was opened to initiate the feed to the header box. The slurry tank inside the separator was filled with fresh water. The machine was fed in a semi-batch operation with enough feed slurry to allow the run to reach a visually established pseudo-steady state some minutes into the run. It is expected that pseudo-steady state was achieved for at least ten minutes of the run, during which time product samples were collected. Recycle to the feed tank was maintained throughout a run to allow mixing of the solids throughout the slurry volume, but it is believed that some segregation of the heavy chrome solids nevertheless occurred across the height of the sump. Samples were collected in clean, plastic-lined drums by taking the full stream for the assumed steady-state duration of the run. There were volumes of holdup of material in the feed sump, feed lines, header box, return lines, and the WHIMS slurry tank that did not report through as products; these difficult-to-collect volumes were not always collected but care was taken to wash this material out in-between test runs. Feed samples were taken as grab samples from the recycle line return to the feed sump (with the agitator) running as reference. Wash water was adjusted to wash the concentrate from the matrix using a supply water valve, but the flow rate was not measured.

Product streams were left to settle until the water header at the top of the sample collection drum was clear, and then carefully decanted. The remainder of the drum was then dewatered in a vacuum suction filter pot through filter paper with sufficient backing to ensure drainage and prevent tears in the paper, thus avoiding any solids losses. No solids were seen to be lost in the filtrate. The vacuum-dried material was then transferred into stainless steel oven pans and dried at 40–60°C. Flakes of material that formed during drying (particularly noted in the nonmagnetic product pan, where higher slimes and silica were present) were manually crushed. The dry solids weights were recorded to allow subsequent mass balance calculations, and the samples handed back to the customers for analysis under agreements that results would be made available.

The nonmagnetics, middlings, and magnetic product weights were carefully recorded and the feed solids weight calculated as the sum of these weights. Mass pull was calculated using the weights recorded and recovery calculated using the product of the product stream weight and mineral grade.

The test runs were all confidential and therefore samples are only labelled as X and Y - where it was possible to conduct tests runs on 0.4, 0.7, 1.0, and 1.3 T; and A and B where test runs were conducted at 0.4 and 1.0 T only. All samples were UG2 material collected from unspecified locations of the tailings dams. Particle size distributions were not analysed as the sources confirmed that all material should have a P_{80} of less than 100 µm.

The Cr:Fe ratio was not tested as this was not a critical parameter for any of the clients.

Chemical analysis was done by potential clients' laboratories and the methods chosen were not within the author's control.

RESULTS AND DISCUSSION

Effect of Magnetic Intensity on Grade and Recovery of Chromite

Figure 7 shows the effect of magnetic background field intensity on chromite grade and recovery for sample X. It can be seen that chromite recovery increases from 50 to 70% as a result of increasing magnetic field intensity from 0.7 to 1.0 T. Conversely, recovery to middlings and tailings decreases. This result is expected, since more chromite particles will be attracted to the matrix elements and be transported out of the slurry bath to be recovered as magnetic concentrate. It can also be seen that the product grade decreases with increasing background intensity, as is expected: more nonmagnetic particles (assumed to be mostly silica) will be entrained with the magnetic concentrate. Furthermore, non-liberated particles will also be recovered at these higher intensities where the magnetic force acting on the interstitial paramagnetic mineral is now large enough to overcome the shear experienced on the surface of the total particle due to the slurry flow. Conversely, the chromite grade in the nonmagnetic tailings decreases as the magnetic field intensity increases, because more chromite is removed to the magnetic concentrate. Recoveries are calculated as the weight of mineral reporting to a stream divided by the total weight of that mineral in the feed stream. It is expected that recovery of chromite to the tailings stream will decrease as more of the chromite reports to the magnetic concentrate with increasing magnetic intensity. Variations in feed grade are noted between runs, possibly resulting from sample source variations.

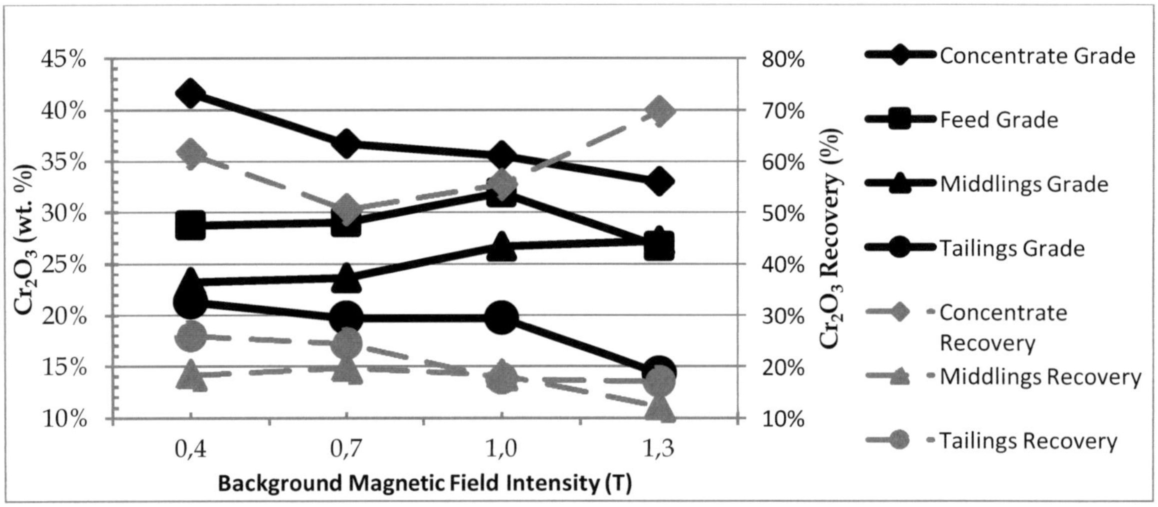

Figure 7. Cr_2O_3 grade and recovery versus background magnetic field intensity for sample X.

Figure 8 compares the chromite grades and recoveries of samples X and Y as a function of background magnetic field intensity. As expected, chromite recovery from sample X increases as the background intensity is increased from 0.7 through 1.3 T, while (surprisingly) that of sample Y decreases very slightly across this range. This could possibly be ascribed to a higher viscosity of the slurry resulting from sample Y, which counters the transport of the increasing amount of chromite particles that are pulled towards the matrix elements. This effect is possibly pronounced as more particles are excited by the magnetic field towards the matrix elements, where hindered settling effects could affect recovery adversely. The presence of flotation chemicals was indicated by the smell and by frothing of the slurry during the test run, and it is suspected that even though sample Y was from a tailings dam, it could have been recently arising material that still contained unreacted flotation chemicals.

It can also be seen that the concentrate grade for UG2 sample Y follows the same trend as for sample X. The upgrade ratio of sample Y (defined here as the ratio between the concentrate grade and the feed grade) is, however, substantially lower (1.39, 1.30, 1.24, and 1.24) for low magnetic field intensities, compared to that of sample X (1.45, 1.27, 1.11, 1.24) at background intensities of 0.4 through 1.3 T. It is proposed that the more viscous fluid discussed in the previous paragraph counters the transport of material to the matrix surface. Secondly, the viscosity around the matrix elements could also prevent displacement of nonmagnetic material from the surface of the element, resulting in more nonmagnetic material being entrapped and thus lowering the concentrate grade. However, as background magnetic field strength is increased it seems possible that the magnetic force is sufficient to overcome the proposed viscosity effects, and upgrade ratios seem to be in line with the results from sample X.

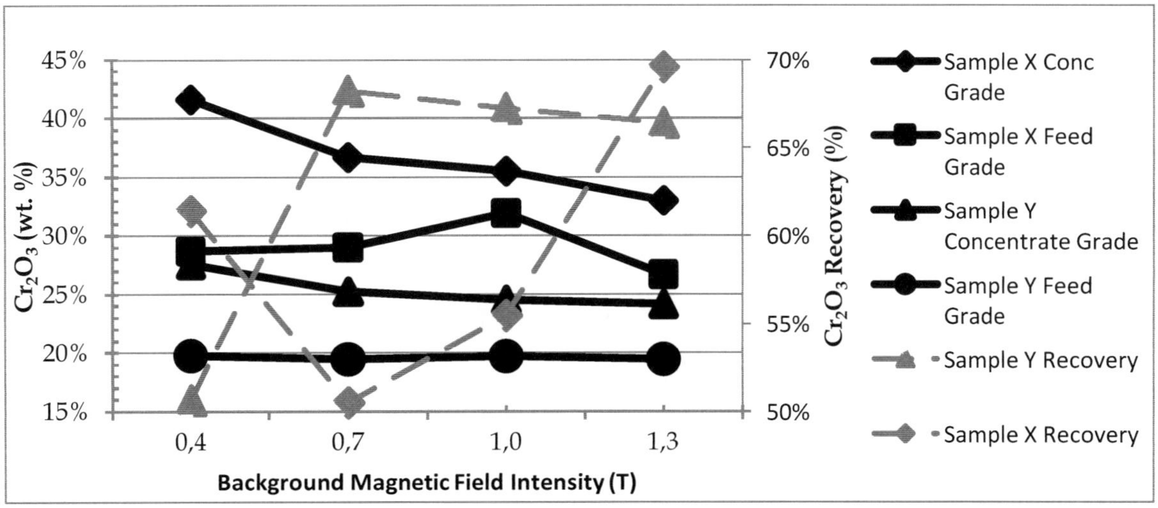

Figure 8. Cr_2O_3 grade and recovery versus background magnetic field intensity for sample X and sample Y.

Figure 9 shows the grade and recovery results for samples A and B, which were only subjected to background field intensities of 0.4 and 1.0 T. Even though the samples were from different sources their feed grades were similar. The concentrate grades produced were also very similar, at 31% and 27% at the two intensities. There is, however, a notable difference in recovery of almost 15% between the two samples. This result cannot be explained.

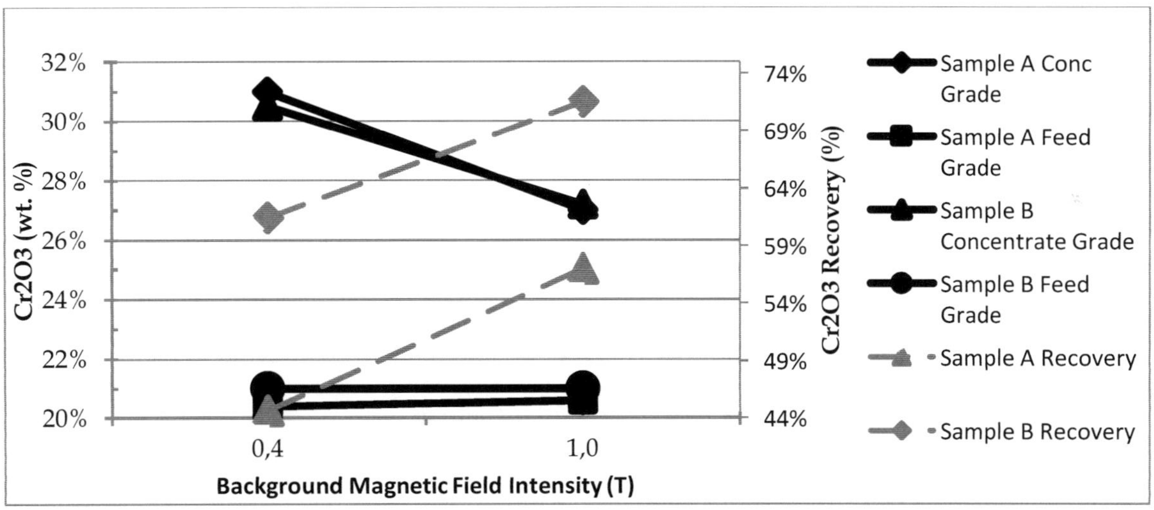

Figure 9. Cr_2O_3 grade and recovery versus background magnetic field intensity for sample A and sample B.

From the results discussed above, deeper investigation into the parameter(s) that affects the upgrade ratio is considered in Figure 10. As can be seen, the upgrade ratios for the various samples follow the same trend when plotted as a function of background magnetic field intensity. Even though the values are of the same order, there are differences that could be ascribed to the viscosity effects discussed before, particularly for sample Y results. The results indicate that the upgrade ratios at higher background field intensity are more consistent, proving that the increase in background magnetic field intensity can be sufficient to overcome these effects, but not without the sacrifice of increasing entrapped nonmagnetics or increased attraction of non-liberated material resulting in lower concentrate grades.

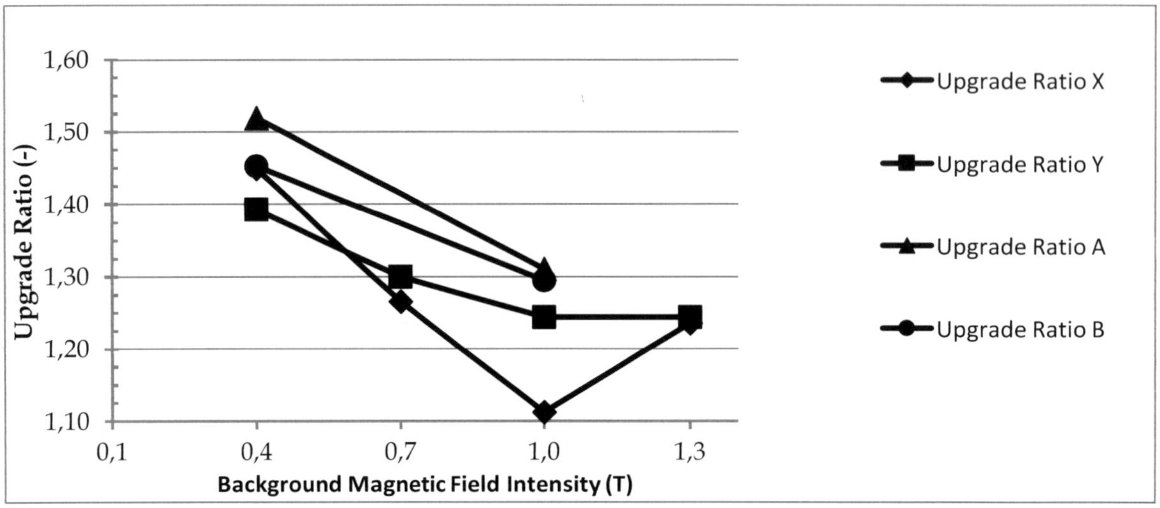

Figure 10. Upgrade ratio of chromite concentrate as a function of background field intensity.

A possible relationship between upgrade ratio and feed grade was investigated, but no correlation was evident (graph not shown here). This means that lower feed grades will be subjected to upgrade ratios that are (within the scope of this paper) only a function of magnetic field intensity.

Figure 11 shows the relationship between concentrate product grade and yield for all test samples. From the graph, an increase in magnetic field intensity generally resulted in an increased yield and a decreased concentrate product grade owing to the factors discussed above. The grade *versus* yield result for sample Y is the lowest, again emphasising the suspected viscosity effects. The grades versus yield values for different samples are significantly different and the response to magnetic separation must be tested to confirm the response for a specific sample. Further investigation into other parameters that affect the separation is recommended.

These discussions above prove that grade and recovery results are very predictable except at lower intensities, where site-specific higher viscosity effects could adversely influence grade and recovery. Inevitably more stages of WHIMS will be required to upgrade material to the required saleable grade of 42%. The number of stages will be determined by the recovery that is required; high overall recovery will require machines to be run at 0.4 T, whereas recoveries are typically reported here over a broad range of 15–32%. Site-specific work on fresh arising and older material, where no active flotation chemicals are present, is strongly recommended. This work should also identify the presence of magnetite in the material to determine the size of the scalping wet drum magnetic separator that must be installed prior to WHIMS.

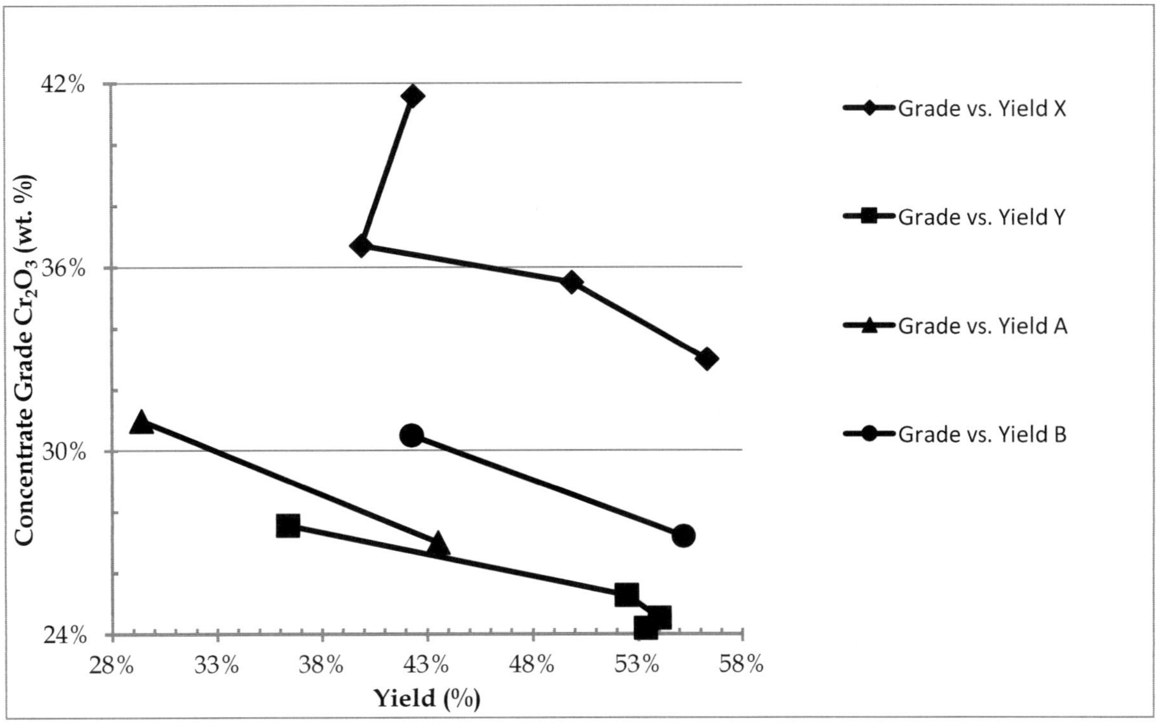

Figure 11. Concentrate grade versus yield for Cr_2O_3 for all four test samples.

Effect of Magnetic Intensity on Grade and Recovery of PGMs

Figure 10 shows the reverse upgrade effect of chromite removal to the magnetic concentrate on the PGM concentration in the middlings and tailings streams of samples X and Y. Reasonable reverse upgrades are possible, and recoveries to the combined middlings and tailings streams are more than 60%, with recoveries as high as 80% achieved. PGM grades were as high as 6.5 ppm, which could indicate that re-floating of these streams would be economically viable, as the threshold is believed to be around 4 ppm. Results are, however, limited and more test work runs are recommended.

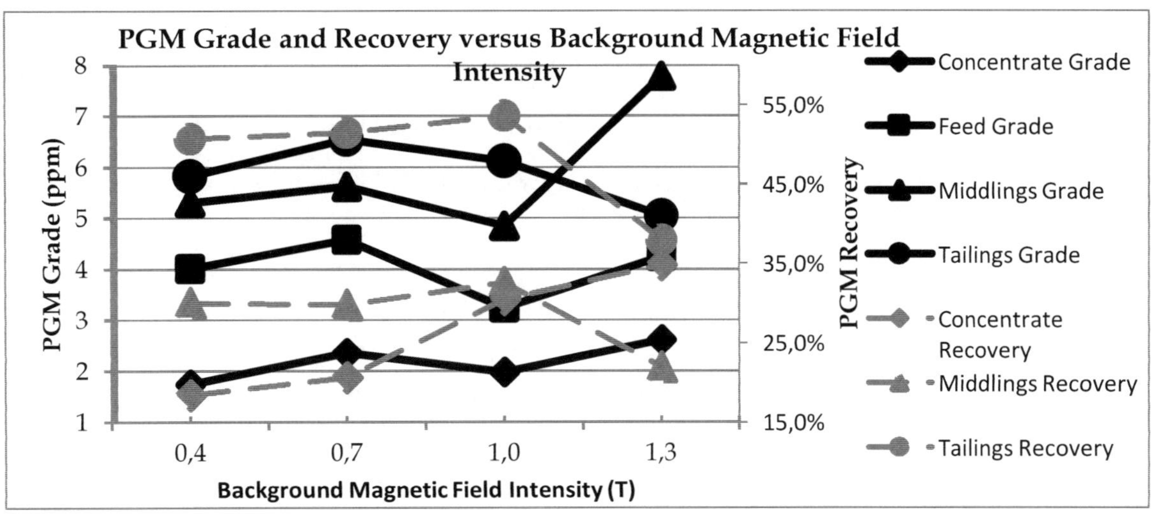

Figure 12. PGM grade and recovery versus background magnetic field intensity for sample X.

CONCLUSIONS AND RECOMMENDATIONS

Effect of Magnetic Intensity on Grade and Recovery of Chromite

Very fine chromite from gravity separation circuit tailings can be upgraded using a vertical ring WHIMS. Chromite recoveries to the magnetic concentrate range between 44–62% at 0.4 T and 55–72% at 1.3 T. Yields varied from 27–42% at 0.4 T and generally decreased with increased magnetic intensity to 1.3 T, where yields were between 24–33%. Recoveries and yield are strongly dependent on background magnetic intensity, but more importantly seems to be site-specific.

Upgrades of 6–14% were achieved at 0.4 T and 4–6% at 1.3 T. Upgrade ratio analysis proved that upgrade is a function of magnetic intensity, but not of chromite feed grade.

The broad ranges of grade and recovery values are suspected to be the result of different viscosities, attributed to the possible presence of unreacted flotation chemicals or ultra-fine material. Differences in mineralogy and liberation of chromite from different source materials could also account for the broad range of grades, recoveries, and yields and should be further investigated.

Further results indicated that at higher background magnetic intensities the proposed viscosity effects are diminished as larger magnetic forces act on particles and overcome these counteracting forces. However, as the background intensity increases, grade decreases and recovery and yield increase, which means inevitably more stages of WHIMS will be required to upgrade material to the required saleable grade of 42% Cr_2O_3.

Site-specific work on fresh arisings and older material, where no active flotation chemicals are present, is strongly recommended. This work should also aim to identify the presence of magnetite in the material to determine the size of the scalping wet drum magnetic separator that must be installed prior to WHIMS.

The reverse upgrade of PGMs in the vertical WHIMS tailings stream through chromite removal to the magnetic concentrate stream seems feasible, with tailings grades of up to 6.5 ppm achieved at a PGM recovery of 50%. It is believed that this stream could be floated economically to recover additional PGM material. The advantage of processing existing tailings streams lies in the cost savings, since mining, conveying to surface, and milling have already been done.

REFERENCES

Engelbrecht, J. (2012). Potential changes in the physical beneficiation process that can improve the recovery grade or costs for the platinum group metals. *Proceedings of the Fifth International Platinum Conference: 'A Catalyst for Change'*, Sun City, South Africa, 17–21 September 2012. Southern African Institute of Mining and Metallurgy, Johannesburg. pp. 343-360.

Forssberg, K.S.E. and Kostkevicius, N.R. (1982). Comparative pilot scale tests with wet high intensity magnetic separators. *Erzmetal*, 35 (6), 284 - 293.

Murthy, Y.R., Tripathy, S.K., and Kumar, C.R. (2011). Chrome ore beneficiation challenges and opportunities - a review. *Minerals Engineering*, 24, 375–380.

Tripathy, K.S., Murthy, Y.R., and Singh, V. (2013). Characterisation and separation studies of Indian chromite beneficiation plant tailing. *International Journal of Mineral Processing*, 122, 47–53.

Willem Pieter Slabbert

Applications and Process Manager
Multotec Process Equipment (Pty) Ltd

Willem is a Master's degree graduate that started his working career in water treatment and industrial filtration before joining Multotec Process Equipment (Pty) Ltd in February 2012. His experience and responsibilities includes magnetic separation, where he serves as Multotec's Product Specialist, centrifuges and filtration as well as wet and dry Sampling, where he serves as Applications and Process Manager.

Crude $TiCl_4$ purification: a review of the current state of the art and future opportunities

L. Hockaday and A. Kale

Mintek, Randburg, South Africa

Raw titanium tetrachloride ($TiCl_4$) is produced by chlorination of titania feedstock. Prior to its utilization in the manufacture of titanium dioxide pigment or titanium metal, the $TiCl_4$ is purified by adding reagents that react with the impurities in the crude $TiCl_4$, followed by distillation of the mixture. The current state of the art in purification of crude $TiCl_4$ was surveyed. Tests were conducted at the laboratory scale to assess the efficiency of various additives used in the purification process to minimize the vanadium content in the distillate. The effects of various reagents, namely oleic acid, sodium oleate, potassium oleate, and copper on the purification of crude $TiCl_4$ were compared. The purified $TiCl_4$ was water-clear in aspect and analysed < 2 ppm vanadium for all reagents tested. Possible reactions between the impurities in the crude $TiCl_4$ and the reagents are investigated thermodynamically with the FactSage program.

INTRODUCTION

Titanium tetrachloride ($TiCl_4$) is obtained by the chlorination of a titanium-bearing mineral such as ilmenite and rutile, as well as from titanium slag. The chlorination process usually takes place in a fluidized bed in the presence of chlorine gas and a reducing agent such as petroleum coke, resulting in crude titanium tetrachloride. $TiCl_4$ is an intermediate product in the production of titanium dioxide (TiO_2) pigment and titanium metal. The pure form of titanium dioxide is used in the pigment industry for paints, food colouring *etc.*, and titanium metal is widely used in the aerospace, medicine, sport, and semiconductor production industries. The crude $TiCl_4$ has many impurities, most of which can be removed through distillation; however, some are difficult to remove *e.g.* vanadium oxychloride ($VOCl_3$). This impurity has a boiling point close to that of the desired product, making separation by distillation ineffective. The removal of impurities is especially important as their presence in TiO_2 pigment and titanium sponge reduces the quality of pigment and increases the hardness of the metal respectively (Xiong *et al.*, 2012).

A chemical treatment is often used to precipitate such impurities or to convert them into high-boiling impurities, ensuring that separation by distillation is effective. Various substances can be used as chemical treatment agents, such as metal powders (iron, aluminium, copper *etc.*), petroleum fractions, alkali metal hydroxides, inorganic gases, and vegetable or animal oils and their derivatives.

The aim of this work was to produce $TiCl_4$ of high purity, containing significantly low levels of vanadium and suitable for use in the TiO_2 pigment and Ti metal industries, utilizing Mintek's laboratory-scale set-up. Various purification experiments were conducted using both organic and inorganic treatment agents to purify crude $TiCl_4$. The agents were compared in terms of their effects and economic viability.

From Table I, it can be seen that Air Products and TOHO high grade have stringent restrictions on vanadium content, at 1 ppb and 1 ppm respectively. Air Products uses the $TiCl_4$ for production of titanium metal and TOHO high grade supplies the semiconductor industry. For use in the pigment industry, the vanadium content must be below 10 ppm and preferably below 5 ppm. The vanadium content in crude $TiCl_4$ is typically between 200 and 500 ppm. Judging from the commercial specifications, purification of $TiCl_4$ is essential. Table I shows some commercial specifications by various industries.

Table I. Commercial specifications of $TiCl_4$.

	$TiCl_4$ (%)	V (ppm)	Fe (ppm)	Al (ppm)	Sn (ppm)	Si (ppm)	Sb (ppm)	As (ppm)
Air products (Koiro, 2013)	>99.7	<0.001	<0.003	<0.002	<0.001	-	<0.001	<0.002
TLT -1 (TOHO Titanium, 2016)	>99.9	<10	<10	-	<30	<5	-	-
High grade (TOHO Titanium, 2016)	-	1	1	1	1	2	0.1	0.1
Osaka Titanium Technologies (, 2013)	99.9	1	1	-	-	1	-	-

LITERATURE REVIEW

Literature sources were reviewed in order to gain a better understanding of the crude $TiCl_4$ purification process. Various reagents, namely alkali metal hydroxides, metal powders (Fe, Zn, Sn, Cu, Sb, Na, Mg, Ti or Al), $TiCl_3.1/3AlCl_3$, the salts of stearic, myristic, palmitic, oleic, and lauric acids, oleic acid, animal fat, vegetable oil, petroleum fractions, hydrocarbon polymers, soap, dichlorobenzene, chlorotoluene, hydrogenated naphtenic oil, white mineral oil (WMO), and chloronaphtolene have been used as treatment agents, alone or in combination, to bind impurities before purified $TiCl_4$ is distilled off. The treatment agents, operating conditions and results are summarized in Table II. Some unique processes are discussed separately as they did not conform to the common procedure of refluxing followed by distillation.

In a paper on the preparation of high-purity $TiCl_4$, Clabaugh, Leslie, and Gilchrist (1955) claim that most metal impurities except V and Sn can be removed by distillation without any treatment agents. They also describe the removal of organic impurities by the addition of Cl_2, aluminium chloride hexahydrate ($AlCl(OH)_6$), and water or by the addition of Br; the use of Cu to remove V is also described.

A patent entitled 'Process for the purification of titanium tetrachloride' (Sironi *et al.*, 1971) describes a method of producing titanium tetrachloride with a vanadium content of less than 1 ppm, using hydrogen sulphide gas (H_2S) to achieve the desired purity levels through precipitation of vanadium impurities. According to the invention, the crude $TiCl_4$ from the chlorination process is passed through a spray quencher where it is injected with the liquid $TiCl_4$ and at the same time treated with H_2S in order to cool the vapour stream from high temperatures to 130–180°C, to remove the vanadium and other low-boiling impurities in the vapour phase.

Winter (1975) proposes the purification of vanadium-containing $TiCl_4$ by heating with $TiCl_3.1/3AlCl_3$. The reaction between the reagent and impurities in the $TiCl_4$ has very fast kinetics, and the reaction products are filterable from liquid $TiCl_4$.

Bonsack (1978) added water to crude $TiCl_4$ to remove impurities of niobium pentachloride and tantalum pentachloride. Water reacts preferentially with the niobium and tantalum compounds to form insoluble hydrates.

Table II. Experimental conditions and results from some reviewed articles.

Reference	Treatment agent	Addition of treatment agent	Initial vanadium content	Final vanadium content	Operating temperature (°C)	Refluxing time
Cole and Meister, 1946	Alkali metal hydroxide	-	-	-	136	1-6 h
Meyers, 1946	Alkali salts of fatty acids	0.1 to 1% wt% crude $TiCl_4$	-	<10 ppm	No heating	No reflux
Stambaugh, 1956	Alkali metal hydroxide + metal	0.15-0.3 wt% crude $TiCl_4$	2500-3500 ppm	1-3 ppm	134-138	1-2 h
Hansley and Schott, 1960	Fine salt and powdered metal	0.5-0.6 wt% metal to crude	-	<10 ppm	136	90 min
Clark and Kolb, 1970	Any organic soap (*e.g.* potassium oleate)	0.05-5 wt% crude $TiCl_4$	200-4000 ppm	5 ppm	90-136	15-50 s in scrubber set-up
Sironi *et al.*, 1971	Hydrogen sulphide	3 mol H_2S/1 mol V	2200 ppm	1 ppm	145	15-50 s
Goddard and Litwin, 2002	Copper	1-2 wt% crude $TiCl_4$	0.1 ppm	0.03 ppm	136	30 min
Xiong *et al.*, 2012	Aluminium powder + WMO	1.6 g/g $VOCl_3$	2860 ppm	4.9 ppm	136	40 min

Bonsack and Schneider (1982) claim that vanadium chlorinated with high-surface-area carbon forms VCl_3, rather than $VOCl_3$ or VCl_4, which can readily be separated as it remains in solid form below 450°C. This alternative process allows the separation of vanadium impurities before $TiCl_4$ condensation.

Ishizuka (1988) removed impurities from crude $TiCl_4$ by heating over a catalytic metal (Cu, Fe, or Cu and Fe alloys) above 300°C to reduce oxygen impurities to substances less volatile than $TiCl_4$, which can then be removed from gaseous $TiCl_4$ or in the subsequent distillation. Lowering the oxygen content in $TiCl_4$ used for metal production improves the ductility of the titanium metal produced.

According to the literature, the problem is clearly not to find a treatment agent, since many has been described, but rather to find a reagent that achieves the desired $TiCl_4$ purity in a cost-effective way with minimal adverse impact on the production process.

THERMODYNAMIC DATA

Table III shows the boiling points of metal chlorides often associated with titanium tetrachloride in the crude mixture. It can be seen that $AlCl_3$, $AsCl_3$, $VOCl_3$, VCl_4, and $SnCl_4$ have boiling points close to that of titanium tetrachloride (136°C), making it difficult to separate them from one another using physical processes such as distillation. Although ferric chloride, which is a very common impurity in titaniferous materials, has a boiling point significantly higher than 136°C, it is known to react with other components in the system to form lower-boiling complexes/compounds that can contaminate the titanium tetrachloride. Antimony chlorides, ferrous chloride, and calcium and magnesium chlorides are high-boiling compounds and do not pose any problems during the distillation process.

Table III. Boiling points of common crude $TiCl_4$ constituents and possible reaction products (Perry and Green, 1997).

Chemical name	Chemical formula	Boiling point (°C)
Titanium tetrachloride	$TiCl_4$	136
Titanium trichloride	$TiCl_3$	Decomposes at 440°C
Vanadium oxychloride	$VOCl_3$, yellow	127
Vanadium tetrachloride	VCl_4, red	149
Vanadium trichloride	VCl_3, pink or violet	Decomposes at 300°C
Vanadium dichloride	VCl_2, green	1506
Antimony pentachloride	$SbCl_5$	Decomposes at 140°C
Antimony trichloride	$SbCl_3$	220
Ferric chloride	$FeCl_3$	315
Tin tetrachloride	$SnCl_4$	114
Arsenic trichloride	$AsCl_3$	130
Silicon tetrachloride	$SiCl_4$	58
Aluminium chloride	$AlCl_3$	183
Ferrous chloride	$FeCl_2$	1023
Copper chloride,	$CuCl_2$, brown-yellow	993
Nantokite	Cu_2Cl_2, white	1366
Calcium chloride	$CaCl_2$	1935
Magnesium chloride	$MgCl_2$	1412

Evaluation of reactions

The crude titanium tetrachloride contains a lot of impurities as a result a number of chemical reactions that can occur between $TiCl_4$ and other metal chlorides present in the crude mixture upon addition of either an organic or an inorganic treating agent. Some chemical reactions between the treatment agent and various chlorides/oxychlorides of vanadium and iron have been thermodynamically evaluated and the results are discussed in the following sections. These reactions are by no means the only reactions that can take place; for instance, no data was available for reactions with VOCl and $VOCl_2$ as products, and these reactions were not investigated. Vanadium is the most detrimental impurity in $TiCl_4$, and iron chlorides are often present in higher concentrations than desirable.

The thermodynamic properties of the reactions, such as changes in Gibbs free energy, enthalpy, and entropy, can be calculated using the FactSage thermodynamic simulation software. This requires an input of possible reactants and products, operating temperature, and pressure.

$Cu + VOCl_3 \rightarrow CuO + VCl_3$ [1]

$Cu + VCl_4 \rightarrow CuCl + VCl_3$ [2]

$Cu + VCl_4 \rightarrow CuCl_2 + VCl_2$ [3]

$FeCl_3 + Cu \rightarrow FeCl_2 + CuCl$ [4]

$FeCl_3 + CuCl \rightarrow FeCl_2 + CuCl_2$ [5]

$VOCl_3 + \frac{1}{2}C \rightarrow VCl_3 + \frac{1}{2}CO_2$ [6]

$VOCl_3 + C \rightarrow VCl_3 + CO$ [7]

$VOCl_3 + \frac{1}{2}\frac{1}{2}C \rightarrow \frac{1}{2}VCl_4 + \frac{1}{2}VCl_2 + \frac{1}{2}CO_2$ [8]

$VOCl_3 + C \rightarrow \frac{1}{2}VCl_4 + \frac{1}{2}VCl_2 + CO$ [9]

$VCl_4 + \frac{1}{2}C \rightarrow VCl_2 + \frac{1}{2}CCl_4$ [10]

$VCl_4 + \frac{1}{2}C \rightarrow VCl_3 + \frac{1}{2}CCl_4$ [11]

$FeCl_3 + \frac{1}{2}\frac{1}{2}C \rightarrow FeCl_2 + \frac{1}{2}CCl_4$ [12]

Solubility data for the iron chlorides and vanadium oxychlorides in $TiCl_4$ was not available.

The thermodynamic evaluation shows that copper may indeed react with vanadium and iron chlorides, producing chlorides with much higher boiling points than $TiCl_4$ and thus ensuring that purified $TiCl_4$ is obtained through distillation.

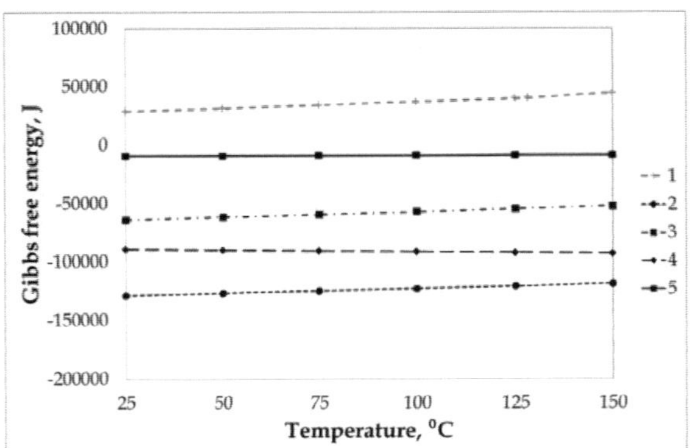

Figure 1. Gibbs free energy evaluation for reactions [1] to [5].

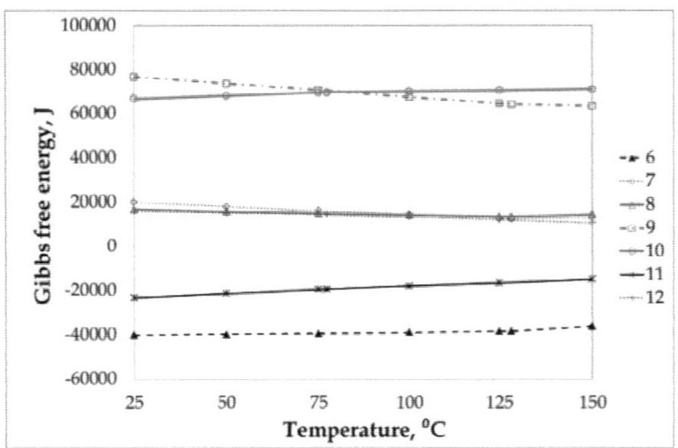

Figure 3. Gibbs free energy evaluation for reactions [6] to [12].

Xiong *et al.* (2012) proposed that $VOCl_3$ was converted to $VOCl_2$ (solid phase) by $TiCl_3$ formed through the reduction of $TiCl_4$ using aluminium powder and carbon, which was obtained from the pyrolytic cracking of WMO. Although pyrolytic cracking of the white mineral oil to carbon seems unlikely at such low temperatures and is contra-indicated by the fouling caused by the polymerized organic residue in industry, thermodynamic investigation confirms that carbon might be an effective treatment agent to pacify vanadium and iron impurities.

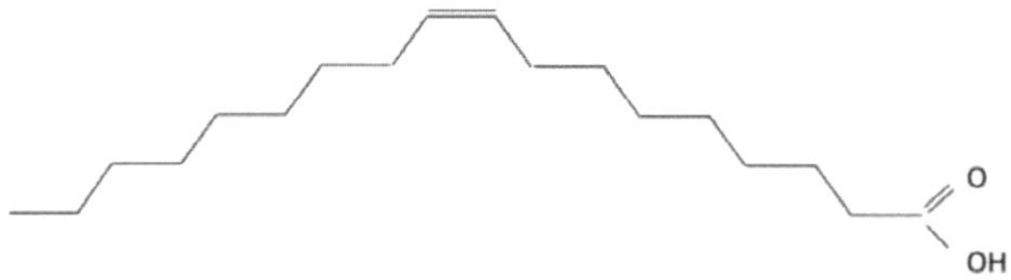

$CH_3(CH_2)_7CH{=}CH(CH_2)_7COOH$
Oleic acid

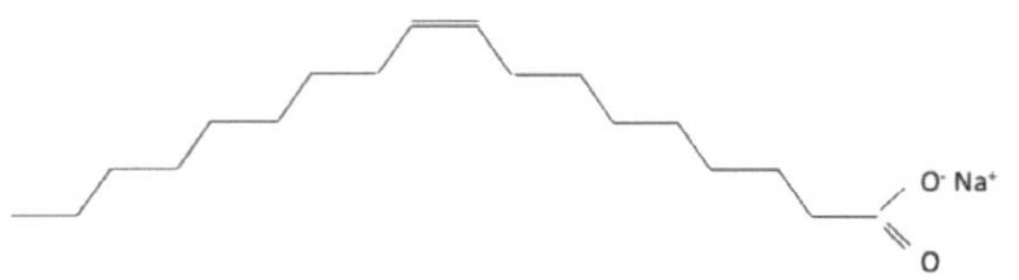

$CH_3(CH_2)_7CH{=}CH(CH_2)_7COONa$
Sodium oleate

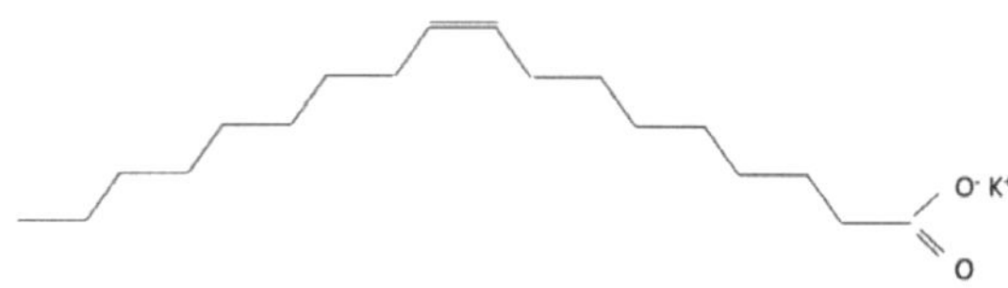

$CH_3(CH_2)_7CH{=}CH(CH_2)_7COOK$
Potassium oleate

Figure 2. Line structure, chemical formulae, and common names for oleic acid and its salts.

Oleic acid ($CH_3(CH_2)_7CH{=}CH(CH_2)_7COOH$) is an unsaturated organic carboxylic acid and therefore likely to undergo the typical reactions of both alkenes and carboxylic acids. These reactions include addition and polymerization of the alkene bond and nucleophilic substitution reactions of the carboxylic acid bonds (McMurry, 1996). In Figure 2 the line bond structures of the organic treatment agents are compared.

Sodium oleate and potassium oleate share the same structural form as oleic acid, with the alkali metals replacing the hydrogen of the -OH group. For ease of discussion we replace $CH_3(CH_2)_7CH{=}CH(CH_2)_7CO$ with placeholder R. We can then write oleic acid as ROH and its salts as RONa and ROK.

$VOCl_3$ reacts with alcohols, especially in the presence of a proton acceptor, to give alkoxides (McMurry, 1996). Writing these reactions for our organic reagents gives:

$$VOCl_3 + 3\ ROH \rightarrow VO(OR)_3 + 3\ HCl \qquad [13]$$

$$VOCl_3 + 3\ RONa \rightarrow VO(OR)_3 + 3\ NaCl \qquad [14]$$

$$VOCl_3 + 3\ ROK \rightarrow VO(OR)_3 + 3\ KCl \qquad [15]$$

Vanadium oxytrichloride ($VOCl_3$) is also strongly Lewis acidic (Wikipedia, 2016) and is used as a catalyst in alkene polymerization. It would therefore promote polymerization of the alkene bond in the carboxylic acid and its salts, leading to the organic residue that fouls reactors in industry. $TiCl_3$ is also known to act as a catalyst for polymerization of alkenes (McMurry, 1996).

EXPERIMENTAL PROCEDURES

Titanium tetrachloride is a highly insoluble, thermally stable metal halide that is liquid at room temperature. It is a clear colourless liquid in pure form, but crude titanium tetrachloride may be yellow or reddish-brown in appearance. It has low melting and boiling temperatures, namely -24.1°C and 136.4°C respectively. $TiCl_4$ is extremely moisture-sensitive and reacts violently with water/air to form toxic HCl gas and titanium dioxide/oxychloride.

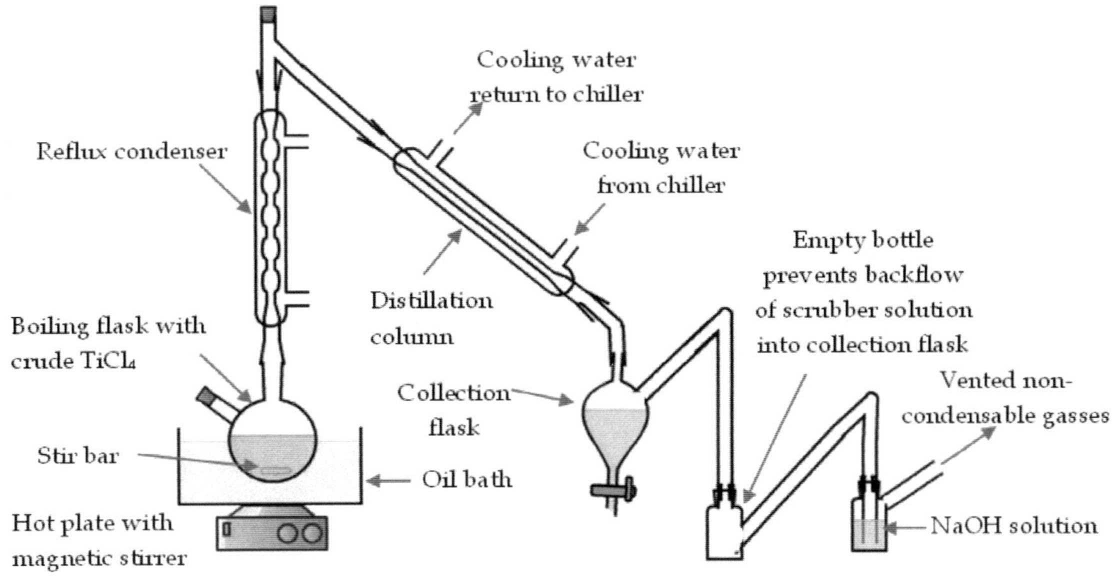

Figure 3. Experimental set-up of reflux column and condenser.

Figure 3 depicts the set-up used for investigation of the chemical treatment and distillation of the crude mixture of $TiCl_4$ using various additives. The set-up consists of a round-bottom flask containing a mixture of crude $TiCl_4$ and treatment agents agitated with a magnetic stir bar; the flask is immersed in a beaker containing heating oil. The oil should be stable at high temperatures as it is heated to up to 200°C to enable distillation. The beaker is placed on a hot plate with a magnetic stirrer. During the experiment, the vapours from the heated crude mixture enter the reflux condenser where they are condensed back into the liquid phase and return to the flask; this process is called refluxing. The distillation column is connected to the reflux column in an inclined position. The collection flask is connected to the bottom of the distillation column. After the crude mixture has undergone chemical pre-treatment, a heating mantle/heating oil bath can be used to provide increased energy to the mixture in the round-bottom flask, causing the vapour to reach the distillation column, condense, and collect in the collection flask. The non-condensable gases from the distillation column are vented after being bubbled through NaOH solution. The distillation column may be cooled with cold water from a chiller. In the present configuration, the cooling water to the distillation column was at room temperature and no cooling was needed on the reflux column. A data logger connected to a B-type thermocouple was used to monitor the temperature of the oil bath around the feed flask.

Table IV. Operating conditions of the purification trials.

	Refluxing agent	Dosage (wt%)	Refluxing T(°C)	Time (min)
Trial 1	Oleic acid	0.15	136	30
Trial 2	Sodium oleate	0.25	136	30
Trial 3	Potassium oleate	0.25	136	30
Trial 4	Copper metal powder	1	136	30

Table IV shows the details of each purification trial, including the type and dosage of the treatment agent and the refluxing time. The dosages selected for the tests were estimated from the literature.

The crude and purified batches of titanium tetrachloride were analysed for metals such as Fe, Al, Cu, and V using inductively coupled plasma-optical emission spectrometry.

RESULTS

Trials for the purification of $TiCl_4$ with oleic acid were successfully executed. The measured oil bath temperatures are given in Figure 4 to Figure 7. The oil bath temperature proved a reliable indicator of the reflux and distillation zones and temperature measurements inside the reflux column were unnecessary. A sticky organic residue formed in the boiling flask. This residue proved impossible to decant and was scrubbed out of the reactor with soapy water. A sticky organic residue also formed in the boiling flask in the trials with sodium oleate and potassium oleate. Although similar in colour to the residue produced by treatment with oleic acid, the residues from sodium oleate and potassium oleate appeared to the eye less stringy and more particulate in nature. This is, however, subjective and would need to be confirmed by analysis before decisions are based on this observation. The residues from the trial purification with Cu appeared much like crude $TiCl_4$ with copper shavings in it. After standing overnight in a sealed bottle, the residue appeared to solidify and turn white. This may be explained by oxidation of residual chlorides in contact with air.

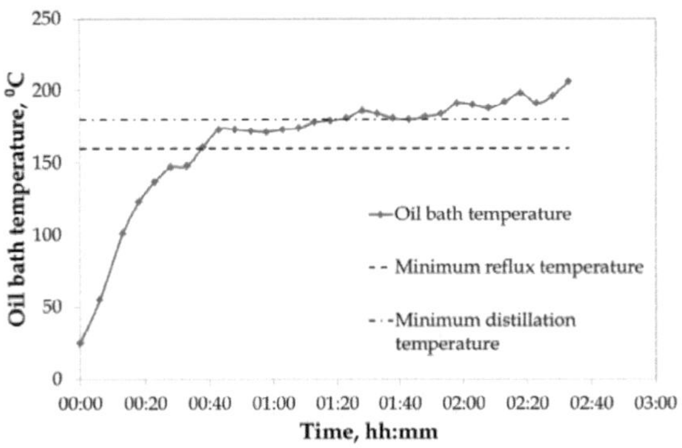

Figure 1. Oil bath temperature for purification of $TiCl_4$ with oleic acid.

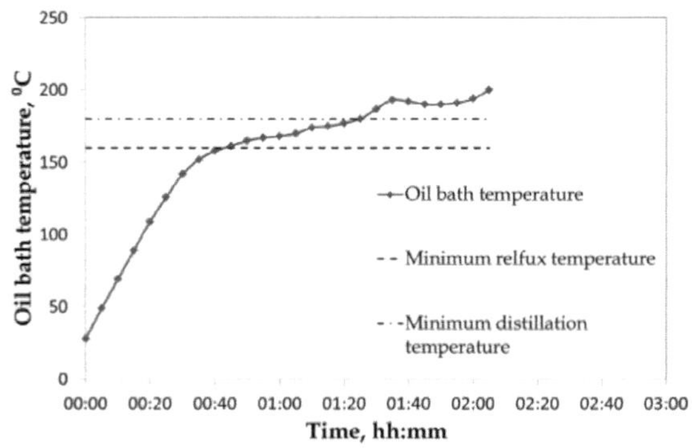

Figure 2. Oil bath temperature for purification of $TiCl_4$ with sodium oleate.

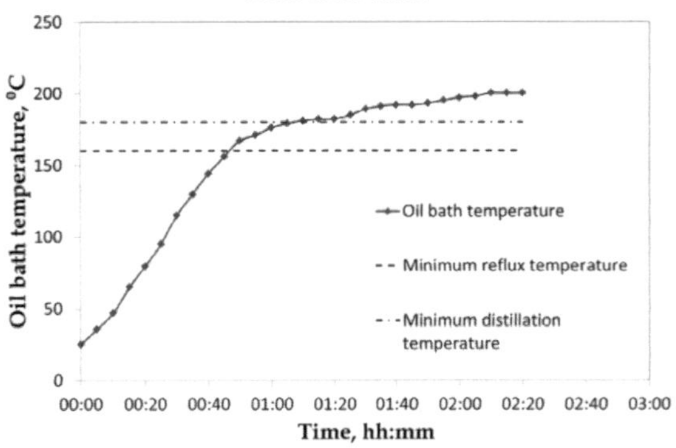

Figure 3. Oil bath temperature for purification of $TiCl_4$ with potassium oleate.

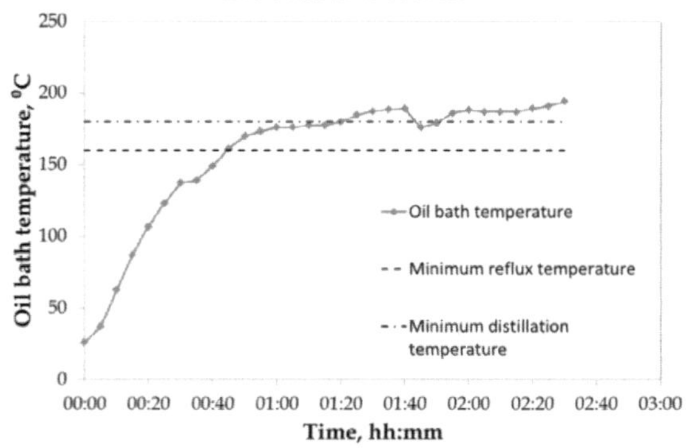

Figure 4. Oil bath temperature for purification of $TiCl_4$ with copper.

When cleaning the boiling flask, the addition of water to residues remaining in the flask produced a mixture purple in colour. Copper reacts with some of the titanium tetrachloride to produce a lavender material that renders some impurities non-volatile. It is believed that the lavender material is probably $CuCl.TiCl_3$ or involves the reduction of some of the titanium tetrachloride to a trivalent titanium chloride compound.

Figure 5. Purple colour on cleaning boiling flask used for copper trial.

Figure 6. Crude $TiCl_4$ (left) and purified $TiCl_4$ (four product bottles on right).

Product analysis

The crude $TiCl_4$ used in the experiments was produced in Mintek's pilot plant chlorination facility using rutile as a titanium feedstock. For all the reagents tested, clear $TiCl_4$ were produced, as shown in Figure 9. The product analyses for impurities are given Table V. Typical V contents in crude $TiCl_4$ in industry are around 200–500 ppm; Al is in the range of 50–200 ppm and Fe below 50 ppm. Our test uses a V level of around 700 ppm. This is much higher than the typical value of 200–500 ppm V in crude $TiCl_4$.

Table V. Analysis of crude and purified $TiCl_4$.

Purified Product	Mass, g	337	304	307	322		
	Crude $TiCl_4$					Ref. actual	Ref. result
Reagent	None	Oleic acid	Sodium oleate	Potassium oleate	Copper	References	
V ppm	726	<2	<2	<2	<2	15	16
Al ppm	4	5.7	5.3	5.1	5.3	15	15
Fe ppm	20	<2	<2	<2	<2	15	14

Analysis by Mintek's Analytical Science Division (ASD)

The analytical method is challenging, as $TiCl_4$ is very reactive with any moisture and the resulting TiO_2 particles and HCl vapour cause blockages in the instruments and interfere with the readings.

Between 88 and 97 wt% of the crude $TiCl_4$ was recovered as purified product.

Economic evaluation

In Table VI the high cost of copper compared to the organic reagents can be seen. Various grades of mineral oils are currently used in the commercial production of $TiCl_4$. The use of metal powders is not favoured due to economic reasons. The use of mineral oil entails some operational challenges as it forms sludge that causes fouling in the condensers.

Table VI. Economic comparison of reagents used in experiments (R15.71 per US dollar).

Reagent	Source	Price US$/t	Dosage wt% of crude $TiCl_4$	Price R/t crude $TiCl_4$
Copper	InvestMine, 2016	4776	1.00	750.31
Oleic acid	Alibaba, 2016a	1200	0.15	28.28
Sodium oleate	Alibaba, 2016b	2000	0.25	78.55
Potassium oleate	Alibaba, 2016c	2000	0.25	78.55

DISCUSSION

The use of various treatment agents for the purification of raw $TiCl_4$ by distillation was investigated. The purity of raw and distilled $TiCl_4$ samples was measured by the amount of impurities present. However, impurity levels below 10 ppm were difficult to detect and required extreme precision in the make-up of reference and control solutions.

The use of organic materials as treatment agents often results in the introduction of organic contaminants into the system and requires an additional step for their removal. Regular shutdowns and maintenance are required in order to clean the equipment, as the residue formed by the organic contaminants reduces the heat transfer efficiencies. If these contaminants are formed by polymerization of the treatment agents, an agent that does not polymerizse easily, such as saturated alkanes or alcohols, should produce less fouling. Such treatment reagents would need to be tested in practise to ensure that they provide the same or better efficiency than those currently used.

The use of inorganic treatment agents such as copper results in chlorides that are non-volatile and do not interfere with heat transfer; however, metals and their powders are expensive and furthermore their recovery from waste before disposal could prove problematic.

According to the literature H_2S is effective in the removal of vanadium impurities, but since it is a poisonous gas its use would be very limited because of the associated safety, transportation, and environmental issues.

CONCLUSION

Experimental trials confirmed the effectiveness of copper and oleic acid and its alkali salts as reagents to bind impurities in crude $TiCl_4$, enabling the distillation of purified $TiCl_4$ suitable for use in the TiO_2 pigment industry.

The literature survey and thermodynamic studies illustrated possible reactions taking place between impurities and reagents. Additional work is planned to confirm the reaction mechanism for carbonaceous reagents by comparing the effectiveness of carbon, and saturated hydrocarbons such as alkanes, alcohols, and esters, with the known carboxylic acids as treatment agents in order to identify a cost-effective non-fouling reagent suitable for industrial use.

REFERENCES

Alibaba. (2016a). Potassium oleate price search. http://www.alibaba.com/trade [accessed 16 Feb. 2016].

Alibaba. (2016b). Sodium oleate price search. http://www.alibaba.com/trade [accessed 19 Feb. 2016].

Alibaba. (2016c). Wholesale oleic acid price search. http://www.alibaba.com/trade [accessed 19 Feb. 2016].

Bonsack, J.P. (1978). Purification of crude titanium tetrachloride. US patent 4070252. SCM Corporation.

Bonsack, J.P. and Schneider, F. (1982). Chlorination of titaniferous material using treated coal for vanadium removal. US patent 4329322. SCM Corporation.

Clabaugh, W., Leslie, R., and Gilchrist, R. (1955). Preparation of titanium tetrachloride of high purity. *Journal of Research of the National Bureau of Standards*, 55 (5), 261 - 264.

Clark, H. and Kolb, K. (1970). Purifying crude $TiCl_4$ vapor. US patent 3533733. E.I. du Pont de Nemours.

Cole, S. and Meister, W.F. (1946). Method for the production of titanium tetrachloride. US patent 2396458. National Lead Company.

Cronin, J., Elkins, T., Helberg, L.E., and Merkle, J.E. (2004). Purification of titanium tetrachloride. EP 1585707B1. E.I. du Pont de Nemours.

Goddard, J. and Litwin, M. (2002). Process for purifying titanium tetrachloride. US patent 2002/0179427 A1. Praxair Technology Inc.

Hansley, V.L. and Schott, S. (1960). Purification of titanium tetrachloride. US Patent 2958574. National Distillers and Chemical Corporation.

InvestMine. (2016). 5 year copper prices and price charts. http://www.infomine.com/investment/metal-prices/copper/5-year/ [accessed 5 April 2016].

Ishizuka, H. (1988). Method of purificating titanium tetrachloride. US patent 4762696. Ebara 6-chrome.

Koiro, K. (2013). Product data sheet titanium-tetrachloride-extrema. http://www.airproducts.com/products/Chemicals/Specialty-Chemicals-Electronics/Metallization.aspx [accessed 17 Feb. 2016].

Krchma, I.J. (1949). Distillation of titanium tetrachloride. US patent 2463396. E.I. du Pont de Nemours.

McMurry, J. (1996). *Organic Chemistry*, 4th edn. Brooks/Cole Publishing Company.

Meister, W. (1947). Method for the purification of titanium tetrachloride. US patent 2416191. National Lead Company.

Meyers, B. (1946). Purification of titanium tetrachloride. US paent 2412349. Pittsburgh Plate Glass Company.

Michigan State University. (2016). Carboxylic acids. https://www2.chemistry.msu.edu/faculty/reusch/virttxtjml/crbacid1.htm [accessed 19 Feb. 2016].

Perry, R. and Green, D. (1997). *Perry's Chemical Engineers' Handbook*, 7th edn. McGraw-Hill.

Sironi, G., Sacaerdote, R. and Ferrero, F. (1971). Process for the purification of titanium tetrachloride. US patent 3627481. Montecatini Edison S.p.A.

Stambaugh, E. (1956). Decolorizing titanium tetrachloride. US patent 2754255. National Lead Company.

TOHO Titanium Co. Ltd. (2016). Chemicals – titanium tetrachloride. http://www.toho-titanium.co.jp/en/products/ticl4_en.html [accessed 17 Feb. 2016].

Wiggington, R. J. (1961). Laporte Titanium Limited. Process for the purification of titanium tetrachloride. US Pat. 3009772.

Wikipedia. (2016). Vanadium oxytrichloride . https://en.wikipedia.org/wiki/Vanadium_oxytrichloride [accessed 24 Feb. 2016].

Winter, G. (1975). Purification of vanadium containing $TiCl_4$ by heating with $TiCl_3.0.33AlCl_3$. US patent 3871874. Bayer Aktiengesellschaft.

Xiong, S., Yuan, Z., Yin, Z., and Yan, W. (2012). Removal of the vanadium impurities from crude TiCl4 with high content of vanadium using a mixture of Al powder and white mineral oil. *Hydrometallurgy*, 119-120, 16-22.

Osaka Titanium Technologies. (2013). $TiCl_4$ (titanium tetrachloride). http://www.osaka-ti.co.jp/e/e_product/pdf/table_e.pdf [accessed 17 Feb. 2012].

Susanna Aletta Carolina Hockaday

Engineer
Mintek

After obtaining her masters of science in engineering degree at the University of Stellenbosch, Lina joined Mintek's Pyrometallurgy Division in 2002 and gained experience in DC arc furnace technology and extractive metallurgy including the recovery of PGMs to iron alloys, ferrochrome smelting and the production of titanium metal in a Kroll reactor. In 2011 she resigned to be a full time mother, but re-joined Mintek in 2015. She is currently involved in $TiCl_4$ production and purification research, as well as research into the application of concentrated solar thermal energy in the minerals processing industry.

The thermodynamic potential and reaction kinetics of synthetic rutile and other feedstocks during chlorination

H. Kotzé[1], N. Bernard[2], A.a Kale[3] and T. Kekana[3]

[1]Consensi Consulting, Mtunzini, South Africa
[2]Iluka Resources, Perth, Australia
[3]Mintek, Johannesburg, South Africa

The thermodynamic potential of six chlorinator feedstocks used in the chloride pigment production process (a higher and lower Ti ilmenite, an oxidized and a reduced slag, natural rutile, and synthetic rutile) was modelled using free energy minimization. With this model the maximum production rates, theoretical coke and chlorine consumptions, and the process temperatures were calculated and compared. Among the high-Ti ores (slags and rutiles), chlorine consumption was found to be the most significant difference between feedstocks: natural and synthetic rutile had the lowest chlorine consumption. Depending on the heat removal capacity of the chlorinator, the higher process temperatures required for slags can result in reduced production rates. Results from experimental work on chlorination kinetics were used to add a kinetic model to the thermodynamic model. For a hypothetical fluidized bed, the bed height for a minimum target residence time for 90% Ti conversion was found to be well within the typical chlorinator bed heights.

INTRODUCTION

A number of papers have been published comparing the kinetic behaviour of different feedstocks during chlorination (den Hoed and Nell, 2003; Burger Bessinger, and Moodley, 2009; Moodley *et al.*, 2012). Notably, data on synthetic rutile is often omitted from the published works. Iluka Resources commissioned a research programme in collaboration with Mintek to address this omission. The primary objective of this programme is to understand the reasons for the different behaviours of the various chlorinator feedstocks, and to place synthetic rutile relative to the other feedstocks for pigment manufacture.

This paper is the first publication arising from this work. We attempt to describe the thermodynamic potentials in chlorination of two ilmenites that differ in titanium content and iron oxidation states, an over-reduced slag, an oxidized slag, a natural rutile, and a synthetic rutile. Acknowledging that thermodynamics and kinetics are two sides of the same coin, we experimentally determined the chlorination kinetic behaviour of these six feedstocks. Finally, to enable comparison between the feedstocks, we applied the thermodynamic calculations and kinetic results to a hypothetical industrial-size chlorinator.

The chemical compositions of the six feedstocks are given in Table I. Ilmenites A and B originated from Iluka Resources' mines and operations; as did the natural and synthetic ilmenites.

- The two ilmenites were selected for their difference in titanium content and iron oxidation states
- The oxidized slag was produced in water granulation trials: the tap stream from the furnace was directed onto water jets which broke the stream into granules sized less than 3 mm. Compared with traditionally tapped slags, this oxidized slag has a higher Ti^{4+}/Ti^{3+} ratio for its FeO content. The level of oxidation increased as the size of the granules decreased (Bungu and Pistorius, 2009). The as-granulated size fraction used in this work is +1 mm; oxidation was not severe enough to form Fe^{3+} and a significant amount of Ti^{3+} is still present, albeit less than with block-cooled slags
- The highly reduced slag originated from a batch smelting process in which the FeO content in the slag was reduced to 5%. As a consequence, the Ti^{3+} content of this slag is higher than that of the usual 9–12% FeO high-titania slags.

For the purposes of the thermodynamic model the forms of iron and titanium were, for the ilmenites and slags, obtained from Rietveld XRD analyses. Synthetic rutile from Iluka typically contains 10–12% Ti_2O_3. It was assumed that this is also true for the synthetic rutile used in this work. The mineralogy of the synthetic rutile was hence assumed to fall within the $FeO.2TiO_2$-$Ti_2O_3.TiO_2$-TiO_2 phase triangle, while the iron and titanium contents were balanced with the analysed numbers. The results are given in Table II.

Table I. Chemical compositions (mass%) of the ilmenites, highly reduced slag, oxidized slag, natural rutile, and synthetic rutile used for thermodynamic and kinetic studies.

	Ilmenite A	Ilmenite B	Oxidized slag	Reduced slag	Natural rutile	Synthetic rutile
TiO_2	50.7	59.1	61.5	54.8	96.0	81.3
Ti_2O_3	-	-	23.4	32.4	-	11.1
Ti_{total} as TiO_2	50.7	59.1	87.5	90.8	96.0	93.7
Fe_{total}	32.8	28.0				3.2
FeO	21.0	9.7	10.0	3.6	-	4.1
Fe_2O_3	22.9	27.8			1.5	-
SiO_2	0.52	0.41	0.84	2.4	1.1	0.86
Al_2O_3	0.58	0.58	0.72	2.2	0.22	0.98
CaO	0.08	0.03	0.05	0.8	0.05	0.05
MgO	2.8	0.33	1.02	0.8	0.05	0.37
MnO	0.9	1.3	1.4	1.1	0.05	0.8
V_2O_5	0.2	0.3	0.4	0.4	0.4	0.4
Cr_2O_3	0.2	0.1	0.1	0.1	0.2	0.1
ZrO_2	0.1	0.5	0.2	0.2	0.4	0.1

Table II. Calculated mineralogical compositions (mass%) of the ilmenites, highly reduced slag, oxidized slag, natural rutile, and synthetic rutile used for thermodynamic and kinetic studies.

	Ilmenite A	Ilmenite B	Oxidized slag	Reduced slag	Natural rutile	Synthetic rutile
$FeO.TiO_2$	13.9	-	-	-	-	-
$FeO.2TiO_2$	36.4	31.4	32.1	11.5	-	9.2
$Fe_2O_3.TiO_2$	43.4	41.7	-	-	-	-
$Fe_2O_3.3TiO_2$	-	-	-	-	3.8	-
TiO_2	-	23.6	26.3	28.9	97.3	68.8
Ti_3O_5	-	-	36.4	50.4	-	17.3

THERMODYNAMIC MODEL

Description and Assumptions

The inputs into the thermodynamic model were the feedstock data in Table I and Table II, the petroleum coke composition shown in Table III, recycled chlorine gas comprising 93% Cl_2, 7% O_2, and 4% N_2 at 200 °C, and make-up chlorine gas at 25°C. Feedstock ore and coke were fed at 25°C. The majority of the Gibbs free energy and enthalpy numbers were obtained from the FactSage and HSC databases. The thermodynamic data for less common species was obtained from literature (Kang and Lei, 1996; Eriksson and Pelton, 1993).

Table III. Short analysis of the petroleum coke used in the chlorination thermodynamic model

Constituent	Mass %
Fixed carbon	98.91
Volatiles	0.15
Moisture	0.11
Ash	0.83
TOTAL	100.0

The heat losses from the chlorinator were quantified by building a hypothetical chlorinator into the mass and energy model, with 5.0 m working diameter, 14 m total height, and a refractory lining of 250 mm throughout. Conduction heat losses were calculated assuming that the refractory provides the greatest resistance to the heat flux. The cold face temperature was set at 100°C, and the hot face equal to the process temperature. The thermal conductivity of the refractories (as a function of average brick temperature) was assumed to be that of a 96% Al_2O_3 brick with 15% porosity, *i.e.* 5.0 W/m°C at 441°C (Shimizu *et al.*, 2013). From these assumptions, the calculated heat losses from the chlorinator ranged between 3.17 MW for a hot face temperature of 750°C to 3.80 MW for a hot face temperature of 1100°C. The process temperature was selected such that the overall energy balance over the process was calculated to within ±10 kW. This is an oversimplified heat loss model, but in view of the comparative objective of the broader model, it is regarded as adequate. It is also noted that the calculated process temperatures are not absolute for the ores modelled, but a result of the hypothetical heat loss assumptions.

As a number of industrial chlorinators are chlorine-constrained, the total chlorine supply rate was set constant at 3.0 Nm^3/h. This supply consisted of recycled and make-up Cl_2. It was assumed that 95% of the free and Ti-associated chlorine in the product gas stream was recycled back to the chlorinator at 200°C. This recycled chlorine stream contained 7% oxygen and 4% nitrogen. Make-up chlorine at 25°C made up the difference required to reach the Cl_2 requirement. Under these assumptions, and for the given chlorinator dimensions, the superficial gas velocity was calculated to be 20.8–23.4 cm/s. In this work the Cl_2 requirement is defined as the total stream Cl_2 fed to the chlorinator, while the Cl_2 consumption is the chlorine lost to impurities and inefficiencies in the recovery process.

The thermodynamic model was built as a Microsoft Excel® spreadsheet, applying free energy minimization via the software's Solver function while maintaining the mass balance at an elemental level. Provision was made for titanium to be present as TiO_2, Ti_3O_5, Ti_2O_3, $TiOCl_2$, and $TiCl_4$. Just enough coke was added to ensure conversion of all titanium to $TiCl_4$ and $TiOCl_2$. Since this is a theoretical model, a carbon efficiency of 100% was assumed, although this assumption is not not realistic (Gazquez *et al.*, 2014.,

Thermodynamic Model Results

The results of the thermodynamic model calculations are summarized in Tables IV and V. The following discussion attempts to verify these results against published literature. In attempting to understand the differences, interesting clarifications were found.

Table IV. Ilmenites A and B: results from the chlorinator thermodynamic model for chlorinator dimensions and operational conditions as described in the text.

	CO/CO_2 ratio = 1		CO_2 >> CO	
	Ilmenite A	Ilmenite B	Ilmenite A	Ilmenite B
Max. ore feed rate (t/h)	14.4	14.4	14.4	14.4
$TiCl_4$ production rate (t/h) (as TiO_2)	7.3	8.5	7.3	8.5
$TiCl_4$ production rate (t/t ore)	1.21	1.40	1.21	1.40
Coke ratio (% of ore)	19.5	20.4	14.7	15.4
Coke consumption (kg/t $TiCl_4$)	162	145	122	110
Cl_2 requirement (kg/t $TiCl_4$)	1689	1439	1689	1430
Cl_2 consumption (kg/t $TiCl_4$)	600	420	591	413
Bed temperature (°C)	792	779	1065	1071
Energy balance (MW)	0.000	0.001	0.003	0.003
Cl_2 slip (%)	16.5	15.0	18.8	16.8
Log P_{O2} (atm.)	-18.6	-19.0	-16.3	-16.5
Log P_{Cl2} (atm.)	-0.78	-0.82	-0.73	-0.77
Raw $TiCl_4$ composition:				
$TiCl_4$ (vol.%)	18.1	20.6	20.1	23.3
$TiOCl_2$ (vol.%)	0.0	0.0	0.0	0.0
CO (vol.%)	22.8	23.4	0.8	0.6
CO_2 (vol.%)	22.7	23.4	37.6	39.4
$FeCl_3$ (vol.%)	15.9	13.5	17.3	14.9
Gas fraction (mass % of products)	96.4	98.8	97.2	98.8
Liquid fraction (mass % of products)	3.6	1.1	2.8	1.2
Bed height	0.66	0.70	0.58	0.62

In the industrial chlorination of ilmenite, the formation of $FeCl_{2(\ell)}$ is limited in favour of $FeCl_{3(g)}$ by running at a Cl_2 partial pressure of around -0.7 log(Cl_2) (Harris, Henderson, and Campbell, 1976). Under these conditions the Cl_2 slip (the unreacted chlorine in the raw titanium-tetrachloride stream) is high. In the thermodynamic model where the Cl_2 supply rate was fixed, the $FeCl_{2(\ell)}$ was limited to around 1% by limiting the ilmenite feed rate. The model calculates the range for Cl_2 slip at 15–19 vol% which translates to -0.73 to -0.82 log(Cl_2) - in agreement with the work of Harris *et al.* (1976).

Setting the CO/CO_2 ratio close to unity (Harris, Henderson, and Campbell, 976; Burger, Bessinger, and Moodley, 2009) for ilmenite results in relative low calculated process temperatures of 779°C and 792°C. Coke ratios are around 20% (Table IV). Thermodynamically, ilmenite can be chlorinated at very low CO/CO_2 ratios of 0.02 by reducing the coke supply to around 15%. Under this condition the process temperatures increase to 1065°C and 1071°C (Table IV). This is explained by the exothermic formation of the additional CO_2.

It is noted in Equations [1a] and [1b] that the production of CO requires twice the amount of carbon required for the production of CO_2 (den Hoed and Nell, 2003), hence the lower coke requirement when running at higher CO_2 concentrations. A higher CO_2 content (and lower CO/CO_2 ratio) is therefore desirable from a carbon consumption point of view.

$$TiO_2 + C + 2Cl_2 = TiCl_4 + CO_2 \qquad [1a]$$

$$TiO_2 + 2C + 2Cl_2 = TiCl_4 + 2CO \qquad [1b]$$

Despite the higher coke consumption, the 1.0 CO/CO_2 case could be favoured for kinetic reasons. It has been shown that the process kinetics are significantly dependent on the CO partial pressure (Rhee and Sohn, 1990). Potentially higher refractory wear and increased heat extraction loads on downstream equipment due to higher process temperatures would also count against running at high CO_2 contents.

All of the above predicted process temperatures are in line with the literature. Chlorination temperatures as low as 700°C were investigated for ilmenite by Rhee and Sohn (1990); while higher temperature ranges from 900–1000°C (Gazquez *et al.*, 2014) to 1000–1050°C (Kale and Bisaka, 2011) are also reported. The process temperatures appear to be equally a function of operational philosophy and feedstock properties.

For the higher Ti ores, two operational philosophies were modelled: (i) maximizing the ore feed rate for a given Cl_2 supply to limit the Cl_2 slip to 1% or lower (Cong, Zhangfu, and Xiaoqiang, 2006; Kang and Okabe, 2014), and (ii) for a given Cl_2 supply rate, controlling the CO/CO_2 ratio around unity (Harris, Henderson, and Campbell, 1976; Burger, Bessinger, and Moodley, 2009). Both conditions are quoted in the literature. From thermodynamic modelling they appear to be mutually exclusive.

When limiting the Cl_2 slip to around 1 vol%, the CO/CO_2 ratios calculate to 1.4–1.5. Process temperatures range from 819°C to 1009°C, and coke feed ratios are 22.5% and 24.5% (Table V). All Ti is converted to $TiCl_4$, with only 0.1–0.2% remaining as $TiOCl_2$. At 1% Cl_2 slip, the $\log(P_{Cl2})$ of the raw product gas calculates to -1.9 to -2.3, which is in agreement with the range given for typical chlorinators (Kang and Okabe, 2014) but adjusted for temperature.

When reducing the coke feed ratio to 20–22%, which enables a CO/CO_2 ratio of around unity, the unconverted $TiOCl_2$ increases to 3–5%. This in turn allows the Cl_2 slip to increase to 3.3–4.2 vol% (Table V). This is expected: from stoichiometry (Equations [2b] and [2c]), since converting TiO_2 to $TiOCl_2$ instead of $TiCl_4$ requires half the amount of Cl_2. The decrease in exothermic CO_2 production contributes to the 60–70 °C decrease in process temperature range: 759°C to 940°C.

Bearing in mind that the model assumes 100% carbon efficiency, the calculated coke ratios of 20 to 25% for both cases are in agreement with literature reports of 25% (Kale and Bisaka, 2011; Nell and den Hoed, 2003).

It is known that Ti^{3+} present in slags chlorinates exothermally, increasing process temperatures (Pistorius and Roux, 2002; Bungu and Pistorius, 2009). In agreement with this, when running the thermodynamic model on the 1% Cl_2 slip criterion, the highly reduced slag with 9% more Ti_2O_3 yields the higher process temperature (Table V). However, when running on the 1.0 CO/CO_2 criterion, the calculated process temperature for the reduced slag is lower than that for the oxidized slag. This seems to be due to the presence of $TiOCl_2$ in the 1.0 CO/CO_2 case. The reaction enthalpy of the TiO_2 to $TiCl_4$ reaction is slightly exothermic, while that of TiO_2 to $TiOCl_2$ is comparatively strongly endothermic (Equations [2a]–[2c]).

$$\tfrac{1}{2}Ti_2O_3 + 1.5C + 2Cl_2 = \tfrac{1}{2}TiCl_4 + 1.5CO \qquad \Delta H° = -186.1\ kJ/mol \qquad [2a]$$

$$TiO_2 + C + 2Cl_2 = TiCl_4 + CO \qquad \Delta H° = -60.2\ kJ/mol \qquad [2b]$$

$$TiO_2 + C + Cl_2 = TiOCl_2 + CO \qquad \Delta H° = +274.4\ kJ/mol \qquad [2c]$$

The presence of $TiOCl_2$ thus also contributes to the lower process temperatures of the 1.0 CO/CO_2 case. In the case of the reduced slag, which contains more Ti^{3+}, more coke is required for complete conversion of Ti to $TiCl_4$; hence the 1.47 CO/CO_2 ratio. The comparable CO/CO_2 ratio for the oxidized slag is 1.39. The negative change in coke supply when moving to 1.0 CO/CO_2 is therefore greater for the reduced slag, resulting in a higher fraction of unconverted $TiOCl_2$: 24 kg-mol/h *vs.* 16 kg-mol/h. This difference is at the endothermic reaction enthalpy of Equation [2c] is sufficient to reverse the effect of Ti^{3+} on the process temperature.

Aside from the coke consumption and the process temperature, the CO/CO_2 ratio does not seem to affect the production rate or chlorine consumption significantly, assuming that $TiOCl_2$ is recovered as successfully as $TiCl_4$ during downstream oxidation to TiO_2.

Table V. Results from the chlorinator thermodynamic model for chlorinator dimensions and operational conditions as described in the text. CO/CO_2 ratio of 1.0.

	CO/CO_2 ratio = 1				Cl_2 = 1 vol%			
	Oxidized slag	Reduced slag	Natural rutile	Synthetic rutile	Oxidized slag	Reduced slag	Natural rutile	Synthetic rutile
Max. ore feed rate (t/h)	15.6	15.6	15.5	15.5	15.5	15.1	15.5	15.5
$TiCl_4$ production rate (t/h) (as TiO_2)	13.7	14.2	14.9	14.5	13.6	13.7	14.9	14.5
$TiCl_4$ production rate (t/t ore)	2.1	2.2	2.3	2.2	2.1	2.2	2.3	2.2
Coke ratio (% of ore)	20.4	20.1	22.3	21.3	22.4	23.0	24.5	23.7
Coke consumption (kg/t $TiCl_4$)	98	93	98	96	108	107	107	107
Cl_2 requirement (kg/t $TiCl_4$)	879	846	797	820	884	874	798	820
Cl_2 consumption (kg/t $TiCl_4$)	125	116	41	69	124	116	41	69
Bed temperature (°C)	940	921	759	833	991	1009	819	894
Energy balance (MW)	-0.002	-0.001	-0.001	0.001	0.007	-0.001	-0.005	0.003
Cl_2 slip (%)	3.8	3.3	4.2	4.0	1.2	1.0	1.0	0.5
Log P_{O2} (atm.)	-15.3	-15.7	-19.5	-17.6	-14.0	-13.6	-17.7	-15.9
Log P_{Cl2} (atm.)	-1.4	-1.5	-1.4	-1.4	-1.9	-2.0	-2.0	-2.3
Raw $TiCl_4$ composition:								
$TiCl_4$ (vol.%)	31.0	30.7	32.1	31.7	33.1	33.7	34.7	34.4
$TiOCl_2$ (vol.%)	3.2	4.9	3.5	3.8	0.1	0.1	0.2	0.3
CO (vol.%)	26.3	25.9	27.2	26.6	32.5	33.5	33.8	33.9
CO_2 (vol.%)	26.0	26.0	27.2	26.6	23.5	22.8	24.7	23.8
$FeCl_3$ (vol.%)	4.6	2.3	0.6	1.8	4.5	2.2	0.6	1.7
Gas fraction (mass% of products)	98.1	97.9	99.5	99.0	98.0	97.9	99.4	98.8
Liquid fraction (mass% of products)	1.9	2.1	0.3	1.1	1.9	2.1	0.5	1.1
Bed height	1.5	2.1	2.5	1.9	1.5	2.1	2.6	2.0

KINETIC MODELLING

Experimental Work

An 80 mm diameter quartz reactor at Mintek was used as a fluid bed chlorinator (Figure 1). Quartz frit at the bottom of the reactor acted as a gas distributor and helped in fluidizing the solids. A thermowell protruding through the frit into the reactor was used to measure the bed temperature. The reactor was placed in a temperature-controlled electric furnace. The furnace temperature controller was connected to the fluid bed thermocouple, thus maintaining the set bed temperature in the reactor. A crossover duct protruded out of the furnace and connected to a condenser.

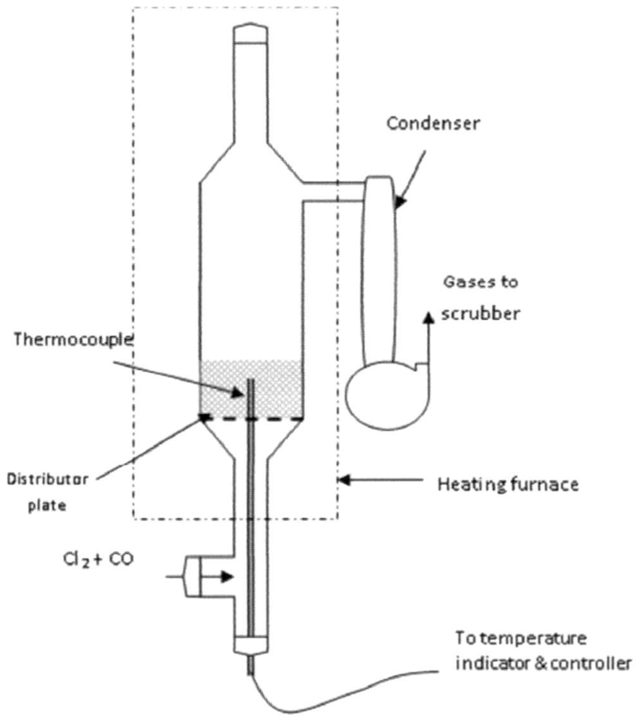

Figure 1. Sketch of the 80 mm experimental chlorinator set-up at Mintek.

Gases entered the bottom of the reactor through a flexible tube suitable for chlorine. All gas flows were measured and controlled using appropriate gas flow rotameters. Cl_2, Ar, and CO gases were supplied from gas cylinders equipped with pressure regulators to maintain constant pressure supply.

In order to limit the effect of particle size on chlorination kinetics, 200 g of feedstock particles in the size range 106–150 µm and 40 g of petroleum coke ranging between 600–850 µm were charged into the quartz reactor. The material was fluidized with argon gas while the reactor set-up and material were heated to 1000°C. Once the reaction temperature was reached, the Ar gas was substituted by an 8 Ndm^3/min Cl_2–4 Ndm^3/min CO gas mixture. The chlorination reaction commenced immediately, with dark brown fumes exiting the reactor. Over time the colour of the fumes changed from dark brown to rusty brown to grey to white. The high-boiling-point metal chlorides were condensed in the condenser and uncondensed gases were vented off to a caustic scrubber. Blown-over particles were collected in the glass condenser.

After the stipulated chlorination time the furnace power was shut off and the CO-Cl_2 gas mixture was replaced with Ar, thus flushing all remaining chlorides from the reactor while cooling the reactor and remaining bed material. When the temperature in the reactor dropped to room temperature, the Ar gas flow was ceased. Material from the reactor was removed and weighed. The petroleum coke and slag were separated using appropriate sieves. The fine petroleum coke in the chlorinated material was burnt off in a muffle furnace maintained at 900°C for 2 hours. The crossover duct, condenser, and venting tubes were washed with water to dissolve all metal chlorides while collecting all blowover particles. The blowover was dried, and weighed.

Chlorination trials on the two ilmenites were terminated at 90 minutes, after it was evident from these tests that the starting bed mass of 200 g was insufficient to last 180 minutes. The bed masses after 90 minutes were also too small for chemical analysis; hence the Ti conversions for the ilmenites were calculated up to 60 minutes only.

Experimental Results

The mass and Ti conversions of the six ore fractions converted (or degree of chlorination) were calculated using Equation [3] and are shown in Figure 2. Among the ilmenites, the mass and Ti conversions are higher for ilmenite B than for ilmenite A. Of the high-Ti ores, oxidized slag, followed closely by synthetic rutile, had marginal higher mass and Ti conversions at 180 minutes. At 90 minutes synthetic rutile and the oxidized slag outperformed natural rutile and reduced slag substantially. Reduced slag consistently had the lowest mass conversion, as well as the lowest Ti conversion at 60 minutes and longer.

The formula for fraction converted (mass or element) is:

$$Fraction\ converted = \frac{m_{in} - m_{bed} - m_{blowover}}{m_{in}} \qquad [3]$$

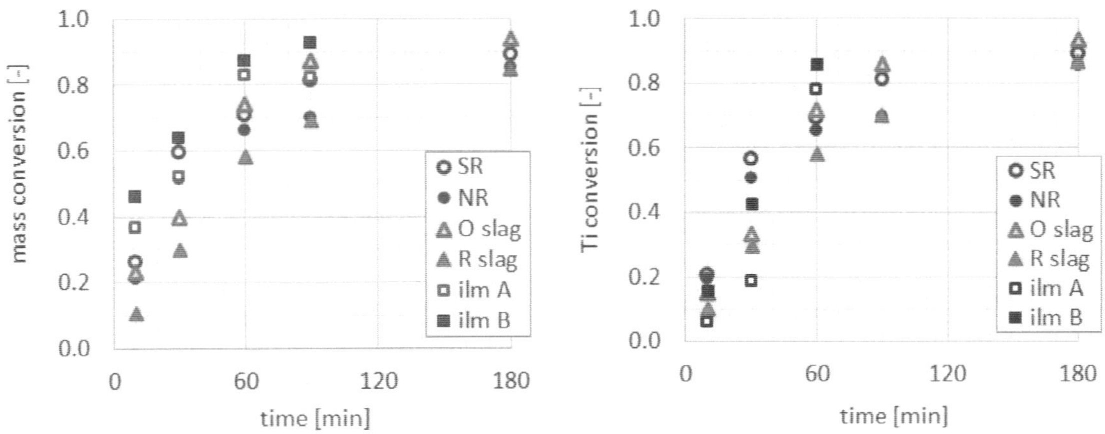

Figure 2. Fractional conversion for the overall mass (left) and Ti (right) over time during experimental carbochlorination at 1000°C. Synthetic rutile – SR, natural rutile – NR, oxidized slag – O slag, reduced slag – R slag, ilmenite A – ilm A, and ilmenite B – ilm B.

The measured conversion kinetics were tested against three kinetic rate laws: a uniform reaction rate (Equation [4a]), shrinking core rate law (Equation [4b]), and pore blocking rate law (Equation [4c]). In all of these equations X_A is the converted fraction of either the total sample mass or Ti (dimensionless); k is the reaction rate constant (min^{-1-}); t is the reaction time (min), and λ is the reaction rate coefficient (dimensionless). The reaction rate coefficient was selected for a best fit with the experimental data. In the present work, the particle size was kept to a narrow range; the gas composition was maintained at a given composition, and the reaction temperature was kept constant. The reaction rate constant k was hence treated as a constant number, although it is dependent on gas composition/partial pressure, reaction temperature and particle size (Sohn and Zhou, 1999; Pistorius and Roux, 2002).

$$-\ln(1 - X_A) = k_1 t \qquad [4a]$$

$$1 - (1 - X_A)^{\frac{1}{3}} = k_2 t \qquad [4b]$$

$$\lambda \left[exp\left(\frac{X_A}{\lambda}\right) - 1 \right] = k_3 t \qquad [4c]$$

The rate plots for above equations are given in Figures 3 to 5. In all cases the linear fits are equally good for all three rate laws; the exceptions being the shrinking core model for synthetic and natural rutile, which have regression coefficients of 0.85 compared to 0.91 and 0.98 for the uniform reaction rate and pore blocking rate respectively.

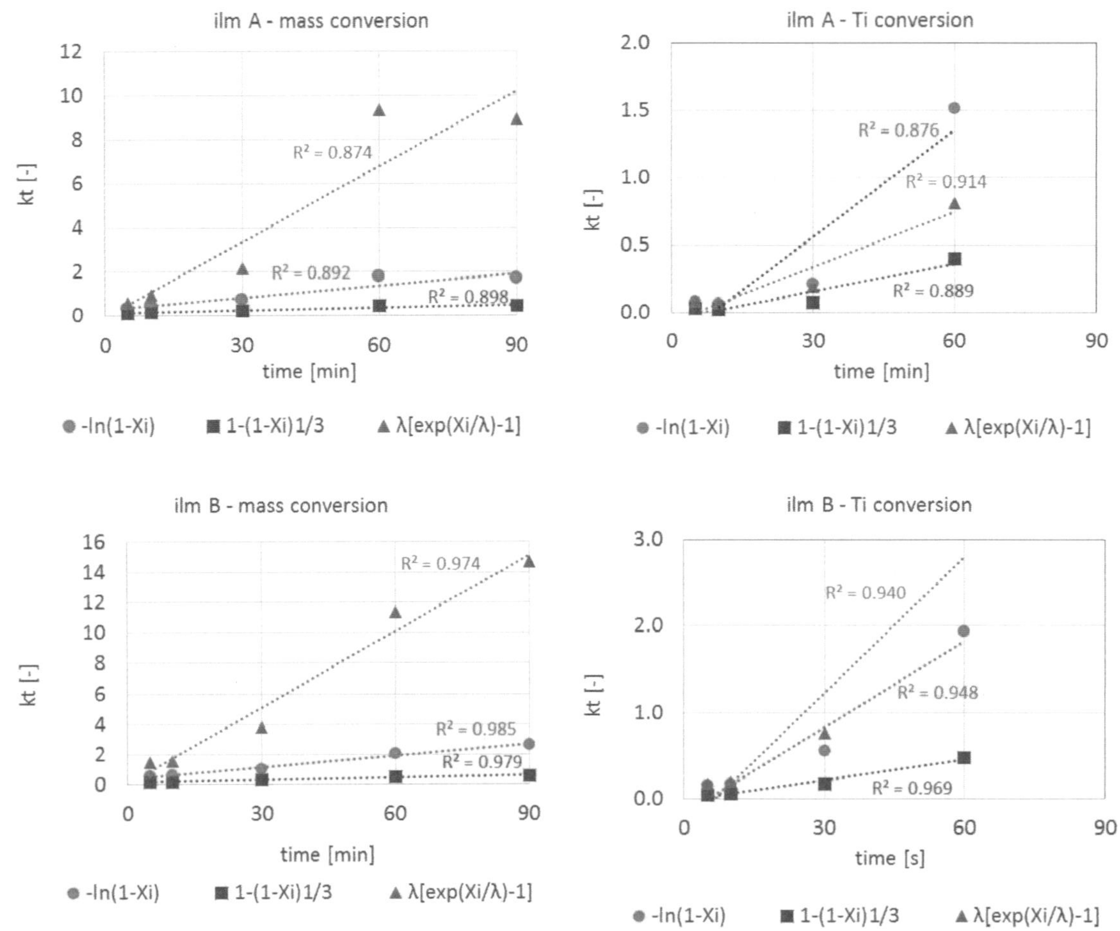

Figure 3. Kinetic rate law plots for ilmenite A (top) and ilmenite B (bottom).

The rate constants k were calculated from the slopes of Figures 3 to 5. Using these rate constants, the required reaction times for 80% mass and Ti conversion were calculated by inserting the k-values into the rate laws with X_A=0.80. The results are given in Tables VI and VII. For the reduced slag and natural rutile, almost all the rate laws fit the (interpolated) experimental data equally well at 80% mass and Ti conversions. Overall, for all six ores at 80% mass and Ti conversion, the pore blocking rate law gives as good a fit or better with the interpolated experimental data.

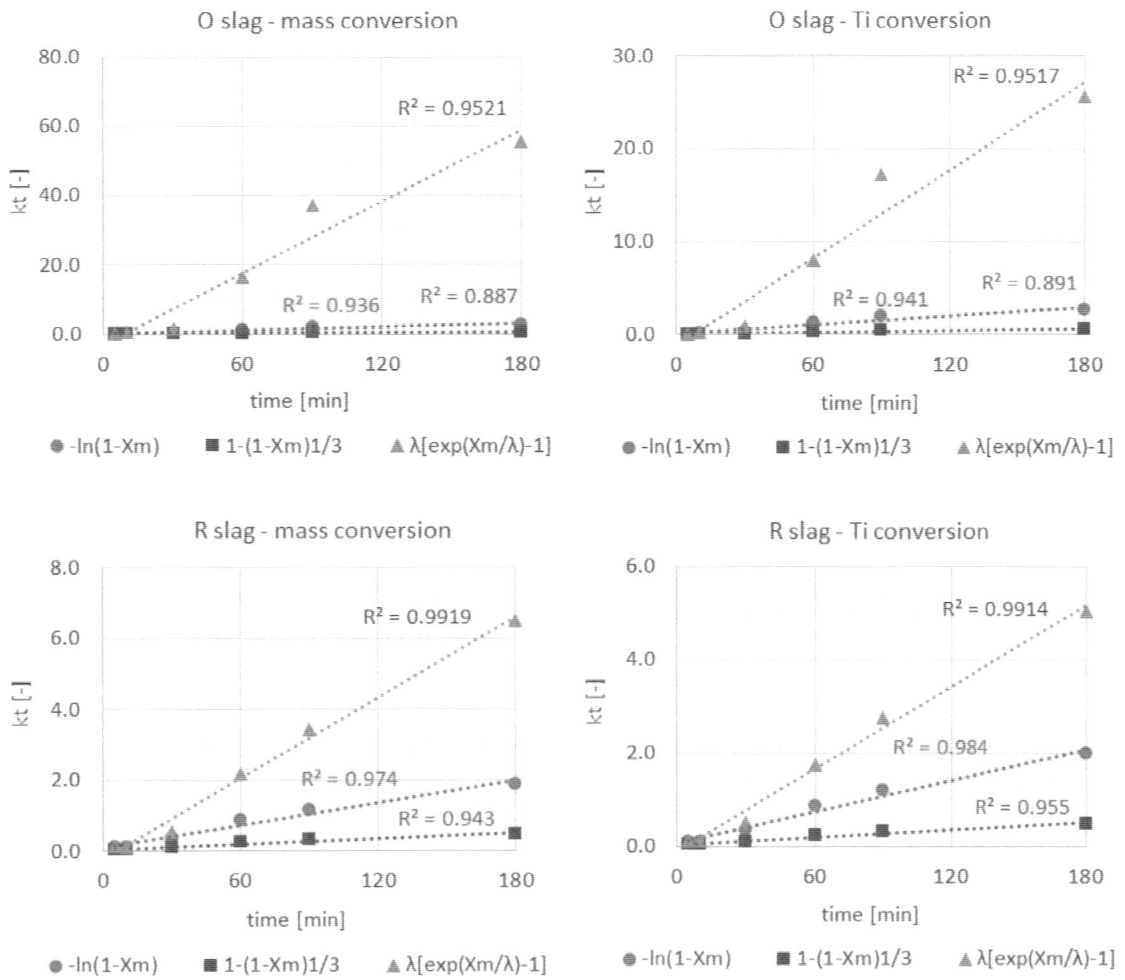

Figure 4. Kinetic rate law plots for oxidized (top) and reduced (bottom) slags.

Table VI. Experimentally vs. rate law calculated times (min) to 80% mass conversion. Rate constants (min^{-1}) and kinetic coefficients (dimensionless) for the pore blocking rate law were selected for a best fit with the experimental data.

	Ilmenite A	Ilmenite B	Oxidized slag	Reduced slag	Synthetic rutile	Natural rutile
Experimental time (interpolated)	57	50	74	154	86	150
Uniform reaction model	88	64	104	155	140	181
Shrinking core	97	82	127	164	162	145
Pore blocking rate	64	47	73	146	92	150
	k =0.1144	k =0.1677	k =0.3415	k =0.0377	k =0.2469	k =0.0780
	λ = 0.22	λ = 0.22	λ = 0.16	λ = 0.26	λ = 0.16	λ = 0.20

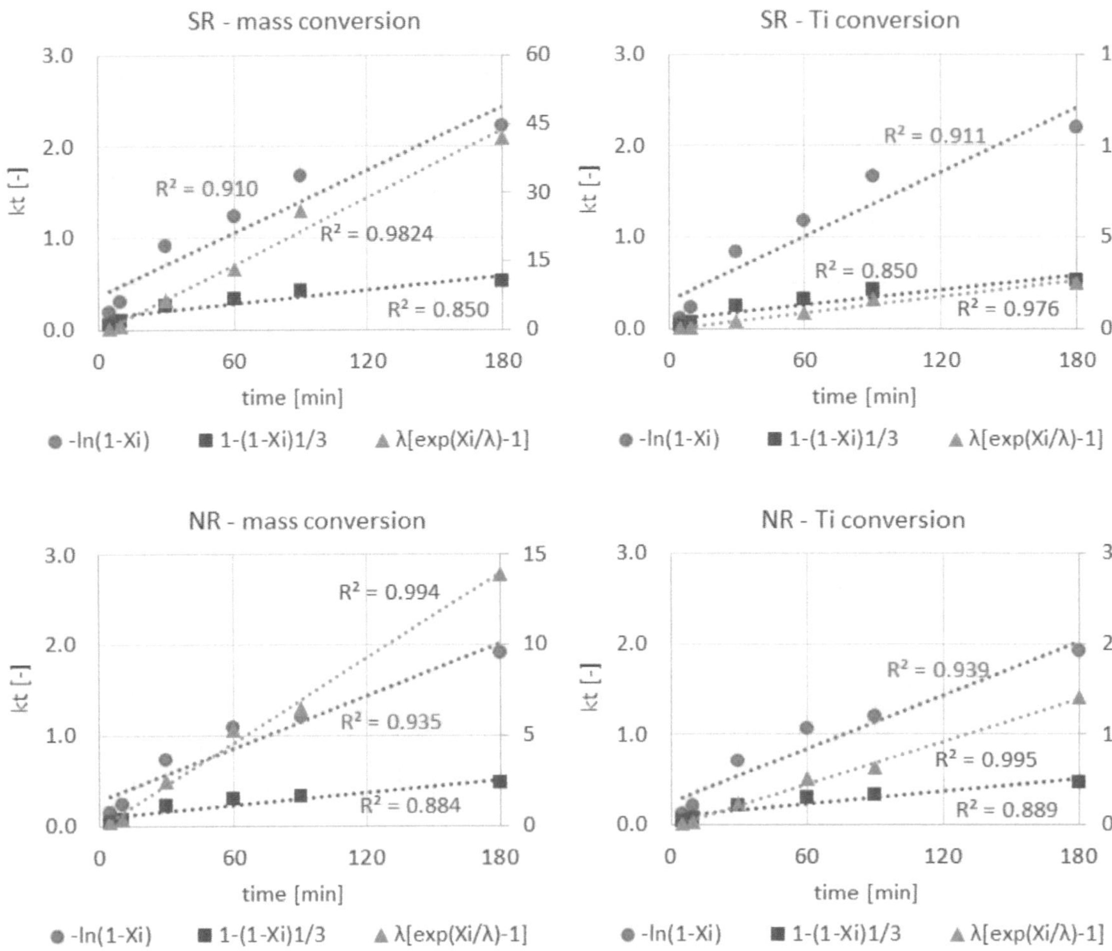

Figure 5. Kinetic rate law plots for synthetic (top), and natural (bottom) rutiles. The pore blocking rate law references to the secondary y-axis.

Table VII. Experimentally vs. rate law calculated times (min) to 80% Ti conversion. Rate constants (min^{-1}) and kinetic coefficients (dimensionless) for the pore blocking rate law were selected for a best fit with the experimental data.

	Ilmenite A	Ilmenite B	Oxidized slag	Reduced slag	Synthetic rutile	Natural rutile
Experimental time (interpolated)	61	58	78	143	90	155
Uniform reaction model	27	61	103	142	139	162
Shrinking core	44	71	123	154	157	176
Pore blocking rate	66 *k* =0.0.0136 λ = 10	71 *k* = 0.0558 λ = 0.35	81 *k* = 0.1577 λ = 0.19	140 *k* = 0.0293 λ = 0.30	91.3 *k* = 0.1466 λ = 0.18	156 *k* = 0.0779 λ = 0.20
Target residence time for 90% Ti conversion (min)	75	88	141	193	187	214
Mass conversion at 90% Ti conversion (%)	78	89	91	88	90	90
Bed height (m)	0.58–0.66	0.62–0.70	1.5	2.1	1.9–2.0	2.5–2.6

The mass and Ti conversions calculated with the pore blocking rate law (using λ and k given in Tables VI and VII) are compared with the experimental data in Figure 6. For the ilmenites, the fit with mass conversion is good; although not so much for lower degrees of Ti conversion. For both slags, good fits are obtained from 60% conversion and upwards - on both mass and Ti. The same is true for synthetic and natural rutiles. Based on these results the pore blocking rate with constants given in Table VII was used for the kinetic modelling.

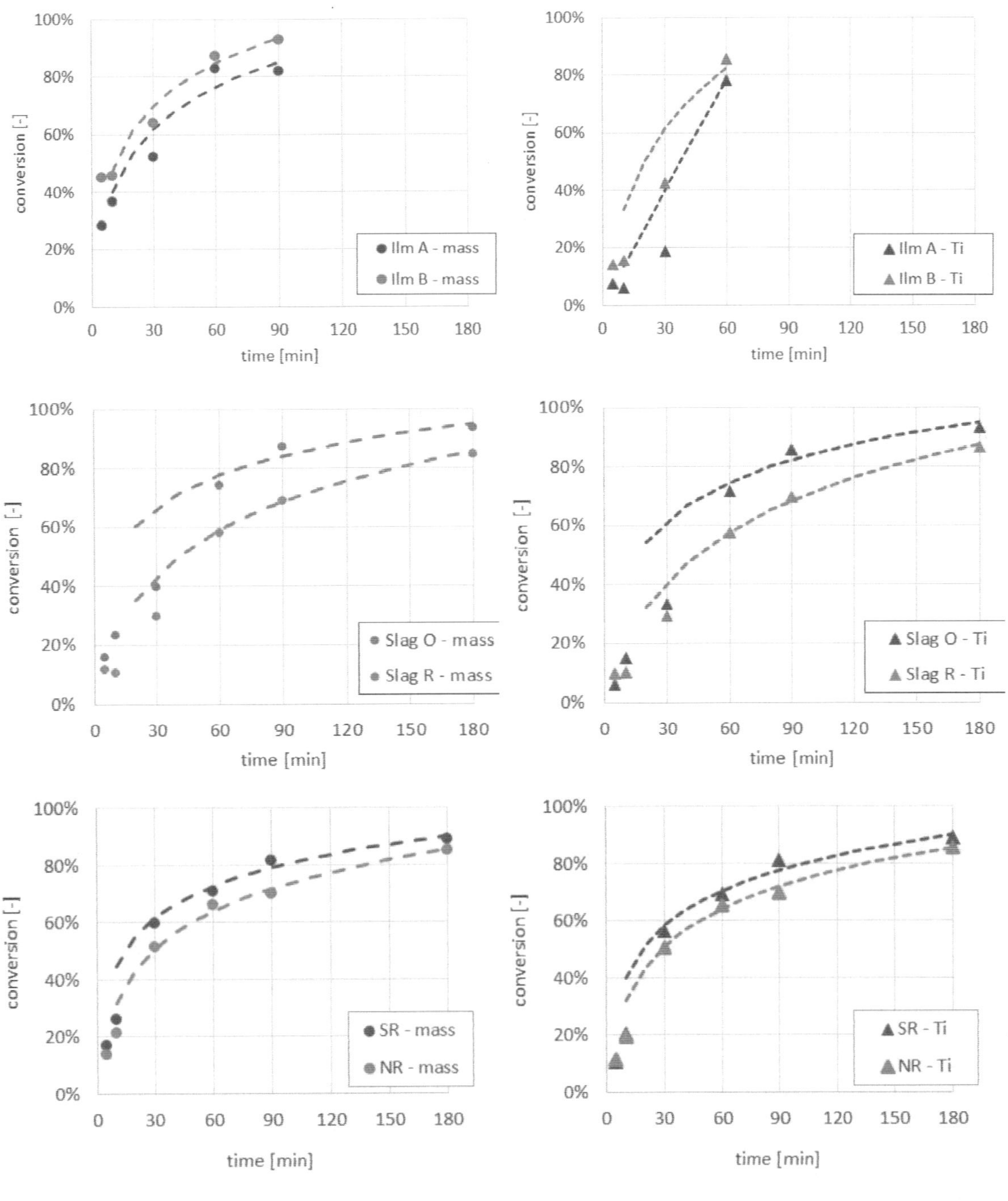

Figure 6. Mass (left) and Ti (right) conversions against time: experimental data (markers), and pore blocking rate law (lines). Top row – ilmenites; middle row – slags, and bottom row – rutiles.

Kinetic Model Results

For all six ore types the average reaction times for 90% Ti conversion were used as the target chlorinator residence times (Table VII). The average residence time for the various ores was calculated from the method given by Kunii and Levenspiel (1991):

$$\bar{t} = \frac{Weight\ of\ a\ particular\ size\ in\ the\ bed}{Flow\ rate\ of\ that\ size\ into\ the\ bed} \qquad [5]$$

Using Equation [5} the average ore residence times were calculated separately from that of the whole bed (where the whole bed includes ore and petroleum coke). The units of the average residence time $\bar{t}$ are determined by the units used on the right-hand side of the equation. With the bed diameter fixed by the dimensions of the hypothetical chlorinator, the ore feed rate as per the mass and energy balance (thermodynamic model), and the target residence time from the chlorination test work, the minimum bed height for each ore was calculated (Table VII).

Typical chlorinator bed heights are between 2.1 and 3.6 m (Hills, 2015). This includes reactive (fresh ore) and unreactive portions. The unreactive portion is made up by the accumulation of unreactive species such as quarts and zirconia and can be of substantial depth. Excessively deep beds are detrimental for elutriation losses due to solids being ejected from bursting bubbles at the bed surface (Kunii and Levenspiel, 1991) and coalescing bubbles resulting in inhomogeneous mixing (Chapadgaonkar and Setty, 1999).

Of the two ilmenites, ilmenite B has a slightly longer residence time and deeper bed. This is against expectation: ilmenite B would be expected to have higher porosity for two reasons: it is a more weathered ilmenite and hence should have a higher initial porosity that ilmenite A, and its initial Fe conversion occurred much faster (more than 90% Fe conversion within 5 minutes compared to ilmenite A which required close to 30 minutes). Nell and den Hoed (2003) correlated the mass chlorination of an ilmenite with a progressive increase in surface area (*i.e.* porosity). In their work, the Fe conversion behaved similar to ilmenite A (den Hoed and Nell, 2003). Initial rapid chlorination of Fe creates higher porosity, allowing access for reactant gases to unreacted zones, thus leading to fast chlorination. From Figure 6 the Ti conversion of ilmenite B is in fact slightly faster than that of ilmenite A. The discrepancy with the calculated residence times results from the poor fit with the kinetic rate law. An improved fit will be possible with shorter time intervals between experimental measurements.

Among the high-Ti ores, the oxidized slag had the lowest target residence time and bed height, followed by synthetic rutile and then the reduced slag. Natural rutile had the longest residence time, and consequently the deepest bed. From surface area measurements (Nell and den Hoed, 2003), slag has a higher surface area than natural rutile (but still lower than ilmenite at mass conversions above 20%). These workers used traditionally block-route cooled slag with 87% TiO_2. Elsewhere it was also shown that in high-titania slags, preferential chlorination of iron and Ti^{3+} creates porous matrixes (Pistorius and Roux, 2002).

Synthetic rutile undergoes an acid wash during production to remove metallic iron and is thus already porous when it enters the chlorinator (Benson, Mellor, and Jackson, 2015). It is possibly this porosity that allows synthetic rutile to chlorinate slightly faster than the reduced slag, despite it having a similar Fe content but substantially less Ti^{3+} than the reduced slag. The increase in actual surface area due to its porosity needs to be confirmed, though (if the pores are too large they will not affect the surface area substantially).

CONCLUSIONS

The results from the thermodynamic model outlined in Tables IV and V are theoretical and specific for a given production rate and chlorinator size. The absolute numbers are therefore not representative of every chlorinator. The value lies in comparing the numbers to a benchmark. In Table VIII, synthetic rutile is set as the benchmark. Ore feed rates, $TiCl_4$ production rates, and coke and chlorine consumptions are expressed as ratios: less than unity is lower than synthetic rutile; greater than unity is higher than synthetic rutile. Bed temperatures and bed heights are deltas: a positive number is higher than synthetic rutile, and a negative number lower. Where two numbers are given, *e.g.* coke consumption as 1.7/2.1, the first applies to the 1.0 CO/CO_2 case, and the second to the 1% Cl_2 slip case. Only one number implies that the outcomes of the two cases were similar relative to the corresponding synthetic rutile number.

Table VIII. Comparative performance data for the various feedstocks expressed relative to those of synthetic rutile. X represents the feedstock and SR represents synthetic rutile. Where two numbers are given, the first applies to the 1.0 CO/CO_2 case; the second to the 1% Cl_2 slip case. One number indicates a similar outcome for both cases.

		Ilmenite A	Ilmenite B	Oxidized slag	Reduced slag	Natural rutile	Synthetic rutile
Max. ore feed rate (t/h)	X/SR	0.92	0.93	1.0	1.0	1.0	1.0
$TiCl_4$ production rate (t/t ore)	X/SR	0.50	0.59	0.93	0.96	1.0	1.0
Coke consumption (kg/t $TiCl_4$)	X/SR	1.7/1.2	1.5/1.0	1.0	0.97/1.0	1.0	1.0
Cl_2 requirement (kg/t $TiCl_4$)	X/SR	2.1	1.8	1.1	1.1	0.97	1.0
Cl_2 consumption (kg/t $TiCl_4$)	X/SR	8.7	6.1	1.8	1.7	0.6	1.0
Bed temperature (°C)	$X-SR$	-41/171	-54/177	107/97	88/115	-75	0
Bed height (m)	$X-SR$	-1.4	-1.3	-0.5	+0.1	+0.6	0.0

The high-Ti ores outperform the ilmenite feedstocks in all aspects: the maximum ilmenite feed rates are 8% lower than those for the high-Ti ores; the $TiCl_4$ per unit ore mass is 41–50% lower than that of synthetic rutile; the chlorine consumption is 6–9 kg/t $TiCl_4$ for the ilmenites, and the coke consumption is 20–70% higher than for synthetic rutile.

Among the high-Ti ores there is no difference between the maximum feed rates. The slags have lower $TiCl_4$ production rates of 4% and 7% – oxidised slag having the smaller difference. The higher Ti^{3+} content of the reduced slag does seem to benefit coke consumption when running 1% Cl_2. This effect is neutralized when running 1.0 CO/CO_2.

The significant difference between the high-Ti ores lies in the chlorine consumption. Though their Cl_2 requirements do not differ much, natural rutile has a gangue content of 4%, whereas synthetic rutile contains 8% and the slags 13–15%. Higher gangue contents result in higher chlorine losses. The slags consequently consume 70–80% more chlorine than synthetic rutile, while natural rutile consumes 40% less.

A second potentially significant difference among the high-Ti ores is the process temperature. The slags result in process temperatures on average 100°C higher than synthetic rutile. In chlorinators where the heat removal is constrained by equipment capabilities (*e.g.* downstream heat exchanging capacity), the production rate may need to be reduced when running on slags (Hills, 2015). This applies even more when chlorinating ilmenite at high CO_2 levels.

With the better understanding from the thermodynamic model, it is clear that no general statement can hold for the performance of any chlorinator or any feedstock. The performance of an ore depends on both its chemical composition (oxidation states of iron and titanium and impurity levels) and chlorinator operating philosophy. The following principles hold:

- The carbon consumption of high-Ti ores can be reduced by running at 1.0 CO/CO_2 – but potentially at the cost of kinetics (lower process temperature and lower CO partial pressure). The opposite is true for ilmenites: a 1.0 CO/CO_2 ratio increases the carbon consumption substantially at lower the process temperatures

- The forms and contents of Fe and Ti in an ore can play a determining role in the process temperature as was shown for the two slags, and slags *vs*. rutiles. However, the effect can be overshadowed by the CO/CO_2 ratio in the chlorinator
- The impurities in an ore largely determine the chlorine consumption – assuming the degree of chlorine recovery remains constant.

From kinetic measurements it was shown that synthetic rutile has faster chlorination kinetics than natural rutile and reduced slag. Oxidized slag chlorinates faster than reduced slag. From the preceding discussions, traditionally block-cooled slag is expected to chlorinate slightly faster than oxidized slag: it has similar levels of FeO with slightly higher Ti_2O_3.

Thus far, the feedstocks were treated as separate feedstocks in the chlorinator. Pigment producers often combine feedstocks to reduce or benefit from the impact of a component and/or effect. In addition, the effect of varying particle size distributions of the different feedstocks needs to be incorporated in the kinetic modelling. In continuing with the research programme initiated by Iluka Resources, these aspects will be further investigated by looking at both thermodynamic and hydrodynamic effects of feed blends.

ACKNOWLEDGEMENTS

We would like to express our gratitude to Iluka Resources for sponsoring this project and giving their permission to publish the results. We are also grateful for the work conducted by the supporting staff of Mintek.

REFERENCES

Benson, L., Mellor, I., and Jackson, M. (2015). Solid state extraction and consolidation of titanium alloys direct from synthetic rutile. *Proceedings of the 13th World Conference on Titanium*. The Minerals, Metals & Mterials Society, Warrendale, PA. pp. 71-75.

Bungu, P. and Pistorius, P. (2009). Mineralogy and initial chlorination of water granulated high titania slag. *Canadian Metallurgical Quarterly*, 48 (1), 45-51.

Burger, H., Bessinger, D., and Moodley, S. (2009). Technical considerations and viability of higher titania slag feedstock for the chloride process. *Proceedings of the 7th International Heavy Minerals Conference 'What Next'*. Southern African Institute for Mining and Metallurgy, Johannesburg. pp. 187-194.

Chapadgaonkar, S.S. and Setty, Y.P. (1999). Residence time distribution of solids in a fluidised bed. *Indian Journal of Chemical Technology*, 6 (March), 100-106.

Cong, X., Zhangfu, Y., and Xiaoqiang, W. (2006). Preparation of $TiCl_4$ with the titanium slag containing magnesia and calcia in a combined fluidized bed. *Chinese Journal of Chemical Engineering*, 14 (3), 281-288.

Eriksson, G. and Pelton, A.D. (1993). Critical evaluation and optimisation of the theermodynamic properties and phase diagrams of the $MnO-TiO_2$, $MgO-TiO_2$, $FeO-TiO_2$, $Ti_2O_3-TiO_2$, Na_2O-TiO_2, and K_2O-TiO_2 systems. *Metallurgical Transactions B*, 24B (October), 795-805.

Gazquez, M., Bolivar, J.P., Garcia-Tenorio, R., and Vaca, F. (2014). A review of the production cycle of titanium dioxide pigment. *Material Sciences and Applications*, 5, 441-448.

Harris, H., Henderson, A., and Campbell, T. (1976). Fluidized coke-bed chlorination of ilmenites. US Bureau of Mines, Washington.

Hills, D. 2015. Private communication [Interview].

Den Hoed, P. and Nell, J. (2003). The behaviour of individual species in the carbochlorination of titaniferous oxides. *Proceedings of Heavy Minerals 2003.* South African Institute of Mining and Metallurgy, Johannesburg. pp. 43-56.

Kale, A. and Bisaka, B. (2011). Fluid bed chlorination pilot plant at Mintek. *Journal of the Southern African Institute of Mining and Metallurgy,* 111 (March), 193-197.

Kang, J. and Okabe, T.H. (2014). Thermodynamic consideration of the removal of iron from titanium ore by selective chlorination. *Metallurgical and Materials Transactions B,* 45B (August), 1260-1271.

Kang, S. and Lei, Z. (1996). Measurement of the thermodynamic properties of $Fe_2Ti_3O_9$ and Fe_2TiO_2. *Transactions of NFsoc,* 6 (1), 25-31.

Kunii, D. and Levenspiel, O. (1991). *Fluidization Engineering.* 2nd edn. Buttherworth-Heinemann Series in Chemical Engineering, Newton.

Moodley, S., Kale, A., Bessinger, D., Kucukkaragoz, C., and Eric, R.H. (2012). Fluidisation behaviour of various titania feedstocks. *Journal of the Southern African Institute of Mining and Metallurgy,* 112 (June), 67-471.

Nell, J. and den Hoed, P. (2003). Carbochlorination of rutile, titania slag and ilmenite in a bubbling fluidised-bed reactor. Document Transformation Technologies, Cape Town, South Africa. pp. 1426-1433.

Pistorius, P. and le Roux, J.F. (2002). Thermal, chemical and structural changes during initial chlorination of titania slag. *Canadian Metallurgical Quarterly,* 41 (3), 289-298.

Rhee, K. and Sohn, H. (1990). The selective chlorination of iron from ilmenite ore by CO-Cl_2 mixtures: Part I. Intrinsic kinetics. *Metallurgical Transactions B,* 21B (April), 321-330.

Shimizu, T., Matsuura, K., Furue, H., and Matsuzak, K. (2013). Thermal conductivity of high porosity alumina refractory bricks made by a slurry gelation and foaming method. *Journal of the European Ceramic Society,* 33, 3429-3435.

Sohn, H. and Zhou, L. (1999). The chlorination kinetics of beneficiated ilmenite particles by $CO+Cl_2$ mixtures. *Chemical Engineering Journal,* 72, 37-42.

Hanlie Kotzé

Consulting Metallurgist
Consensi

Hanlie Kotzé qualified as a metallurgical engineer from the University of Pretoria in 1994 and immediately commenced with post graduate studies on high titania slags. Beginning of 1996 she commenced working with Iscor Heavy Minerals as a process engineer and was privileged to be part of the ilmenite smelting commissioning teams. Following commissioning she lead successful business improvement teams, and later also the Process Development section. In 2007 she completed her PhD on the cooling and processing of high titania slags - also at the University of Pretoria. During 2007 Hanlie was Technical Manager for the KZN Sands operations. She opted for life as a consultant and is since 2008 consulting on pyrometallurgical questions for the heavy minerals and furnace industries.

Monitoring lining wear on submerged arc furnaces

P. Schmidt[1], M. Hopf[1], and D. Zeelie

[1]Saveway GmbH & Co. KG, Germany

High-temperature electric arc furnaces, including submerged arc furnaces, are frequently equipped with water-cooling. The proximity of liquid metal/matte/slag and water results in inherent explosion risks, with potential consequential loss of production, extensive repairs, and possible injuries or fatalities. SAVEWAY provides practical refractory wear management solutions for induction furnaces, ladles, open arc furnaces, and submerged arc furnaces. This paper will discuss two practical applications for the monitoring of (a) bulk refractory in a furnace vessel with spray-cooled sidewalls and (b) intensively water-cooled copper tap-blocks. Fibre-optic sensors are installed in specially arranged piping on the inside surface of the spray-cooled shell and inside the copper tap-blocks. Multiple temperature readings can be obtained over the entire length of the fibre-optic sensors. Using the changes in cooling water temperature, dimensional design, and thermal conductivity of the refractory materials, the lining thickness can be calculated on a continuous basis. The practical installation of the sensors will be discussed and the results of the wear measurement compared to the theoretical FEM (finite element method) calculations. This technique offers the efficient and safe monitoring of tap-holes, sidewalls, and other high-risk components of pyrometallurgical smelters.

INTRODUCTION

Furnace operators have identified lining wear as the primary factor limiting increased furnace campaign life. Developments in refractory design and refractory cooling techniques during the last 50 years have contributed significantly to extended lining life. Cooling water systems offer the most efficient heat exchange, at the risk of water ingress, explosions, costly repairs, and personnel safety. Rapid expansion during the phase transition from liquid to water vapour can have catastrophic results, especially when liquid water and metal/slag come into contact inside a furnace vessel.

Thermomechanical and chemical erosion consume furnace refractory linings and water-cooled parts (*e.g.* high-intensity water-cooled sidewall copper panels and copper tap-blocks), reducing the thickness of the physical barrier between the hot metal/slag and the cooling water. At a certain 'wear point' furnace production must be halted to replace the lining or the water-cooled parts (Figure 1).

Figure 1. Critical wear of a copper sidewall cooler panel. The red arrow indicates an existing damaged cooling tube (Hopf, 2014).

Furnace operators are therefore challenged with a tradeoff between safety and maximizing the furnace campaign life. From a safety perspective, comprises cannot be made, but the furnace campaign must be extended to ensure maximum financial benefit.

All possible methods must be used to ascertain the thickness of the 'barrier between the metal/slag and water; surface analysis by laser scanner, hot-spot detection by thermocouples, or by studying return water temperatures.

This paper presents an overview of the high-accuracy OPTISAVE lining monitoring system.

OPTISAVE FUNDAMENTALS

Thermal Gradients and Lining Thickness

The condition of the refractory lining can be derived if the temperature is measured at an appropriate position, and if the material properties (heat conductivity and heat transfer coefficient) of the materials used are known. The location of the temperature measurement has to satisfy the demand for a detectable temperature increase in the case of wear. To determine whether the location of the temperature measurement is suitable, finite element method (FEM) calculations are utilized to obtain the solid-state temperature profiles for different wear situations. The boundary conditions of the heat equation need to obtained: (a) firstly, the temperature of the hot surface (matte or slag temperature), and (b) secondly, the temperature of the outer shell of the furnace (cooling water temperature). With these two boundary conditions and the material properties for heat transfer, it is possible to calculate the temperature at every point between the outer shell of the furnace and the matte or slag for different lining wear cases.

Under steady-state conditions, the lining design properties are used to define the 'new lining' condition. Design properties (or material properties) will be thermal conductivity, heat transfer coefficient(s), dimensions, and orientation. Boundary conditions will include the hot-face temperature (metal/slag) and outer shell temperature (determined by the cooling water). The thermal profile is calculated to provide an indication of the temperature at any location in the lining between the outer shell and the hot face.

The calculations are then extended for different 'simulated wear' conditions (Figure 2).

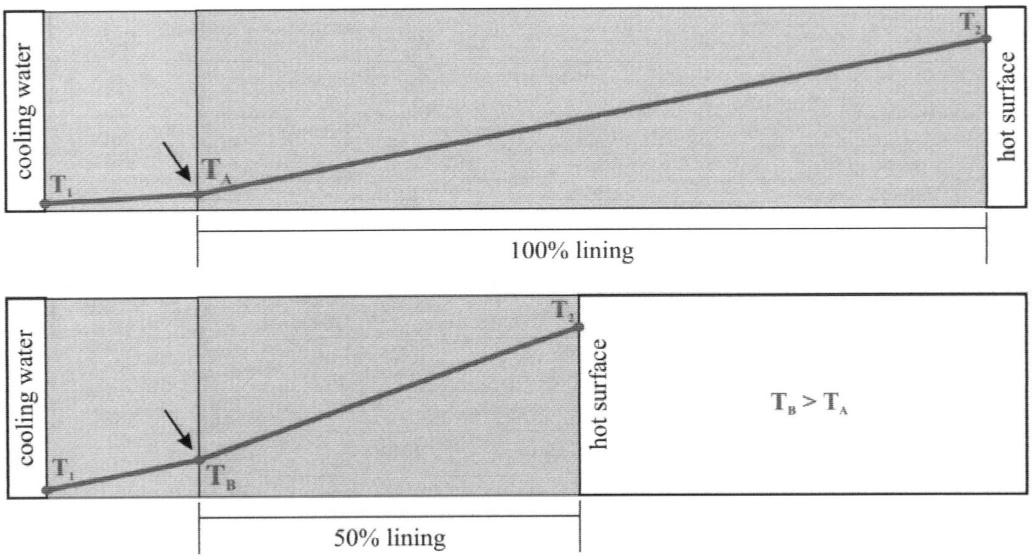

Figure 2. Temperature profile for lining without wear and with 50% wear. Note the temperature increase from T_A to T_B.

In the simulation for the 'new lining' with no wear, the temperature T_A is lower than the temperature T_B in the case with 50% wear at the same measurement position. It therefore possible to predict the condition of the lining based on predetermined thermal profiles, if T_A or T_B can be obtained. All things being equal, an increase in T_B can correspond with a reduction in lining thickness. A reduction in T_B can be caused by re-establishing the freeze lining in a particular area. Because material properties show variances, practical calibration will be required after commissioning and based on on-site experimental work.

This fundamental principle can add value to practice only if T_B can be accurately measured.

Fibre Optic Thermal Measurement

OPTISAVE utilizes the Raman scattering theory in fibre optics to measure temperature. Temperature measurement with optical fibre is a well-established technique and provides high levels of accuracy and repeatability. Different measuring methods are available and have to be evaluated for specific applications. Optical time domain reflectometry (OTDR) and optical frequency domain reflectometry (OFDR) are most commonly used. Both measuring methods allow a localized temperature measurement along an optical fibre.

Relationship between Temperature and Raman Scattering

According to quantum physics, light is considered to consist of small particles (photons). Photons are inserted in a solid body like an optical fibre and some of the photons are scattered (reflected) from atoms or molecules. Most photons are elastically scattered without a change in energy. A minor fraction of the photons is inelastically scattered, resulting in a small change in energy.

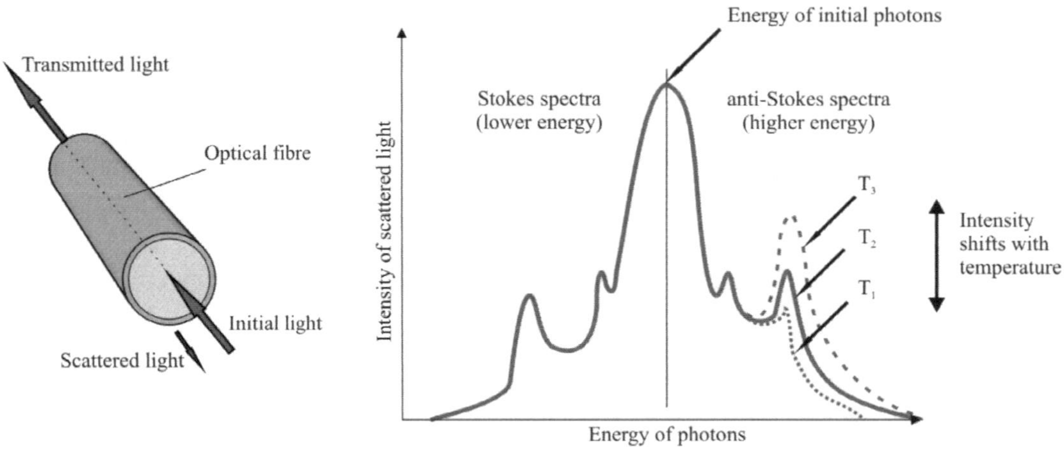

Figure 3. Light scattering in an optical fibre.

Figure 3 shows two spectra on the horizontal axis; the Stokes spectrum and the anti-Stokes spectrum. Photons in the Stokes spectrum retain lower levels of energy and photons in the anti-Stokes spectrum, higher levels of energy. The anti-Stokes spectrum is strongly temperature-dependent compared with the Stokes spectrum for a given temperatures in a solid body.

Determination of Local Temperature Distribution by OTDR

SAVEWAY applies the OTDR technique to the OPTISAVE systems. Using OTDR the location of the temperature measurement is determined by the time required by the light (photons) to travel to the location of scattering and back again to a detector. The location of scattering is the location where the temperature measurement takes place. A pulsed laser is used to avoid superposition of the scattered light from different locations.

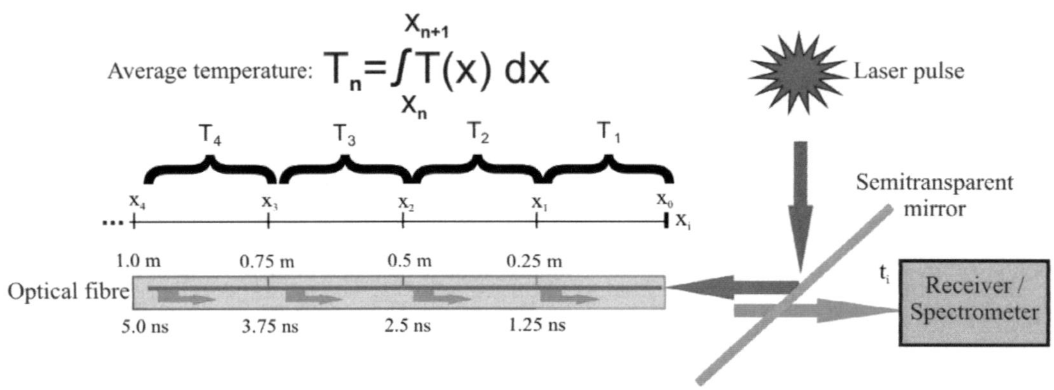

Figure 4. Principle of OTDR method to determine the location of scattering.

Figure 4 presents the principle of OTDR with a laser pulse delivered into the optical fibre via a semitransparent mirror. Scattering takes place at different locations (x_i) along the optical fibre. The scattered light travels back to the semitransparent mirror and is evaluated by a spectrometer (receiver of photons). To distinguish between different locations the spectrometer captures the spectra in short time intervals. The location of scattering is calculated by Equation [1].

$$2 \cdot x_i = c \cdot (t_i - t_0) \qquad [1]$$

where x_i is the distance to the location of scattering, c is the speed of light in the optical fibre, t_i is the time of the detection in the spectrometer, and t_0 is the time when the initial laser pulse was presented into the optical fibre. Due to limitations of the spectrometer, the minimum interval for the time measurement is 1.25 ns. This leads to a maximum spatial resolution of 250 mm. On each 250 mm section the average (integral) temperature is calculated with a repeatability of 0.1 K and an absolute accuracy of ±0.5 K for optical fibres up to 10 km length.

OPTISAVE SENSOR INSTALLATION

Ideally, OPTISAVE sensors are installed on the inner surface of a furnace shell. In hazardous environments, or if it is impossible to fix an OPTISAVE sensor on a surface, embedding technology has to be applied.

OPTISAVE Sensor on the Inner Side of a Water-cooled Steel Shell

OPTISAVE sensors are installed on the inside of a furnace shell before relining of the furnace.

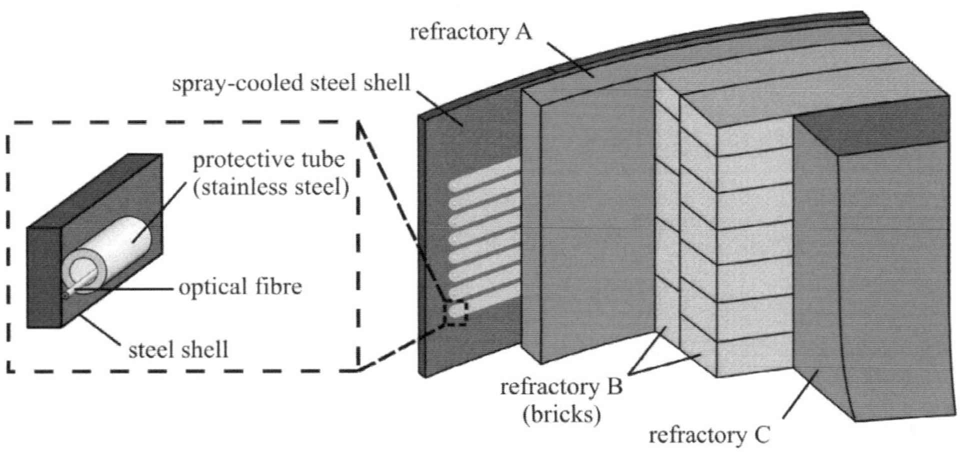

Figure 5. OPTISAVE installation on the inner side of a spray-cooled steel shell .

In Figure 5 the layout of an OPTISAVE installation on a water-cooled steel shell of a submerged arc furnace (SAF) is illustrated. The optical fibre is housed in a stainless steel protective tube that is fixed on the steel shell of the furnace. To monitor an area of the water-cooled steel shell, OPTISAVE sensors are installed at different levels. Figure 6 illustrates an OPTISAVE installation on the water-cooled steel shell of a submerged arc furnace.

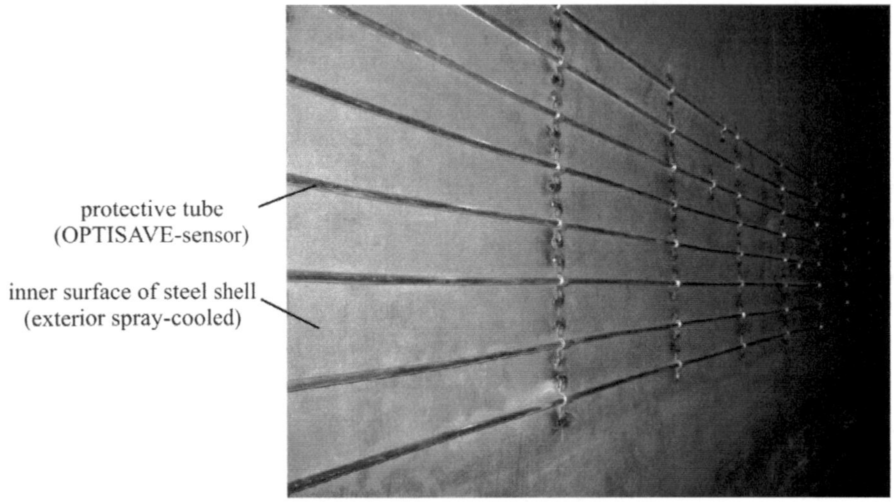

Figure 6.OP'TISAVE sensors on the inside of a water-cooled steel shell of a submerged arc furnace.

The FEM results are presented in Figure 7. OPTISAVE sensor readings range from approximately 50°C (no wear) to about 100°C (90% wear). With the given temperature resolution of 0.1 K it is possible to detect wear in a sufficient number of steps (a 50 K temperature increase equals about 500 steps).

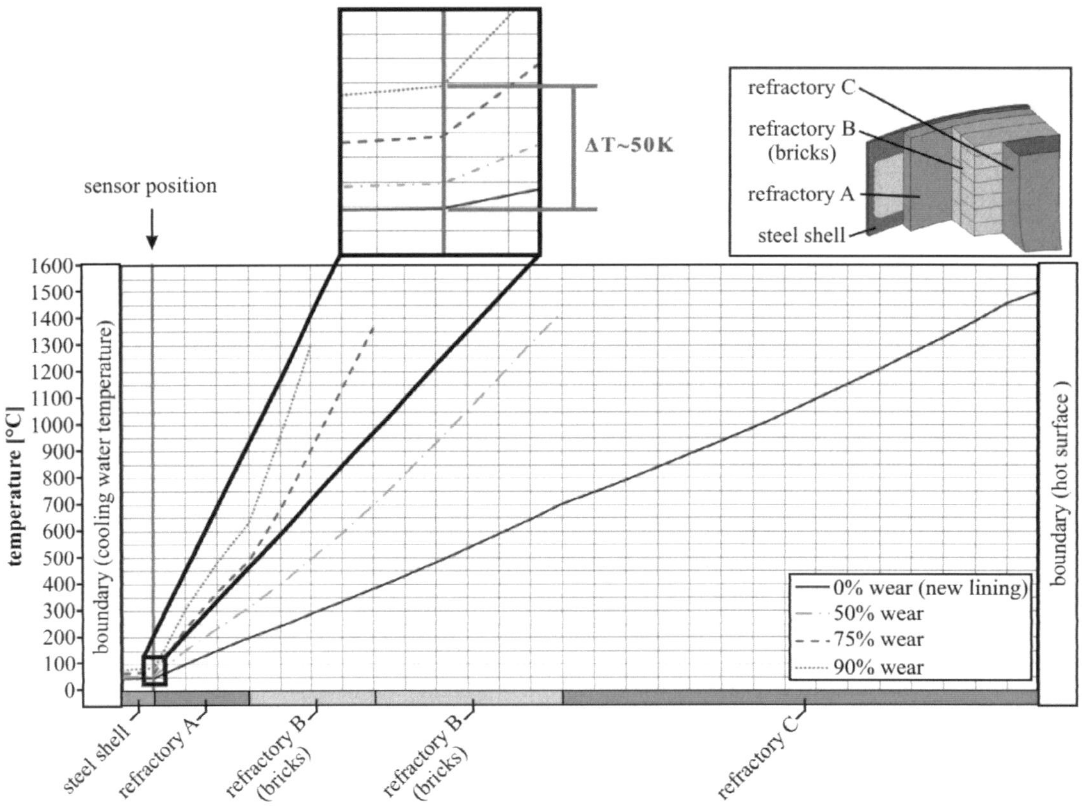

Figure 7. Temperature profiles measured by an OP'TISAVE installation on the inner side of a water-cooled steel shell for different cases of wear obtained by FEM calculations.

Embedding Technology of an OPTISAVE Sensor into Water-cooled Copper Sidewall Panels and Copper Tapbblocks.

If the environmental conditions are too severe for the sensor, *e.g.* on cooling blocks (compare Figure 1 - even the copper block is corroded), the sensor cannot be placed outside of the block. To ensure a long lifetime the OPTISAVE sensor has to be placed inside the copper block. Therefore SAVEWAY, together with the Thos Begbie company, developed a specialized embedding technology allowing the sensor to be placed inside the copper block at the required position during casting of the block. The embedding technology was patented by SAVEWAY and Thos Begbie in 2005 (Hopf and Rossouw, 2007). With this embedding technology, two major advantages arise. Firstly, the sensors are well protected from the environmental conditions by the surrounding copper. Secondly, since the sensors are already pre-installed the on-site effort is greatly reduced. Furthermore, the embedding technology allows the optical fibre to be changed without damaging the copper block or panel.

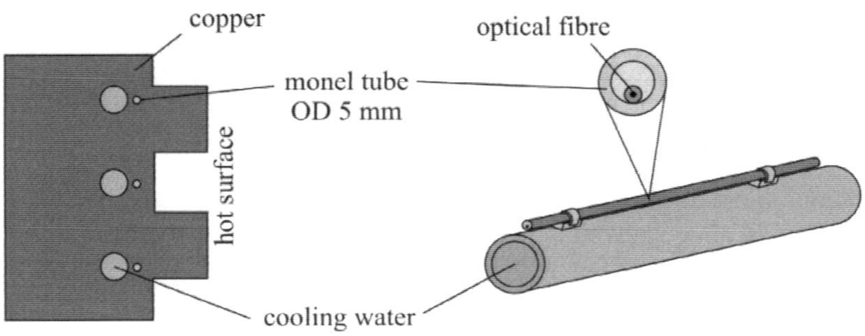

Figure 8. Principle of the embedding technology. Copper is cast around a pre-manufactured cooling tube with an additional monel tubing system (Hopf, 2014).

Figure 8 depicts the principle of the embedding technology. A tubing system (serving as a cooling water channel), with an additional host tube system for the optical fibre, is pre-manufactured and placed in a mould before casting the copper around the system. The host tube system is arranged according to the required position between the cooling channels (tubing system) and the hot surface of the copper block or copper panel. For this reason the host pipe usually follows the structure of the cooling channels (see Figure 9). After casting and solidification of the copper, the system provides safe monitoring of the endangered areas. The first application of the embedding technology on a submerged arc furnace was brought on-line in a slag tap-hole in 2010.

To ensure that the OPTISAVE sensor is capable of detecting the change of temperature at the intended sensor position, a theoretical FEM calculation of the temperature distribution was performed. The results are shown in Figure 10. The major decrease in temperature takes place in the refractory lining (bricks) due to its low thermal conductivity compared to copper. Although copper has a thermal conductivity approximately 100 times higher than the refractory lining, one can still see a detectable change of temperature at the sensor position. If the maximum temperature increase at the position of measurement is calculated to be 16 K, assume a change of wear from zero to 100%. Thus, by taking into account the given temperature resolution of 0.1 K, the remaining lining thickness can be divided into a maximum of 160 steps. Table I shows the calculated temperatures at the OPTISAVE sensor position for the depicted cases of wear.

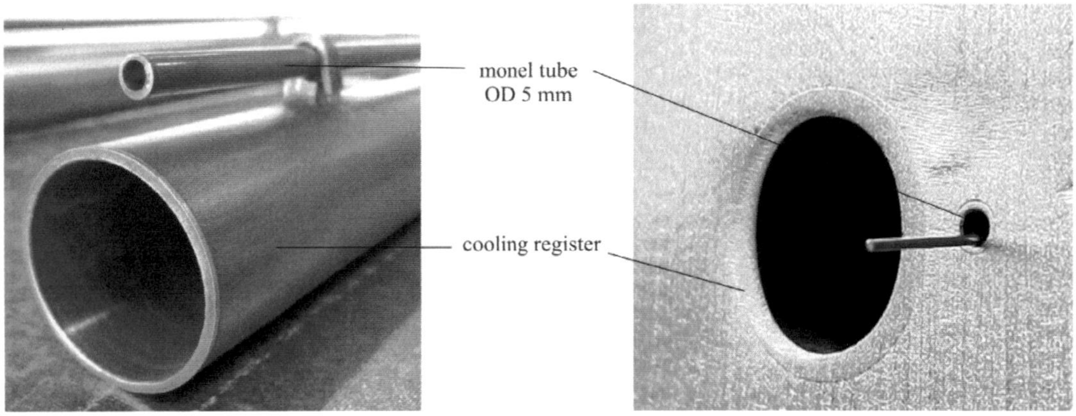

Figure 9. Photograph of a pre-manufactured tubing system with additional monel host tube, before and after casting of the copper, using the embedding technology (Hopf, 2014).

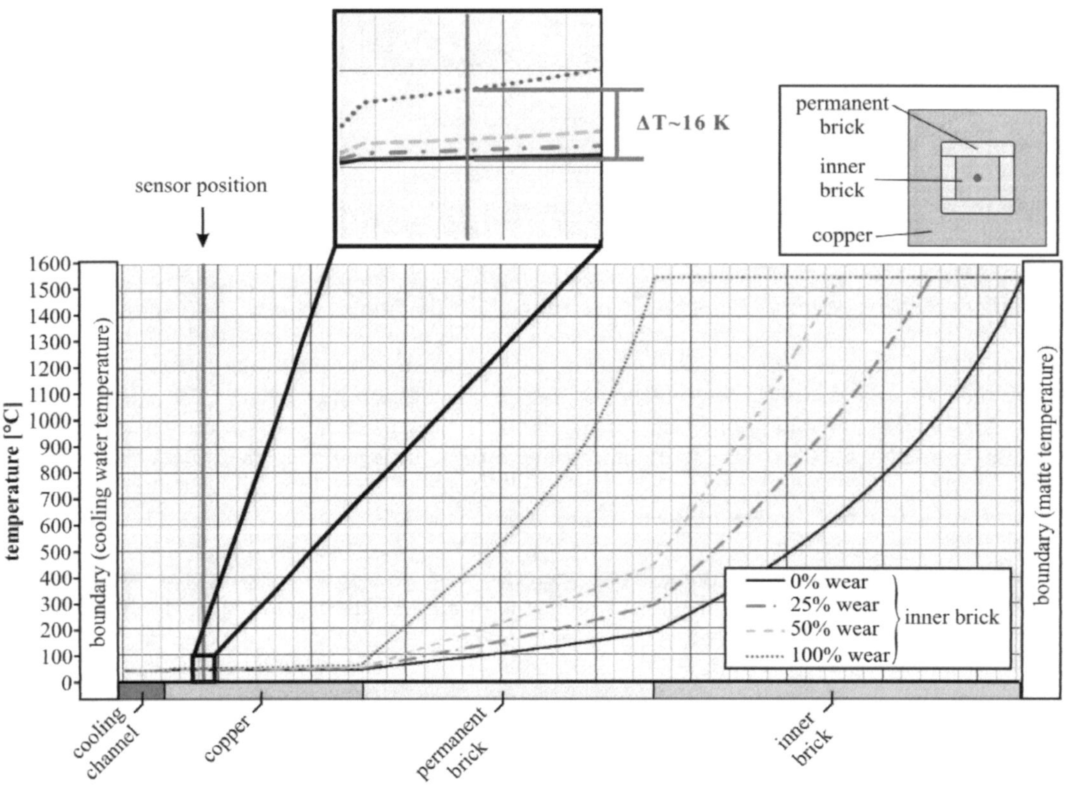

Figure 10. Temperature profiles for different cases of wear obtained by FEM calculations on a OPTISAVE installation in a matte tap-block manufactured using the embedding technology.

Table I. Temperature distribution at sensor location obtained by FEM calculations (Hopf, 2014).

	0% wear	25% wear	50% wear	100% wear
Temperature [°C]	42.36	44.13	46.82	58.79
Temp. increase [K]	-	1.77	4.46	16.43

The absolute change of temperature in Table I indicates that the small temperature increase of 1.77 K in the case of 25% wear, can still be divided into a sufficient number of steps. With the given temperature resolution of 0.1 K it is possible to detect a change in wear between zero and 25% in 17 steps. As the temperature increase becomes steeper with advancing wear, it is certainly possible to detect dangerous wear-levels. In cases of high wear, the temperature increase at the sensor location will be up to 16.43 K. As these are only theoretical values, a calibration is performed to calculate the actual remaining lining thickness from a change in temperature.

PRACTICAL APPLICATIONS

OPTISAVE System on the Inner Side of a Water-cooled Steel Shell

In this particular example the OPTISAVE system was used to monitor the spray-cooled steel shell of a submerged arc furnace with a diameter of about 20 m. The OPTISAVE sensors monitor a height of 1 m in eight levels, leading to more than 2000 measurement values. Figure 11 depicts the temperature readings of one measurement section of an OPTISAVE sensor on a water-cooled steel shell between two maintenance cycles. At the furnace start-up we observe a high temperature increase due to the heating up of the furnace when power is turned on. After this first high increase, the temperature increases at a lower rate with a constant slope. The constant slope indicates ongoing wear of the refractory material. The constant wear of the refractory material was measured and confirmed during the removal of the old lining

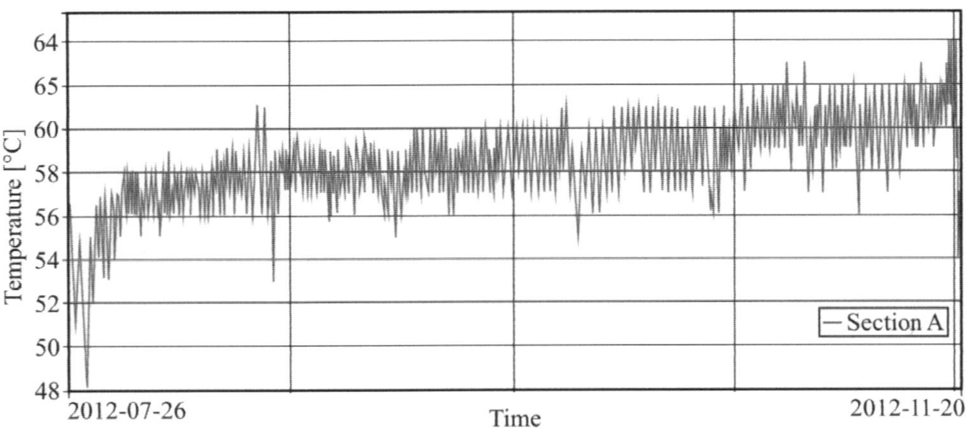

Figure 11. Temperature reading on one measurement section of an OPTISAVE sensor on the water-cooled steel shell of a submerged arc furnace.

As the OPTISAVE sensors are mounted at different levels on the steel shell inside the submerged arc furnace, the system is capable of locating temperature hot-spots. A screenshot example of the OPTISAVE user interface indicating temperature hot-spots is presented in Figure 12. The highest temperatures can be observed on two measurement sections, reading about 90°C. The surrounding measurement sections show only a slight temperature increase according to the remote measurement sections. The remote measurement sections show a temperature reading of about 50°C. In this particular case we displayed a calculated remaining lining thickness of 25% at the hot-spot location, whereas the thickness at the remote measurement sections is close to 100%.

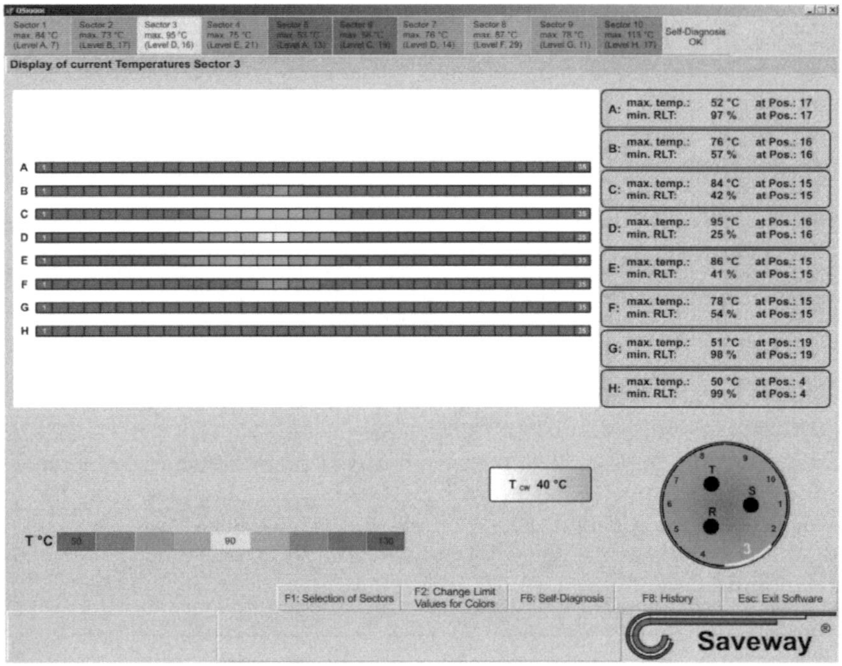

Figure 12. Screenshot of the operator screen of an OPTISAVE system indicating localized wear and displaying the minimum remaining lining thickness (RLT) of every level measured (A-H).

OPTISAVE System on a Matte Tap-hole.

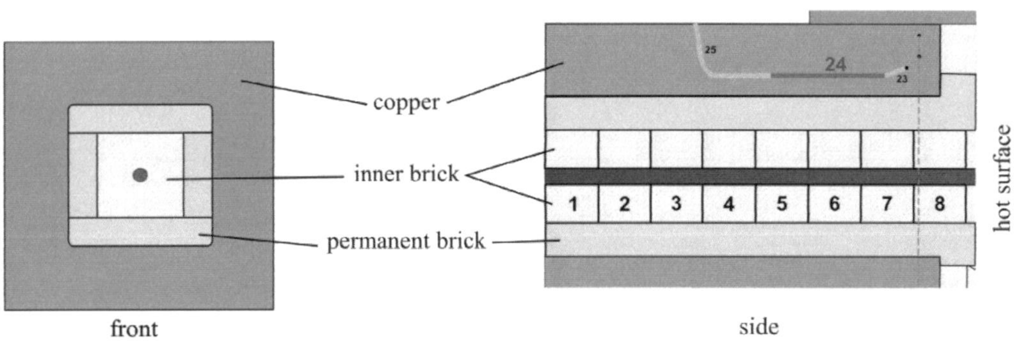

Figure 13. Schematic layout of the monitored matte tap-hole of a submerged arc furnace (Hopf, 2014).

Figure 13 depicts the monitored matte tap-hole, the lining of which consists of several permanent and inner bricks. The bricks numbered from 1 to 8 are changed during regular maintenance cycles. The OPTISAVE sensors are embedded in the water-cooled copper hosting block. The position of the measurement sections inside the hosting block is shown in Figure 14.

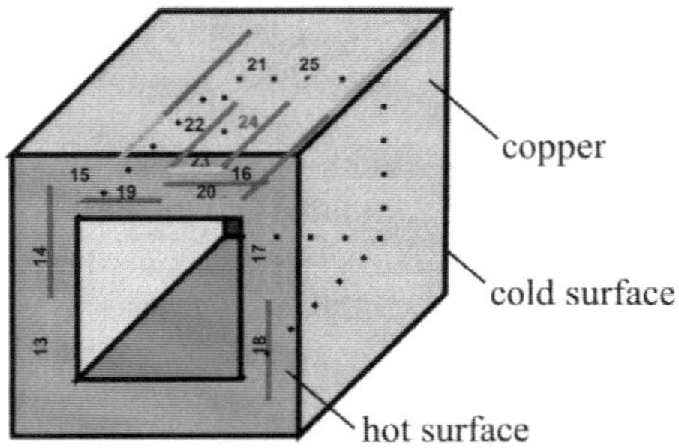

Figure 14. Position of the measurement sections inside the copper hosting block of a matte tap-hole block manufactured using the embedding technology (Hopf, 2014).

In addition to the measurement positions depicted in Figure 14, the temperature of the cooling water (T27) was measured. Figure 15 shows clearly that all measured temperatures vary over time. As all the measured temperatures follow the trend of the temperature T27 (cooling water), we conclude that this is the main impact on the temperature readings. The temperature T27 is influenced by unknown factors like the status of cooling fans, the weather, tapping or other furnaces on the same cooling circuit. This leads to the unpredictable fluctuations in all measured temperatures (see Figure 15).

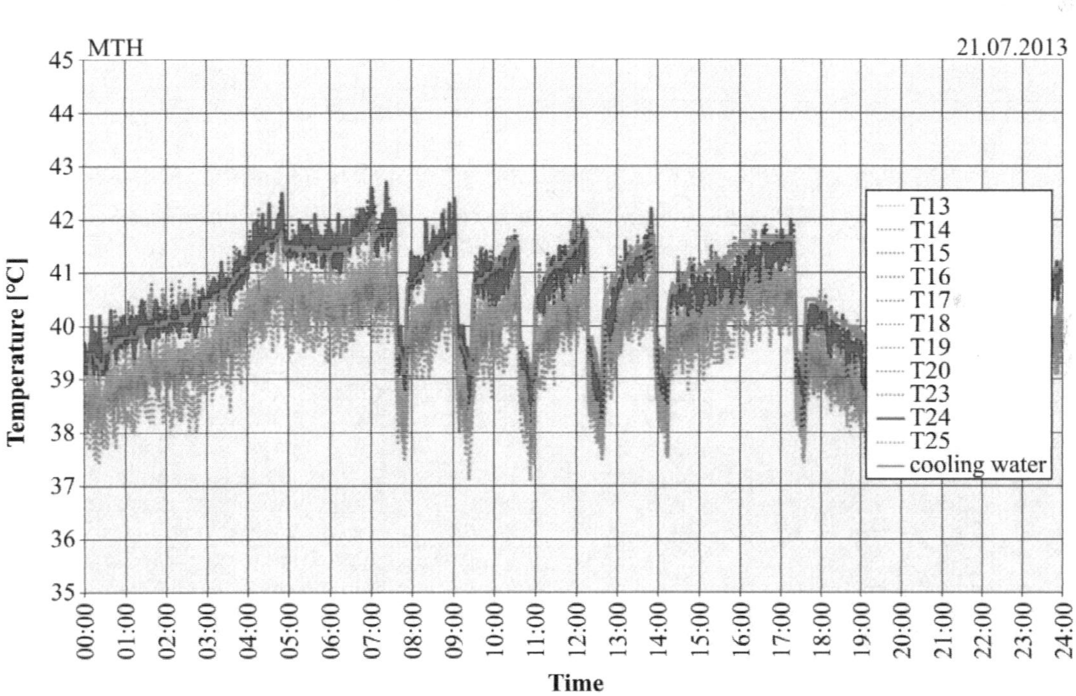

Figure 15. Variation in temperatures of the measurement sections inside the copper hosting block of a matte tap-hole and the temperature of the cooling water during the course of a day (Hopf, 2014).

Therefore the impact of the cooling water has to be eliminated from the measured temperatures. To do this we subtracted the temperature T27 (cooling water) from each temperature signal and, most recently, added the mean temperature of T27 (calculated to a fixed value of 41.5°C). The corrected temperature value is calculated by:

$$T_{i\,corrected} = T_i - T27 + 41.5°C \quad [2]$$

where $T_{i\,corrected}$ is the corrected temperature, T_i is the measured temperature of the measurement section I, and T_{27} is the temperature of the cooling water. The influence of the temperature of the cooling water is thus eliminated for improved analyses and interpretation. In the subsequent discussion, the influence of tapping on the measurement section 24 is investigated using the corrected temperature values (see Figure 14 for position of measurement section 24).

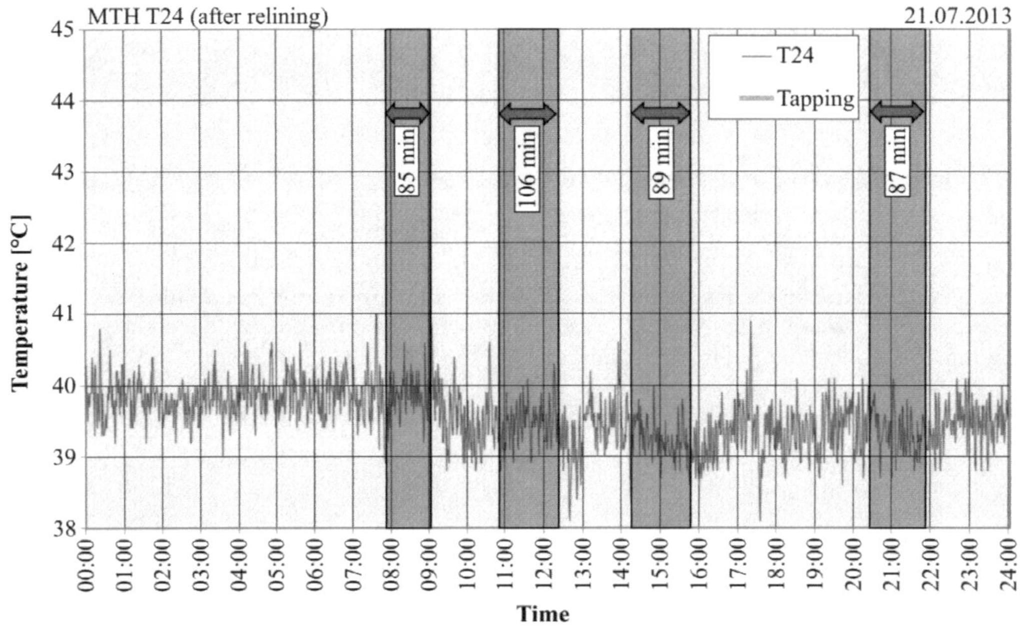

Figure 16. No significant influence from tapping on the corrected temperature of measurement section 24 after relining (no wear) (Hopf, 2014).

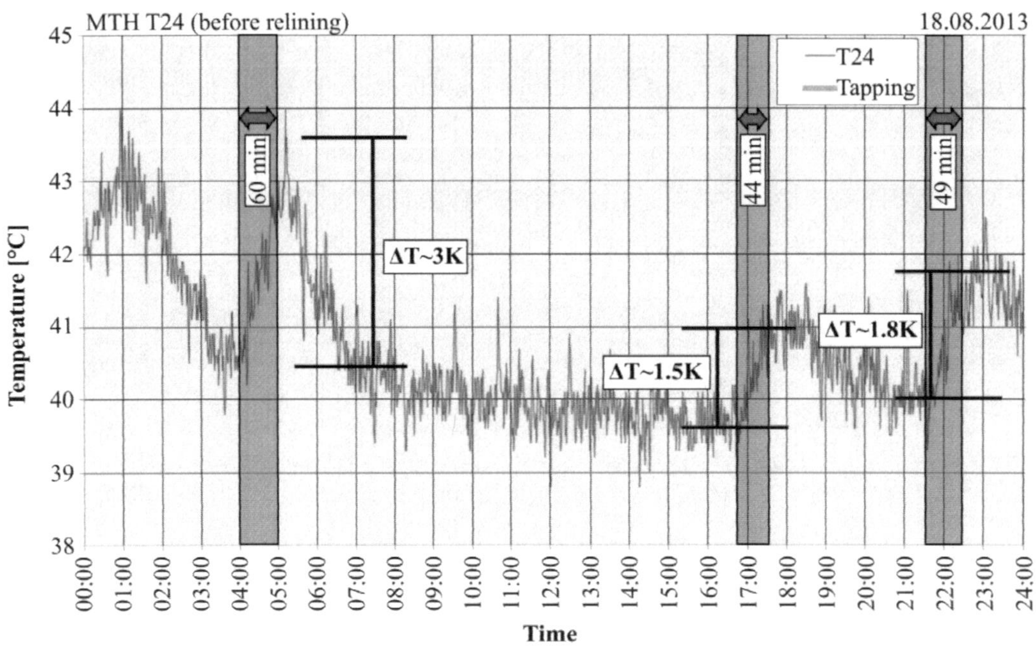

Figure 17. Significant influence from tapping on the corrected temperature of measurement section 24 before relining (wear) (Hopf, 2014).

Figures 16 and 17 show the corrected (in terms of cooling water influence) temperature of measurement section 24 (T24) immediately after a reline, and before the next reline. After a fresh relining, no influence of tapping on the temperature T24 can be observed. However, immediately before the next maintenance cycle took place, the influence of tapping on the temperature T24 is significant. During tapping the temperature T24 increases between about 1.5 K and 3 K, which we relate to less remaining refractory thickness. During the maintenance cycle a breakout was performed, showing visible wear of several bricks.

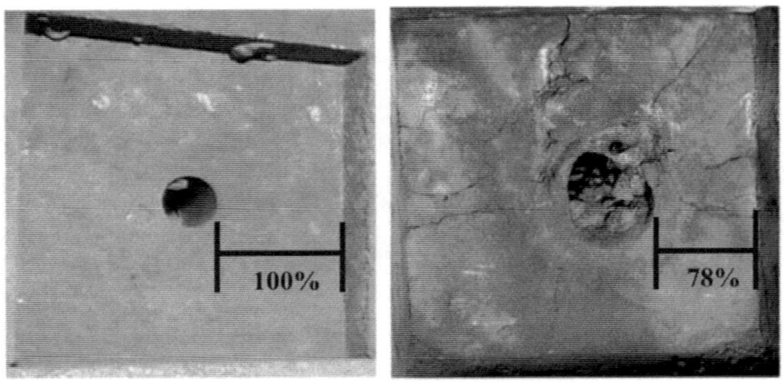

Figure 18. Remaining thickness of the inner brick number 6 of a matte tape-hole without wear (100% after relining) and with wear (78% before maintenance cycle) (Hopf, 2014).

The difference between the unused and the worn-out brick number 6 (located close to measurement position 24) is shown in Figure 18. For brick number 6 the wear was calculated at 22% with reference to the original refractory thickness. The increase in temperature at measurement position 24 of about 1.5 K to 3 K for 22% wear is in the range of the values predicted by our FEM calculation (an increase of 1.77 K in case of 25% wear). The exact temperature increases during tapping were used for the calibration of the OPTISAVE system. Breakouts carried out later during maintenance cycles confirmed the temperature increase in the case of 25% wear.

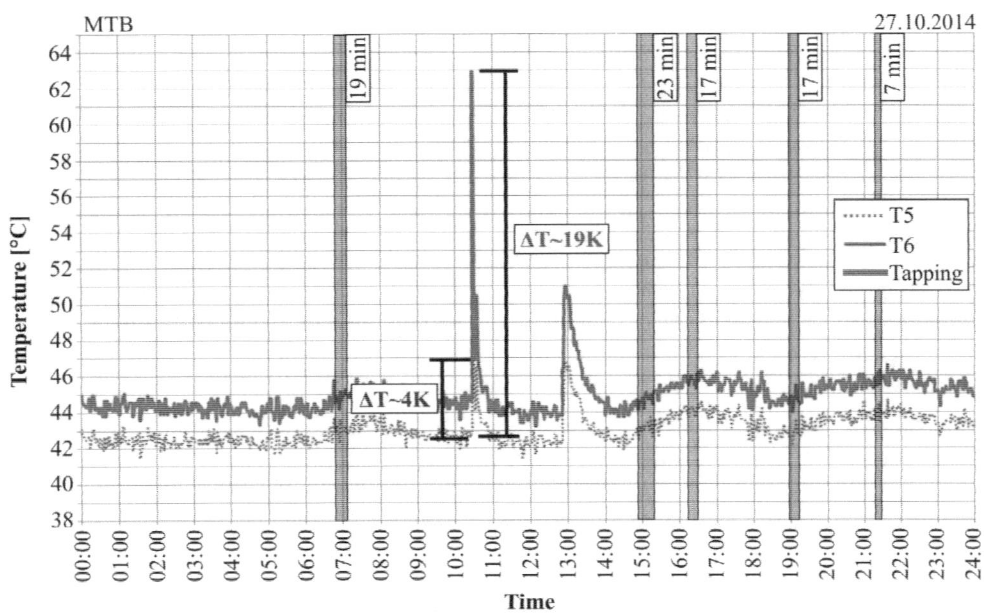

Figure 19. Temperature readings on two adjacent measurement sections while using an oxygen lance (at 10h30) on the matte tap-hole.

The temperature readings on two adjacent measurement sections while an oxygen lance was being used on the matte tap-hole are depicted in Figure 19. While the temperature T5 shows only a small temperature increase of 4 K during oxygen lance use, the temperature of measurement section 6 increases by 19 K. Consultation with the furnace service staff indicated that the oxygen lance was used at the wrong angle, resulting in damage to some bricks. As only measurement section 6 shows this high temperature increase, we conclude that only the bricks close to this measurement section were damaged. Therefore this incident indicates that the OPTISAVE system is capable of detecting localized wear.

At present all sensors are still functioning properly, providing plausible temperature measurements after 3 years of operation. During the period of operation the data captured confirms that the OPTISAVE system can be used to determine the wear situation on the refractory inlays of the tap-hole. As we have shown, the OPTISAVE system is capable of detecting wear accurately. Due to the increasing temperature gradient in case of higher wear (see Figure 10), SAVEWAY claims to be able to safely identify all critical wear situations and constant wear progress.

CONCLUSIONS

The OPTISAVE system was developed to monitor lining wear in water-cooled components on submerged arc furnaces to satisfy the need for the greatest possible safety and cost efficiency. A comparison between the temperature profiles of the lining for different cases of wear leads to a direct correlation between the measured temperature and the remaining lining thickness. Hence it is concluded that it is possible to use temperature measurement to calculate the remaining lining thickness, by including appropriate processing of the measured data to provide a convenient and user-friendly indication of wear. Monitoring of the lining wear can avoid unplanned shutdowns and allow for materials optimization. A technology for embedding of OPTISAVE sensors into cast copper parts was introduced to ensure a long product lifetime and a low installation effort on site. As shown, the embedding technology allows the monitoring of the remaining lining thickness and the location thereof. This results in safer operation of intensively water-cooled parts and allows for a predictable maintenance plan.

REFERENCES

Braun, W., Gebski, P., MacRosty, R., Pula, R., Sadri, A., and Ying, W. (2014). Tap-hole monitoring technologies. *Proceedings of the Furnace Tapping Conference 2014*, Muldersdrift, South Africa. Southern African Institute of Mining and Metallurgy, Johannesburg. pp. 169-182.

Hopf, M. and Rossouw, E. K. (2007). Wandpaneel für Schmelzöfen. German patent DE 10 2005 013 924.

Hopf, M. (2014). Monitoring the wear of water-cooled tap-hole blocks by the OPTISAVE fibre-optic method. *Proceedings of the Furnace Tapping Conference 2014*, Muldersdrift, South Africa. Southern African Institute of Mining and Metallurgy, Johannesburg. pp. 33-49.

Daniel Zeelie

Manager
Saveway Furnace Monitoring Africa

Daniel Zeelie completed his Industrial Engineering degree at the University of Pretoria in 1990 and his Masters degree in Business Leadership in 2001. During the last 26 years Daniel gained experience in the Pyrometallurgical, Mining and Automotive industries, both locally and in selected foreign countries.

Daniel's passion is the design and operation of high temperature furnaces and job creation in South Africa.

Environment-friendly harvesting of heavy mineral deposits in India

V. Subramanian, V. Velmurugan, and C. Swamydas

V.V. Mineral

India is the proud possessor of 35% of the world reserves of ilmenite along with other associated minerals. V.V. Mineral is the largest producer and marketer of garnet in the world and is the largest heavy mineral producer in India. This paper presents the details of the company's operations, with particular emphasis on environmentally responsible mining and processing. Most of the company's deposits occur either on the beach or in the adjacent inland areas. The sands deposited in the beach are collected manually using agricultural tools like spades and baskets and stockpiled manually near the beach before being transported to the wet concentration plants. The adjacent inland deposits are mined by mechanized means using loaders and transported using tipper trucks. The preliminary upgrading of the heavy minerals is carried out by a wet process using seawater. The concentrate is then dried manually using solar power. The dried product is then separated in magnetic separators initially to produce the bulk minerals, namely ilmenite and garnet. The nonmagnetics are then processed with fresh water and subjected to electrostatic and magnetic separation to produce zircon, rutile, and sillimanite. By this process, depletion of groundwater and carbon emissions are to a large extent avoided. The eco-friendly flow sheet eliminates the usage of chemicals either in the wet operation or the dry operations, ensures minimum air, water, and sound pollution, and embodies a zero discharge closed-circuit water circulation system. The company uses its own power generated through wind energy. Gasifiers using waste wood as feed material reduce reliance on fossil fuels.

In view of the importance of water in mineral separation, V.V. Mineral ensures that this valuable resource is not exploited. Instead a smart, pragmatic, and studied approach is employed in water utilization by replacing fresh water by seawater in the operations, thereby saving the extraction of 4 ML of groundwater per day.

By resorting to manual drying using the solar energy instead of conventional dryers using fossil fuels, the company avoids 0.04 Mt of carbon emission to the atmosphere. V.V. Mineral provides employment to around 3000 unskilled labourers every day.

INTRODUCTION

V.V. Mineral (VVM), a partnership company engaged in mining and separation of heavy mineral deposits, is part of the V.V. Group. The company started operations in mining and production of garnet, and subsequently extended its activities to all the beach sand minerals - ilmenite, garnet, zircon, rutile, leucoxene, and sillimanite. VVM is the largest producer and exporter of garnet in the world and is the largest producer of ilmenite in India. From its inception, the company was sincere about the environmental aspects of its activities as well as conservation of energy. Accordingly, every effort is taken to minimize energy consumption as well as reduce or avoid effluents from the operating plants. VVM produces around 0.8 Mt of heavy minerals every year.

The Tenth International Heavy Minerals Conference
Sun City, 16–17 August 2016
Southern African Institute of Mining and Metallurgy

HEAVY MINERALS BEACH PLACERS

The heavy minerals which normally co-exist in the beach placer and adjacent inland areas are ilmenite, zircon, rutile, leucoxene, garnet, and sillimanite. The deposits containing these minerals are also found as teri-sand and teri-soil within a few kilometres of the beaches, in addition to the replenishing deposits along the beach and the deposits being formed due to wind action.

Beach Placer Characteristics

In India, the beach placer deposits are found along the shore are deposited in trap zones (Figure 1). The size and position of the heavy mineral (HM) deposits are a function of the wave energy reaching the beach, the mean grain size of the beach sediments, and the current and height of the ocean. The major trap zones where economic extraction is currently carried out are Chavara and Manavalakurichi on the Arabian coast and from Kanyakumari to Periasamypuram in Tamilnadu, the Srikakulam deposit in Andhra Pradesh, and Chatrapur deposit in Orissa on the Bay of Bengal coast. There are appreciable differences in the mineral assemblage, grain size distribution, and the characteristics of the individual minerals between these deposits. Figure 2 illustrates the concentrated HM deposits on the beach.

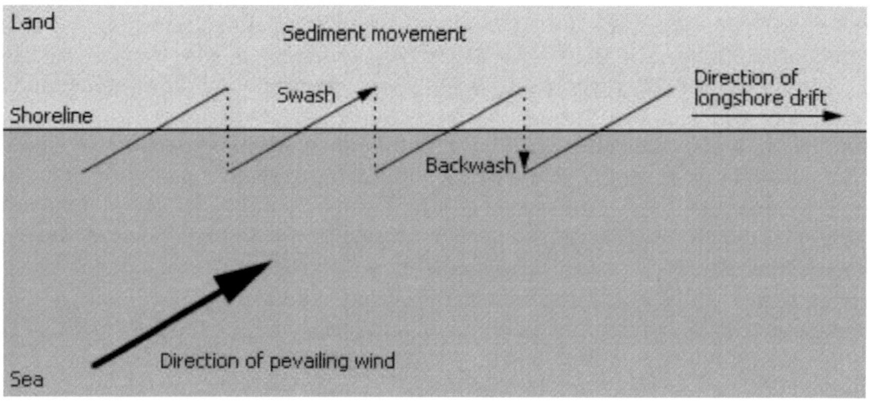

Figure 1. Trap zone processes.

Figure 2. Concentrated beach deposits.

Eco-friendly Mining

The placer deposit is collected by selective manual mining along the beach. VVM relies on the 150-strong unskilled manpower resources available, thereby avoiding the usage of heavy machinery, which disturbs the coastline. Manual mining is carried out using simple agricultural tools such as garden hoes, baskets *etc.* (Figure 3) – hence the method is termed 'harvesting' rather than 'mining'. The advantage of selective mining is that it reduces the collection of gangue materials, thereby improving the efficiency of downstream processing. The inland HM deposit adjacent to the beach placer is mined using mechanized tools like loaders due to the nature of these deposits. This method results in minimum disturbance to the environment and ecology, and maximizes the recovery of all the minerals. VVM provides employment to around 3000 unskilled persons in an area where employment opportunities are scarce. The entire operation has a positive impact on socio-economic development of the area.

Figure 3. Manual mining activities.

Gravity beneficiation

The HMs collected are transported to the gravity separation plants, which increase the HM content to around 90%. These plants are in-house designed, with spiral concentrators and upstream classifiers. Seawater is used as a medium, which reduces the intake of fresh groundwater in an arid area where groundwater availability is scarce. The plant floors and the storage areas are built so as to avoid seepage of saline water to the ground and avoid groundwater contamination. VVM also has some mobile beneficiation plants, which reduces the transportation of the harvested material, resulting in less air pollution. Figure 4 shows one of the wet beneficiation plants. VVM employs a smart, pragmatic, and studied approach in water utilization by replacing fresh water with seawater in the wet beneficiation operations, thereby saving the extraction of 4 ML of groundwater per day.

Figure 4. Wet beneficiation plant.

Manual Drying of Concentrate

The upgraded mineral concentrates are stockpiled and spread in pre-determined drying areas using rakes, spades, and baskets (Figure 5). The dried concentrate is transported using trucks or by conveyor systems to the mineral separation plants (MSPd). The MSPs are located in rain shadow regions with more than 300 bright sunny days per year. VVM is drying around 3000 t of concentrates per day using solar power, thereby avoiding the usage of huge amounts of fossil fuels in the conventional dryers used elsewhere in the world. This process also avoids the emission of 0.02 Mt CO_2 to the atmosphere each year, in addition to providing employment for about 700 unskilled workers in the region. A total of about 2.16 Mt of concentrate are dried using manual labour. Conventional dryers, at the normal rate of 6 L of diesel per ton of concentrate, would consume 12.9 ML of diesel, which could result in the emission of 38 400 t of CO_2. Hence, by adopting solar drying, the company is avoiding the emission of around 0.04 Mt of CO_2. In addition, this reduces the necessity to import fossil fuels. This type of manual drying also avoids coating of the mineral surfaces with fossil fuel combustion products as well as sulphur, which helps improve the separation efficiency in magnetic and electrostatic separation. Dryers are installed in all the plants as standby only for operation during the monsoon seasons.

Mineral Separation Plants

VVM designed their own process for the separation of the individual minerals and constructed the MSPs using in-house expertise. The major separation steps include vibrating screens, rare-earth drum separators, rare-earth roll separators, induced roll separators, high-tension roll separators, and electrostatic plate separators. Figure 6 shows the view of one of the MSPs. VVM produces the following minerals in their operations: ilmenite, garnet, zircon, rutile, and sillimanite.

Figure 5.- Manual drying of concentrate.

Research and Development Facilities

VVM has its own in-house research and development (R&D) facilities and expertise for process design, as well as engineering design for construction of the wet concentration plants as well as MSPs. The facilities include pilot-scale equipment such as various types of spirals, attritioning machines, up-current classifiers, wet tables, rare-earth drums, rare-earth rolls, induced roll magnetic separators, high-tension roll separators, and electrostatic plate separators (Figures 7 and 8). Laboratory facilities are available for carryimg out all analytical work required for daily quality and process control as well as for research and development work (Figure 9)

Figure 6.- Mineral separation plant.

Figure 7 - Pilot-scale wet concentrator plant.

Figure 8. Pilot-scale equipment for dry separation.

ECO-FRIENDLY ACTIVITIES

Mining and mineral separation are carried out in an eco-friendly manner, with partially manual mining, mobile concentrators, solar drying, and the use of seawater as the medium for wet concentration.

Water

Water usage is minimized by a zero discharge closed-circuit water recirculation system. There is no chemical process involved either in the wet concentration plants or the mineral separation plants. Of a total water consumption of 5 ML/d, 4 ML is sea water.

Power Generation

A good quantum of the power required for the V.V. Group of companies is produced using wind energy. The total requirement for the Group is 4.5 million kWh/a. Wind generators meet 30% of this requirement.

Fossil Fuels

The usage of fossil fuels is greatly reduced by solar drying. If all the upgraded heavy minerals were dried using fossil fuels, 12.90 ML of diesel would be used. This would result in the emission of around 0.04 Mt of CO_2 to the atmosphere. In addition, whenever and wherever the dryers are used, gasifiers are used to produce fuel gas by controlled combustion of waste wood to reduce the usage of the fossil fuels.

Upliftment of local communities

VVM employs about 4600 persons, 3000 of whom are local unskilled labourers. The area is semi-arid and no other major employment opportunities are available. Hence, this employment opportunity has helped the local inhabitants by catalysing socio-economic development. The establishment of the mining operations led to the development of all infrastructural facilities, and the standard of living of the local communities has improved considerably. Around 1200 million Indian rupees (US$20 million) are ploughed into the local community through salaries and other benefits to the unskilled labour. Furthermore, other corporate social responsibility activities improve the facilities for the local communities.

Reclamation

The tailings of the wet concentration plants, which contain less than 1% of the valuable minerals, are utilized for refilling the pits in the mining areas and the reclamation of inland mining areas to restore them to the original contour. The reclaimed areas are re-vegetated with various species (Figures 10 to 13).

Figure 9. Analytical laboratory.

Figure 10. Reclamation and afforestation.

Figure 11.- Reclamation and afforestation.

Figure 12. Reclamation and afforestation.

Figure 13. Reclamation and afforestation.

Standards Certification
All VVM units are certified under the integrated management systems comprising ISO 9001, ISO 14001, and OHSAS 18801 by TUV, SUD, Germany. VVM is also certified to supply standards by Saudi Aramco, Qatar Petroleum, the US Defense Department, and other international buyers.

Dedicated In-house Environmental Unit
The V.V. Group has a dedicated in-house environmental unit with qualified personnel to monitor the groundwater in and around the active areas and also to monitor the ambient air quality. Solid, liquid, and gaseous effluents (if any) are regularly monitored to ensure that they are within the specified limits as per the licenses. As monazite, one of the radioactive minerals, is also present in the suite of minerals mined, the monazite tailings are safely stockpiled as per the norms stipulated by Atomic Energy Regulatory Board (AERB). VVM also has 30 Radiological Safety Officers (ISO) trained by the Government of India for constant monitoring of radiation safety of employees.

VALUE ADDITION

The V.V. Group of companies has facilities for value addition to ilmenite and zircon, and produces rutile-grade TiO_2 pigment and various zirconium chemicals. V.V. Group is the only entity in the world to have a fully integrated value chain from mining to paints.

CONCLUSION

The operation of VVM is unique in nature, and is matched by no other HM industry in the world. This is best suited at present to our nation, where human resources are abundantly available, and provides employment to the needy. By successfully implementing selective mining and solar drying methods, the carbon footprint is minimal compared to other mining activities. By eco-friendly activities, air pollution, water pollution, and environmental degradation are kept to a minimum in VVM's operations.

ACKNOWLEDGEMENTS

The authors express their sincere thanks to the management and directors of V.V. Mineral for their encouragement, provision of all facilities to carry out the test work for the preparation of the paper, and for permission to present the paper in this conference. The authors also thank all the executives and staff of VVM involved in carrying out the test work for their sincere co-operation

Vaikundarajan Subramanian

Managing Partner
V V Mineral

Began my career in family's mining business after completing my degree in 2002. Gained experience heading various departments like Accounts, Projects, Production, Marketing, Human Resources, one at a time. Also played crucial role in implementing key projects like separation of Zircon and Rutile and acquiring a value addition project which produced Tio2 Pigment. The group currently owns various business in Mining, Pigments, Paints, Cements, Sugar and Educational Institutions.

Design, construction, and commissioning of a heavy mineral sands concentrator pilot plant

D. Nel

Hatch Africa (Pty) Ltd, Johannesburg, South Africa

Heavy mineral sands processing flow sheets are generally developed in laboratory environments using representative bulk samples of between 10–30 t, depending on the valuable mineral grade and intermediate density non-economic heavy minerals present in the assemblage. Laboratory environments are essential for the development of process flow sheets as these environments provide the stable and controlled conditions necessary for testing process equipment under varying conditions in a systematical and controlled fashion. Pilot plants, on the other hand, perform an equally important function in the development of process flow sheets. When processing flow sheets are developed in a laboratory environment, it is not always possible to adequately evaluate the impact of circulating loads on overall design. Pilot plants, when engineered correctly, provide this capability.

INTRODUCTION

In 2011 Rio Tinto embarked on a Pre-feasibility Study to evaluate options to expand production capacity of chloride and sulphate slag. At the time the Group foresaw that demand in the titanium feedstock market would expand significantly in the coming years, both in the chloride and sulphate sectors. Supply/demand forecasts then suggested medium/long-term opportunities for the chloride feedstock market and short-term opportunities for the sulphate feedstock market. The potential existed for Rio Tinto to meet a portion of the forecast demand increase for TiO_2 pigment-grade feedstock. The Group investigated the possibility of benefiting from that opportunity through increased capacity to meet expanding demand. Unfortunately, due to a downturn in the industry, the expansion project did not materialize or proceed to the next phase (i.e. Feasibility Study).

The project included evaluating expansion options in southern Africa as well as expanding current operations at Quebec Madagascar Minerals (QMM). The scope of work for pilot plant development included developing suitable pilot plants for both locations. Since the *in-situ* grades and the mineralogy of the ore differed significantly, pilot plant work took these differences into account and resulted in different designs. However, this paper deals only with the design, construction, and commissioning of the pilot plant for the QMM operation.

The Pre-feasibility Study scope of work included process design, engineering design, infrastructure design, environmental permitting, as well as logistics definition. Process flow sheet development test work using bulk samples was contracted to a laboratory in Australia. The process engineering scope of work included the design and implementation of a pilot plant to validate laboratory bulk sample flow sheet design. Therefore the objectives of the pilot plant were:

- To validate the process design using bulk samples
- To confirm recirculating loads during the treatment of different ROM samples from various geological ore unit.
- To generate pilot plant survey data to confirm expected performance (grade and recovery)
- To generate bulk HMC samples for further downstream processing.

A particular challenge was to design the pilot plant before the laboratory-developed flow sheet was available. The Mandena plant flow sheet at the time consisted of a 4-stage gravity circuit and work was underway to expand the circuit to five stages therefore it was reasonable to assume that any future process plant circuits would most likely require a 5-stage gravity circuit.

PILOT PLANT DESIGN

Overall Design Scope of Work

The pilot plant design and supply scope of work included the following aspects:

- Design and supply a five-stage gravity movable pilot plant sufficient for the purposes of the project
- Pilot plant to be self-sufficient and independent of the Mandena operations
- Pilot plant to be designed to be potentially set up in a remote area
- Development of an operations plan detailing test work procedures and analytical requirements to validate laboratory-developed flow sheets formed part of the scope.

General Design Requirements for the Pilot Plant

Plant Mobility

The pilot plant was designed to be mobile and capable of being established in the mining area (if required) or of operation at remote sites away from established infrastructure. Therefore the plant was designed to ensure minimal engineering and operational support was required for the efficient operation. The plant was designed in a modular fashion to enable transport in standard 20- and 40-foot (open top) containers. The design allowed for electrical supply from a diesel generator to provide power to the pilot plant as well as other supporting services such as the process water pumps and peripheral services water management. The process water design included two pumps to draw water from a suitable source in close proximity to the pilot plant and to operate the plant for periods of up to 8 hours (daylight hours). A water management plan was developed to ensure the maximum amount of process water from the pilot plant was returned to the process or process water reservoir to minimize the impact on the environment.

Plant Safety

During the design phase, specific emphasis was placed on ensuring compliance with the owner's engineering and safety specifications. Access stairs were included into the design instead of cat ladders to ensure safe access of personnel carrying small tools to equipment at higher elevations. Crawl beams for maintenance of electrical motors and pumps were included in the design as well as for rigging in process equipment when configuration changes to the process circuit were required. Lighting was included in the design in the form of fixed light fittings in appropriate locations to ensure maintenance, modifications, and catch-up operations were possible after daylight hours.

Specific Design Requirements for the Pilot Plant

The following specific design requirements were included after consultation with the client. In this particular case the client was QMM current operations and not the project team, since the pilot plant would become the property of the current operations once the project requirements were fulfilled.

Metallurgical Design Considerations

Since the laboratory-developed process was likely to be a 5-stage gravity circuit, a decision was taken to design the pilot plant to enable simulation of a 5-stage spiral circuit operating under steady-state conditions at a feed rate of 20 t/h ROM. A key requirement was to ensure the plant was designed with sufficient vertical space to accommodate 7-turn spirals from all vendors. No automatic samplers were included in the design and manual sampling of the feed, concentrate, and final tails was made possible. Composites of the grab samples would be generated and submitted to the laboratory for THM determination. Provisions for manual sampling of the feed and products from the unit processes were also included into the design. The plant design also ensured that large-diameter high-capacity spirals could be accommodated in the rougher and scavenger duties, *i.e.* supporting of the spiral centre columns at the product discharge and feed ends of the spirals was catered for. The pilot plant design also had to cater for configuration changes such that (1) direct spiral comparisons are possible (*i.e.* two spirals are fed in closed loop from a single sump with the same feed material mix, (2) other process equipment such as de-sliming cyclones, pilot-scale up- current classifiers, and low-intensity magnetic separators can be added to the circuit with minimal modifications, and (3) product launders from the spirals/other process equipment must consist of light flexible rubber hose to allow easy changes to the pilot plant configuration.

Control Considerations
It was recommended that no process control instrumentation was to be included into the design to ensure a high level of simplicity and fewer issues during commissioning and operations. Therefore the design had to allow accurate manual measurement of all feed and product slurry densities and flows. However one exception was the inclusion of instrumentation on the feed system, *i.e.* an allowance was made for a belt scale on the feed conveyor and a flow meter on the water line to the screen to ensure a constant solids feed rate and density reporting to the rougher spirals. Local MCC panels for the electrical motors were specified to have standard instruments such as volt- and ampmeters. Sump level controls were to be via mechanical float valve arrangement.

Operational Considerations
Pilot plants are known to be difficult to operate as mass balancing of solids can be difficult to achieve, therefore operational design considerations are vital to ensure good operability of the plant. If steady-state operations are difficult, then it is likely that sampling campaigns during interrupted operations will yield poor results and the objectives of the pilot plant runs will not be met. All distributors were designed to have overflow lines back to their respective feed sumps to allow for circulation of excess slurry and to enable balancing of the circuit. All spiral feed pumps were supplied with a second delivery line of smaller or larger internal diameter to allow for flow rate flexibility depending on the situation. The design of the make-up water manifold included two 25NB wash water points with quick-release hose connections and two 30 m lengths of hose. This ensured that the operators had wash water to de-sand pump suction lines after emergency shutdowns. All slurry sump suction lines were fitted with wedge lock dump ports for easy de-sanding. The design of the make-up water system included two supply pumps capable of delivering the required flow to the plant sump manifold at 3 bar to ensure sufficient flow to the bins when required. This water also supplied the vibrating screen spray bars as well as water for general cleaning after daily operations. A 50 m^3 water tank was included in the design to be allowed for in the costing as a takeout price. Due to low volumetric flow rates of the cleaner circuit stages, dilute flows had to be generated to satisfy minimum pumping volumes. Cyclones above the cleaner, re-cleaner, and cleaner tails spirals were included for this purpose to ensure a minimum amount of slurry was pumped to the distributor, and the cyclone overflow had a bleed valve to route some of the over flow water back to the underflow to ensure adequate density control.

Mass and Water Balance
The mass and water balance was based on a dry solids feed rate of 20 t/h. Average, low, and high ROM feed grades of the Sainte Luce ore deposit were used to model the mass and water balance based on expected grades and recoveries of a typical 5-stage gravity circuit. This model proved invaluable as an accurate estimate of the mass balance contributed greatly to the overall success of the project and could be used to determine the extent of any design changes, such as additional spiral starts in any duty or changes in the slurry flow rate of a particular pump. Should this have been required, no simple solution would have been possible as the structure of the pilot plant did not cater for the addition of additional spiral starts. The correct numbers of starts is crucial to ensure loadings on the spirals are in the correct range so that meaningful results are delivered from pilot plant runs.

Once the mass balance was developed, typical slurry densities for feed, concentrate, middling and tailings were applied to calculate volumetric flow rates and slurry densities. It was noted from the mass and water balance that some spirals feed volumetric flows were well below the minimum flows required for the smallest slurry pumps used in this industry. As a result, diluted slurry flows had to be specified for these spiral duties. Since low solids concentration could not be tolerated on the spirals, dewatering cyclones had to be included to ensure the correct solids concentration could be maintained on the spirals. The dewatering cyclone overflows reported to a tundish that allowed some of the overflow water to be directed back to the underflow to ensure optimal control of feed solids concentration.

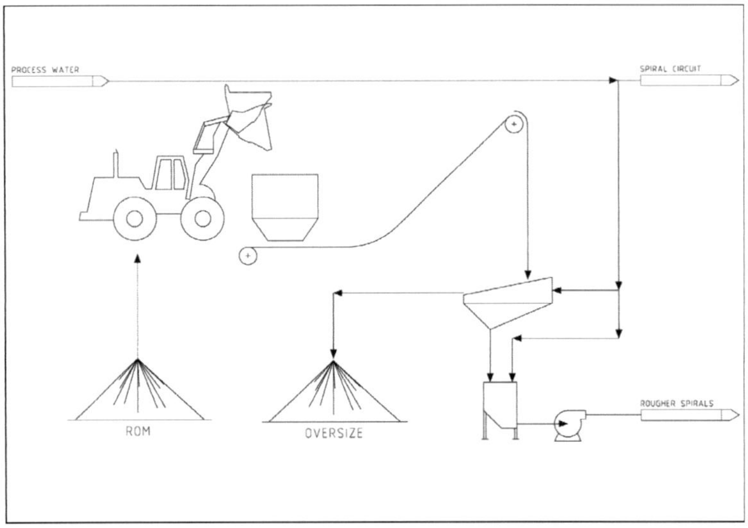

Figure 1. Pilot plant feed preparation process flow diagram.

The pilot plant process flow was based on a typical 5-stage gravity circuit as depicted in the block flow diagram in Figure 1. Upfront of the gravity circuit is a standard feed-preparation section (Figure 2) consisting of feed hopper with static grizzly screen, vibrating feeder to draw moist ROM from the hopper and discharge onto the conveyor belt feeding a vibrating screen with approximately 2.5 mm aperture. The design allowed for sufficient water to be added to the screen deck to wash the mineral particles through the screen as well as spray water to the screen. The feed conveyor was fitted with a belt weigher to ensure accurate feed rate control via the variable speed drive on the vibrating feeder. A flow meter was included on the water supply to the screen to ensure accurate density control to the rougher spirals via control loop.

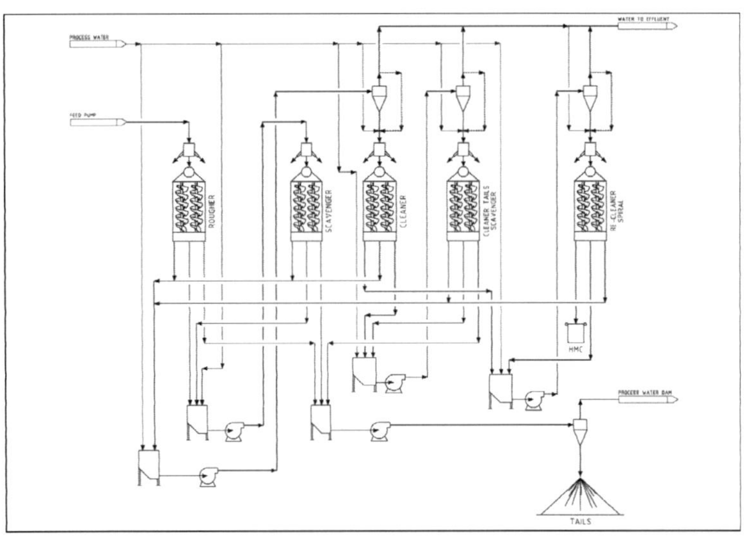

Figure 2. Pilot plant gravity circuit process flow diagram.

However, the plant was designed such that deviations from this process flow can be made by simply changing the light flexible piping between the spiral discharge and the transfer sumps.

Products from the pilot plant consisted of final heavy mineral concentrate and gravity circuit tailings. The HMC gravitated directly into bulk bags while the tails were pumped to the tails stacker dewatering cyclone. Two stackers were included in the design to allow uninterrupted operation while tails from the previous day's operations are being cleared.

Make-up water to the process was provided via a ring main system to ensure adequate pressure to all systems. The pilot plant water supply system included a 50 m^3 buffer tank to ensure controlled startup and shutdown of the plant was possible. All excess water from the process was routed back to the process water tank to minimize consumption of process water.

Detailed Engineering, Pre-assembly, and Pre-commissioning

In order to meet tight schedule targets, the strategy adopted was to complete the basic design package in-house and include detailed engineering in the scope of work of the successful vendor. The selected vendor was a local process equipment supplier with a track record of constructing and commissioning similar pilot plants. The project schedule required the pilot plant to be established and commissioned once the laboratory flow sheet development work was complete to enable pilot plant validation test work to begin.

Apart from the detailed engineering being outsourced to the vendor, it was decided to pre-assemble the pilot plant at the manufacturer's site before shipment to the client's site. This would allow the construction team to identify and change any design issues locally and reduce on-site modifications, further mitigating the risk of not making schedule dates. A further significant risk identified was the potential of poor material flow properties of the ROM in the feed hopper system. It was known that QMM ROM material varies from free-flowing sand to combinations of sand and fairly competent indurated material with variable moisture contents. As a result, it was decided to pre-commission the hopper, vibrating feeder, and conveyor system prior to shipment to the client's site. The commissioning was undertaken using moist building sand with a similar particle size distribution (PSD) to the ROM material. This qualitative test confirmed that the design was sound, and no materials handling issues were experienced during pre-commissioning. Included in the pre-commissioning activities was the commissioning of the one idler belt weigher., Feed rates of up to 32 t/h were recorded on the instrument and verified by manual belt cut sampling and measurement.

Detailed Engineering

Hatch awarded the detailed engineering, supply, and construction to a local process equipment supply company with a track record of supplying similar pilot plants to the industry. All designs were forwarded to Hatch for review prior to the start of shop detailing. Hatch worked closely with the vendor to ensure a smooth and seamless transition from detailed engineering to shop detailing and fabrication.

A HAZOP study for the pilot plant was completed during detailed engineering. Any modifications required to mitigate the hazards identified during this study were incorporated in the design and fabrication of the plant proceeded accordingly.

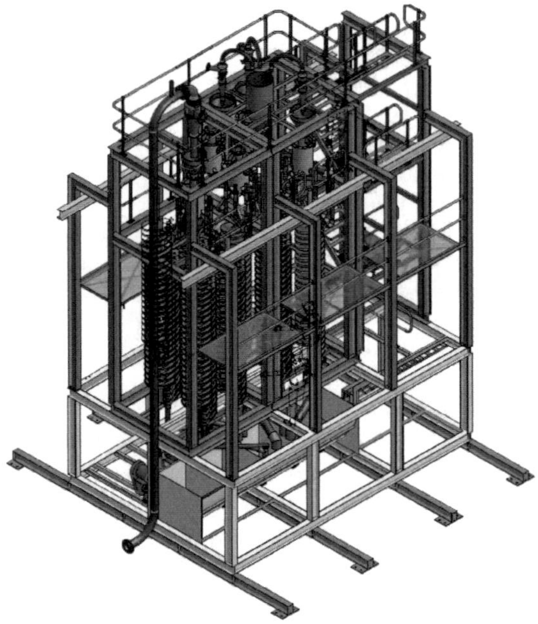

Figure 3. Isometric view of the pilot plant structure.

Pre-assembly

The pilot plant was pre-assembled at the vendor's site to ensure all design and fabrication issues could be addressed before the plant was shipped to site. Even though the plant destination was an operating site with materials and labour, the strategy was to minimize site modifications potentially required as the cost and schedule impact of site modification would be larger. As components were completed in the fabrication facility, equipment and components were sent to the assembly yard for pre-assembly (Figure 4).

Figure 4. Pilot plant at pre-assembly yard.

To assist with reducing the construction time required to re-erect the plant on site, all equipment and structural steel was match-marked before disassembly at the assembly yard. Detailed packaging lists were provided for each shipping container to easily identify and track the equipment stored. The containers were loaded in such a way that the order of unpacking on site would assist with the construction sequence of the plant.

Pre-commissioning

Pre-commissioning activities on the pilot plant (Figure 5) were limited to the front-end feed preparation section due to the potential risk associated with solids flow properties in the feed hopper. This risk was known at the time of basic engineering, and therefore particular emphasis was placed on the design of the ROM hopper and vibrating feeder assembly. Care was taken to ensure maximum wall angles of the hopper as well as non-slip linings to aid solids flow in the hopper. Prior to testing the front-end section of the pilot plant with solids, the belt weigher was commissioned and calibrated. Testing of the hopper design with solids was undertaken using building sand with a similar particle size distribution as the ROM material. Conveyor belt profiles were used to accurately cut and sample a 1 m length of solids on the conveyor belt. This was not only to check the belt weigher reading, but also calculate the actual tonnage on the belt using the loading and speed data.

Figure 5. Pre-commissioning of the pilot plant feed system.

Once the ROM hopper was filled with solids, the conveyor belt was started with the vibrating feeder set in manual mode in the 'off' position. Once the belt weigher functionality was confirmed, the vibrating feeder was set to 50% output and started. Once the vibrating feeder delivered a steady flow of solids to the belt, the belt was stopped. Output from the belt weigher was recorded and a manual 'belt cut' sample was taken using the belt profiles over a 1 m section of the belt. The mass of the solids of this 1 m section was recorded, as was the manual measurement of the belt speed using a handheld tachometer. This procedure was repeated at 75 and 100% output set-points of the vibrating feeder. Manual measurements of the tonnage on the belt at different set-points compared well with the reading of the belt weigher. The pre-commissioning test was considered successful as consistent solids flow was observed during the test and manual measurements compared well with the output from the belt weigher.

Both of the plant electrical control panels were also pre-commissioned and all on-plant pumps rotation-tested.

Construction

Ground Works and Civil Construction

Although the pilot plant was designed to be self-sufficient and could be constructed in remote parts of the ore deposit, a decision was taken to erect it on the Mandena Mineral Separation Plant (MSP) site close to engineering support services, accommodation, laboratory services, and most importantly power and water.

The final construction site selected was in close proximity to power and water and was remote enough not to interfere with regular routes frequented by personnel as well as mobile equipment. A simple raft foundation for the plant as well as the feed preparation circuit was provided to support both structures. The plant foundation was provided with a basic bund wall and trench to ensure all spillage from the pilot plant could be washed into the current MSP drainage trench for final disposal.

Figure 6. Site construction of the feed preparation circuit in foreground and gravity circuit in the background.

Along with the civil works for the feed preparation circuit, a ramp was constructed to ensure that a small loader could access the ROM feed hopper. This ramp was constructed using suitable road construction material and once the basic shape and size was completed, the ramp was further supported by securing the sides with local stone and cement. Once the sides were secured, final compaction of the ramp surface was done before sealing of the surface. This proved to be a low-cost fit-for-purpose solution, since the Mandena plant is situated in a high rainfall area.

Mechanical Construction

A local construction contractor assisted with the erection and installation of all pilot plant structures and mechanical equipment. The same construction team, consisting of Hatch and contractor personnel, had recently completed some modification work at the current QMM operation. Having this team benefited the plant construction, as they were well established before the plant was delivered to site and were fully conversant with all the prescribed construction, health, and safety standards required by the owner. The construction period of the plant was reasonable, and the process was greatly assisted by detailed packing lists and the packing sequence of the plant containers prior to shipment. The plant was inspected on a regular basis by QMM operations personnel (the final owner of the plant), and minor site modifications were implemented to satisfy their requirements. A list of deficiencies (or punch list) was compiled that identified all outstanding rectification work and assigned a priority to each item. On completion of all the safety-critical and cold-commissioning priority items, the plant was handed over to the commissioning team.

Electrical Construction

The electrical equip,ent and instrumentation used on this project is simplistic and elemental. and was selected to ensure that minimal effort was required to commission and run the plant. All the float switches were hardwired interlocked into the different MCC buckets to protect the bins from constant overflow. The float switches were wired to the motor control buckets to stop feed pumps when low levels were reached and so ensured that equipment was protected at all times and that minimal spillage resulted from normal operations. Two MCC's were built and installed to control motors. The primary MCC was installed in the process area from where the cluster of bins was controlled. These bins all have float switches that control the high and low levels of the bins. All the level control valve arrangements were on/off control and designed to open or close depending on the levels in the bins. This was subsequently changed due to the fact that the level switches failed and were not correctly specified for this slurry application. The bins were then manually controlled, which is not ideal but nevertheless worked well in the end.

The secondary MCC was used for controlling the ROM conveyors and vibratory feeders and the weigh scale on the feed conveyor. This was controlled by a local controller that was able to set the speed of the feed conveyor. This was a manual operation where the operator had to start the conveyor in manual control. Once the feed rate was constant and the desired feed was achieved, the controller was changed to auto control and the set-point set to the required feed rate.

Commissioning of the electrical and instrumentation was quick and all the systems were pre-commissioned in 2 weeks.

COMMISSIONING

All electric motors were bump-tested successfully prior to the start of cold commissioning. Rectification of items on the list of deficiencies/punch list continued for the duration of commissioning and beyond. By the time cold commissioning was due to begin, all construction work enabling commissioning was in place.

During the bump start tests most control instruments were inspected to determine their correct functioning. It was found during these inspections that all the level switches that maintain the slurry level in each spiral feed bin had failed. Given the poor availability of replacement spares and the amount of time required to rectify the failure, it was decided to replace the control valves supplying process water to the slurry bins with manual globe valves. It followed from this decision that for the foreseeable future, and for the remainder of activities prior to the handover, the slurry level in each slurry bin would have to be manually controlled.

For all commissioning and post-commissioning activities (up to handover) Hatch provided personnel responsible for the management and execution of all tests and plant operations. These resources included a commissioning manager and process engineer who were on-site from the start of the commissioning until the handover of the pilot plant. The client provided a process engineer to assist in verifying mostly operations-related activities. Personnel from the site-wide construction contractor were employed in rectifying items on the deficiency list/punch list.

Hatch's safety officer was present on-site on a daily basis and assisted in maintaining general safety and safety awareness. Daily toolbox talks, which were translated to French for the benefit of the

French-speaking team members, were held during which the activities for the day were communicated to everyone. This assisted greatly in aligning all team members to a common purpose such that the programme for every day was completed consistently.

Commissioning included quantified cold commissioning of all slurry pumps while all motors installed on the vibrating screen, feeder and feed hopper were checked for the correct direction of rotation. Slurry pumps were commissioned by using the tank drop-test method; all pumps were found to be within the original design parameters. Test data was captured and formal documentation was issued for client/EPCM signatures.

Hot commissioning generally follows on the completion of cold commissioning, but in the case of this project it was not possible. While this was planned as part of the original pilot plant test campaign, the need to bring the project to an end given the new project conditions could not be ignored. It was decided, owing to the relatively simple nature and scale of the pilot plant and the successful outcome of cold commissioning, to forego a full hot commissioning sequence in order to improve on the completion/handover schedule.

By the time cold commissioning was completed only a few hours remained in the day. The team was keen to commence with the operation/handover phase of the project but it was clear that some of the operation/hand-over objectives would not be achievable in the time remaining. Instead, opportunities for doing valuable work in terms of checking system readiness prior to operation were identified. It was decided to operate the pilot plant for a short period (approx. 2 hours) with live solids feed - at a significantly reduced capacity - to determine the system's readiness in terms of the requirements set by the full operational programme. During this pre-operation system check the biggest operational issue was that the plant water supply could not be maintained as the pilot plant water tank kept running empty. Upon investigation it was found that the water supply pump providing water to the pilot plant water tank required a change in its pulley set; this change was effected shortly afterwards. Apart from this issue, the pilot plant's operation during the system check generally went smoothly.

Post-commissioning Operations (Handover)

Original Pilot Plant Operations Plan

In preparation for the pilot plant operations, Hatch developed key documentation that details various practical aspects crucial to the success and reliability of the programme. This documentation included the technical objectives of the programme, resource/personnel requirements, and standard operating procedures (SOPs). Other detail that was further developed included lock-out and isolation procedure.

Technical Objectives

The primary objective for the pilot plant operations was to validate and confirm the suitability of the primary concentration plant process flow sheet as developed by the test work provider by treating bulk samples from deposits containing similar ore. In addition, the HMC produced by the pilot plant would be used to develop the process flow sheet required to produce final products (ilmenite, zircon, and possibly rutile) through a mineral separation plant.

The pilot plant was to be set up to simulate the primary gravity circuit process flow as developed by the test work. Samples from the different ore types were to be treated through the pilot plant to verify the HMC product quality and mineral recovery from each ore type. The bulk HMC produced was to be used for downstream processing test work such as mineral separation circuit flow sheet development.

In completing the pilot plant programme, output of the following deliverables was expected:

- Confirmation of the selected process technology
- Confirmation of optimal operating conditions for process stages

- Validation of the process flow design completed by the flow sheet development test work laboratory by the generation of release curves for the different spiral duties
- Confirmation of recirculating loads during the treatment of different ROM samples from various geological ore units
- Generation of pilot plant survey data to provide operational data to support the release curve data and to confirm expected performance (grade and recovery)
- Generation of bulk HMC samples for further downstream processing.

After commissioning was completed, two bulk samples were to be treated as part of the pilot plant programme. Each sample would require a week of processing with an additional week to allow for startup/ramp-up, and ramp-down/shutdown activities. Ideally, three weeks would be sufficient to complete the programme. This schedule did not allow for unplanned maintenance, although an unplanned maintenance event was not expected to have an major impact on the programme schedule.

The pilot plant's operational programme consisted of normal operation and planned and unplanned downtime. No allowance was made for unplanned downtime within normal operations and planned downtime activities, although its occurrence and impact was expected to be minimal. Normal operations were specified as the period between 8am - 5pm Monday to Friday and 5 hours between 8am and 1pm on Saturday. The rest of the time outside normal operation was scheduled as planned downtime, during which planned maintenance could be carried out in accordance with the relevant SOPs and lock-out procedures.

The test work programme would generate mineral samples that mostly require analysis on a weekly basis. Some samples would require daily analysis as these samples would provide direction for plant management in terms of process set-up and plant operation.

Personnel, Resource, and Utility Requirements
A list of personnel requirements was developed that would enable successful operation of the pilot plant. The list categorizes the main functions as follows:

- Management personnel (client team leader and safety officer)
- Operations personnel (three operators and two loaders)
- Technical personnel (lead process engineer, project metallurgist, and two process engineering assistants)
- Maintenance personnel (millwright).

To successfully operate the pilot plant a complement of 12 persons is required.

For each personnel function a set of roles and responsibilities was identified. Furthermore, responsible, accountable, consulted, and informed (RACI) matrices and organizational structures were developed for the various commissioning, operations, and handover activities.

In order to successfully complete the operations programme, various resources/utilities will be required. Process water and electrical supply were identified and specified as essential utilities; similarly, site access/security, maintenance/operational support, and a mineral laboratory were identified as essential resources. As the pilot plant was to be erected within an existing site's boundaries and the pilot plant's demands on these resources and utilities were minimal, access to and availability of these resources and utilities were deemed secure.

Standard Operating Procedures
As the pilot plant was manufactured as a greenfield plant, new operating procedures tailored to this specific plant were required. Standard operating procedures (SOPs) were developed that detail the sequence of events should the operating status of the pilot plant change or plant management decide to change the operating status of the pilot plant. In general, these SOPs describe the normal and abnormal startup and shutdown and general maintenance of the pilot plant operations.

Pilot Plant Operations

Prior to commencing with post-commissioning activities, the main commissioning activities were completed. Construction work identified as per the list of deficiencies or 'punch list', although almost complete, continued into the early part of the post-commissioning activities. The daily 'toolbox talk' continued during the post-commissioning activities and personnel from the on-site construction contractor were also used again to complete items to be rectified as per the 'punch list' and assist with menial plant operation tasks.

The vendor that supplied the pilot plant also provided extensive documentation in terms of equipment and plant manuals. These manuals provide much reference information with regard to the technical aspects of the pilot plant. In addition to these equipment manuals, Hatch began developing a detailed manual (mentioned earlier) that describes the operational programme objectives for the pilot test campaign and the associated resource requirements, RACI/duty allocation, and organizational structures. The manual structure and some of the sections were well developed, but the revised objectives set for the pilot plant project led to the discontinuation of work on the rest of the manual.

Figure 7. Tails stackers with scaffold structures to provide safe access to cyclones.

Due to the full pilot test programme being put on hold, post-commissioning activities were limited to continuous operation of the pilot plant at full capacity for a pre-specified amount of time as part of the handover requirements. Ore similar to the ore that would have been processed during the pilot test campaign was delivered to the pilot plant's ore storage area for the purpose of completing the required continuous operation programme.

Figure 8. Final HMC gravitating into bulk bags.

The pilot plant's ability to maintain full feed capacity for a specified amount of time was identified as the main deliverable to be met, as the original intent was to operate the pilot plant for a considerable amount of time to complete the full pilot programme. If the pilot plant can maintain full capacity for the duration of a normal shift it will be able to do so repeatedly.

At the start of the continuous operation programme the solids flow rate fed into the pilot plant was increased from a relatively low value to full capacity over the course of a few hours. This enabled the commissioning team to check and confirm equipment capacity under conditions where solids were present in the process slurry. After reaching full solids capacity the continuous operation programme commenced.

The continuous operation of the pilot plant passed without too many incidents. After one or two restarts - caused by minor issues but enough to reset the endurance timer - the pilot plant was operating at full capacity. At or below full feed capacity no capacity constraints were observed and well into the programme it was decided to increase the rougher spiral feed rate to check the capacity of the rest of the slurry pumps at overcapacity. At 110% capacity most of the slurry pumps remained capable of transferring the higher slurry volumes, although the slurry pumps transporting high-grade HM slurries struggled to maintain throughput. After a few spiral feed and product box blockages on the high-grade spirals it was decided to reduce the feed rate to the rougher spiral back to full capacity.

CONCLUSION

As part of this Pre-feasibility Study Hatch, together with the owner's team, successfully designed, constructed and commissioned a 20 t/h wet gravity concentrator pilot plant at Rio Tinto's QMM operations. The decision to pre-assemble the plant as well as to hot commission the feed system before shipment to site resulted in significantly reduced site work. It is believed that this strategic decision made possible the completion of this project within the tight schedule.

Deon Nel

Senior Technical Consultant
Hatch Africa

Deon has 25 years' experience in the heavy mineral sands industry firstly in operations (13 years) followed by a career within the EPCM project environment. During his career in operations he gained in depth experience in both primary concentration and mineral separation circuits including peripheral services such as slimes handling, water treatment and tails management. Deon has developed and managed numerous metallurgical test work programs both locally and abroad to develop process flow sheets for primary concentration and mineral separation plants.

Sillimanite production from placer deposits: a new process

C. Swamydas, A.D. Rajarethnam, and V. Velmurugan

V.V. Mineral, India.

Sillimanite is a refractory mineral having the chemical composition $Al_2O_3.SiO_2$. It is mined from both hard-rock and placer deposits. Hard-rock sillimanite resources have been depleted in many parts of the world. In placer deposits, sillimanite occurs in association with heavy minerals such as ilmenite, rutile, zircon, garnet, and monazite. The placer deposits are of two types, *viz.* beach placers and inland placers. Sillimanite from placer deposits is normally concentrated mostly by wet gravity concentration processesto increase the heavy mineral content to around 90%, the remainder being quartz. The concentrates are processed to separate the individual minerals using physical properties such as electrical conductivity, magnetic susceptibility, grain size, and specific gravity, mostly by dry separation. Quartz and sillimanite report as tailings in the wet concentration process and also as nonmagnetic tailings in the dry separation. The wet concentration tailings are processed to obtain a concentrate containing more than 50% sillimanite, the major contaminant being quartz. The tailings of the dry separation process also contain around 50% sillimanite with quartz as the contaminant. In India, for production of sillimanite from beach sand deposits, the sillimanite is separated from quartz by froth flotation using different types of flotation equipment. This paper deals with a new process for separation of sillimanite from the sillimanite concentrate by a dry method without adopting the conventional froth floatation process. The relative advantages and disadvantages of both these processes are discussed. This process was developed and test work carried out on the plant scale using sillimanite samples from placer deposits in Tamilnadu and Andhra Pradesh in India.

INTRODUCTION

V.V. Mineral (VVM) is a partnership company, an entity in V.V. Group, and is engaged in mining, mineral processing and separation, and value addition of heavy minerals like, ilmenite, garnet, zircon, rutile, leucoxene, and sillimanite. The company was started in 1988 and is currently the world'slargest producer and exporter of garnet and the largest producer of ilmenite in India. The V.V. Group is engaged in value addition to minerals like ilmenite and zircon. VVM has one of the best research and development facilities for beach sand minerals in India and carries out studies for the improvement of processes, quality, and recovery. VVM entered into the production of sillimanite recently and extensive studies were required to finalize the process due to the presence of shell ($CaCO_3$) and CaO in the blend of minerals. The shell and the CaO respond the same way as sillimanite in the various beneficiation processes and it was difficult to separate these impurities from sillimanite. The current study has improved the separation, resulting in reduction of the contamination, if not fully eliminating it.

SILLIMANITE

Sillimanite is one of the three allotropic varieties of minerals corresponding to the formula $Al_2O_3.SiO_2$. The other two are andulasite and kyanite. They are distinguished from one another by their modes of occurrence and physical and optical properties. Sillimanite is white in colour, having a white streak as well as a vitreous to dull lustre. The diaphaneity is transparent to translucent. The specific gravity is 3.3 and the hardness is between 6.5 and 7.5 on the Moh's scale. It has good cleavage and fractures unevenly. It is found as long prismatic crystals, fibrous aggregates, or in granular form. Sillimanite is a nonconducting and nonmagnetic mineral. The crystalline structure is orthorhombic. Sillimanite has many applications:

- Refractory industry
- Foundry
- Precision investment casting
- Electrical porcelain
- Sanitary ware
- Abrasives.

Sillimanite occurs in hard-rock deposits and also in placer deposits together with heavy minerals like ilmenite, rutile, zircon, garnet, and leucoxene.

BACKGROUND TO THE STUDY

Placer deposits normally occur on the shore as well as within a few kilometers from the shore as inland dunes as well as teri-soil and teri-sands. In most of these deposits, calcium in the form of shell ($CaCO_3$) or calcite (CaO) is very common. These calcium minerals have the same physical properties as sillimanite. The individual heavy minerals in placer deposits are separated by utilizing their physical properties and in practice, the sillimanite product is always contaminated by these calcium minerals. The conventional method of separating and producing sillimanite is by flotation using its surface chemistry properties. The presence of these impurities in the sillimanite product is very detrimental in the various applications mentioned above and it became imperative to limit these impurities. However, these calcium impurities are not uniform in all these deposits and hence, only sillimanite produced from certain locations in India had impurity levels within the limit, making the other deposits practically redundant. The only way to utilize these deposits was to resort to acid leaching of the sillimanite product to remove the calcium contamination. This process generatesacid effluents, the disposal of which is an environmental problem as per the prevailing laws in India. Furthermore, the flotation processes also utilizes chemicals like oleic acid, sodium carbonate, and soda ash, resulting in chemical effluents. In order to avoid some of these difficulties, an attempt was made to separate the sillimanite utilizing a dry process instead of flotation.

PRINCIPLE APPLIED

The principle of'tribo-electrification' is used for separating the sillimanite from quartz. This is a type of electrostatic separation in which two nonconductive mineral species acquire opposite charges by contact with each other. The oppositely charged particles can then be separated under the influence of an electric field. This process uses the difference in the electronic surface structure of the particles. The surface phenomenon that comes into play is the work function, which may be defined as the energy required to remove electrons from any surface. The particle that is positively charged after particle–particle charging has a lower work function than the particle that is negatively charged.

STUDIES CONDUCTED

The sillimanite normally reports in the tailings of the dry mill in the heavy mineral separation process, along with other nonconducting, nonmagnetic minerals like quartz, shell *etc.* As these dry plant tails contain some heavy minerals, it is necessary to carry out feed preparation prior to the production of sillimanite. This step is required for the flotation process as well as the proposed new process. Also, the feed preparation removes the surface coating from the minerals.

Feed Preparation

The feed is first subjected to magnetic separation to remove all the magnetic contaminants. The nonmagnetic fraction is subjected to gravity concentration in spirals and on wet tables to remove all the heavier minerals like ilmenite, zircon, rutile, and monazite. The sillimanite concentrate is taken as the middling fraction and contains quartz as the major contaminant. During the feed preparation stage, other unit operations like screening, attritioning, and crossflow separation are also used to improve the feed quality. The feed preparation flow diagram is shown in Figure 1. During this process, the sillimanite content in the feed material is enriched from around 37% to 50%.

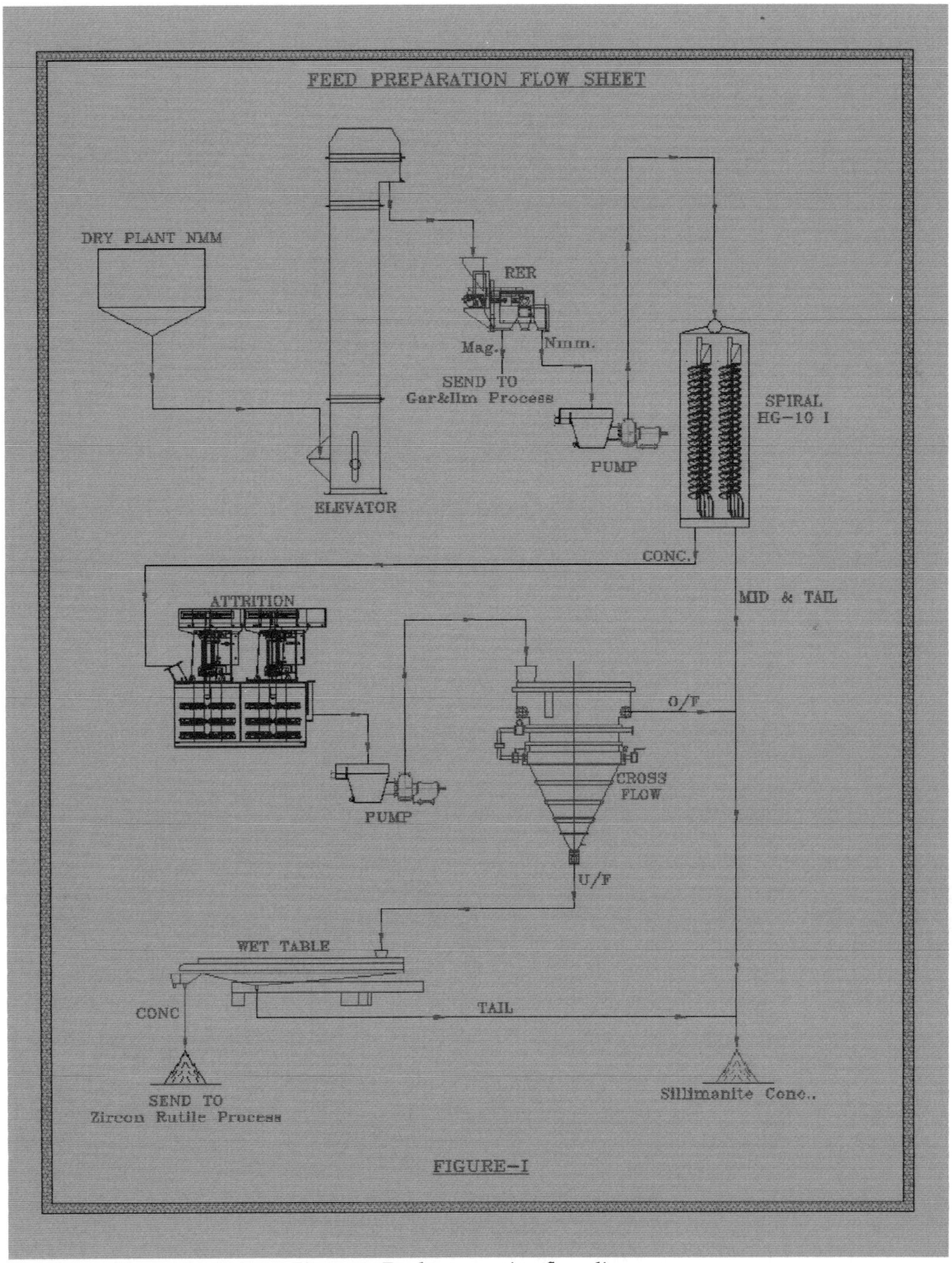

Figure 1. Feed preparation flow diagram.

Test Work for Production of Sillimanite

Flotation

The prepared feed is conditioned in the conditioners, where oleic acid is added for improvement of the surface properties, and then subjected to flotation in a column to produce sillimanite as the float fraction. The float fraction is then dewatered, dried, cooled, and subjected to magnetic separation to remove any magnetic minerals and impurities. The flotation process flow diagram is shown in Figure 2.

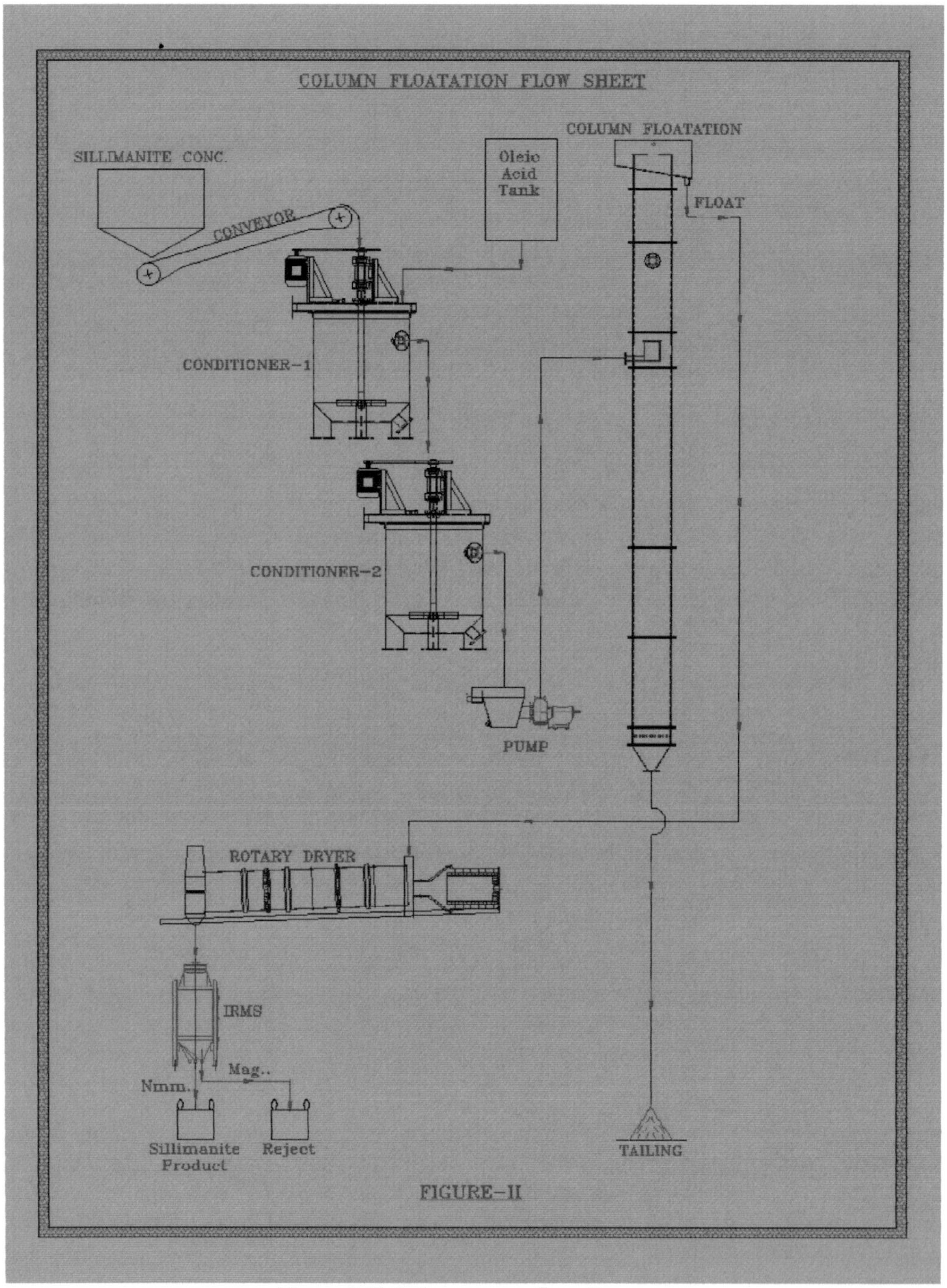

Figure 2.Flotation process flow diagram.

The New Process

The prepared feed is further processed in a crossflow separator to remove the fine materials as overflow, and the underflow is dried and then processed in the electrostatic plate separator (ESPS) supplied by Mineral Technologies, Australia. Using the tribo-electrification principle, the quartz contamination is removed to produce a clean sillimanite concentrate. The feed temperature is a maximum of 110°C, the HT voltage is kept between 24 and 27 kV, and the air gap is varied between 45 and 50 mm. These figures may change depending on the type of feed and nature of the impurities.The concentrate is further cleaned using an induced roll magnetic separator (IRMS) to remove any residual magnetic minerals. The calcium contamination also marginally reduced, although it is not fully removed. This widens the opportunities for usage of this product. The flow diagram of the new process is shown in Figure 3.

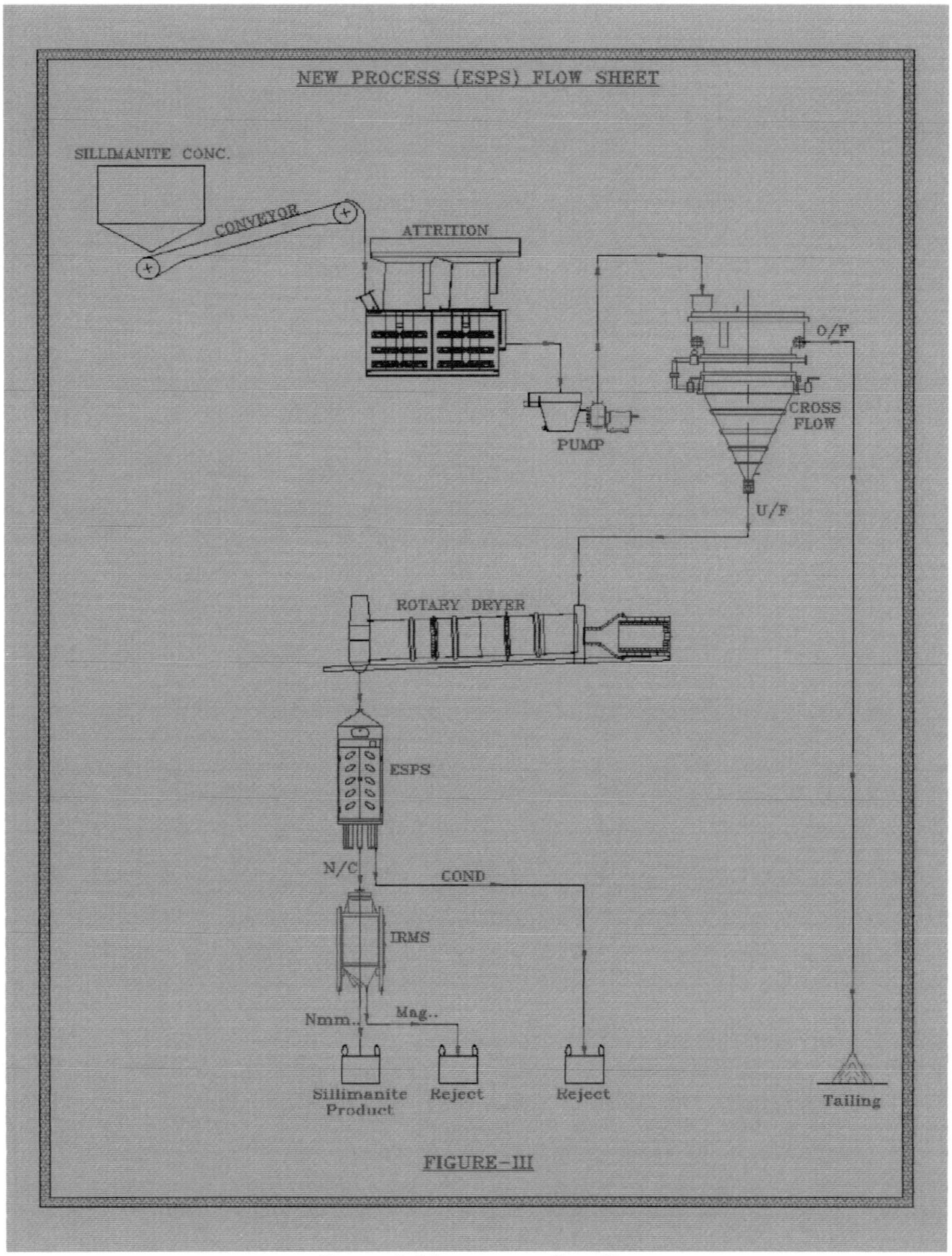

Figure 3. New process flow diagram.

PRODUCT QUALITY

The sillimanite product produced by this process is superior to that from the conventional flotation method. A comparison of the mineralogical analysis of the products of each process is given in Table I. The chemical analyses of the products from both are shown in Table II. The metallurgical data at various upgrading stages is given in Table III.

Table I. Mineralogical analysis of the sillimanite product from both processes.

Sl. no.	Mineral	Unit	Flotation process	New process
1	Garnet	%	*	*
2	Ilmenite	%	*	*
3	Leucoxene	%	*	*
4	Zircon	%	2.05	0.10
5	Rutile	%	0.33	0.06
6	Monazite	%	*	*
7	Sillimanite	%	96.0	96.33
8	Quartz	%	1.39	3.31
9	Shell	%	0.23	0.20
10	Haematite	%	*	*

Table II. XRF analysis of sillimanite product from both processes.

Sl. no.	Constituent	Unit	Floatation process	New process
1	TiO_2	%	0.45	0.06
2	Fe_2O_3	%	0.35	0.86
3	SiO_2	%	37.1	38.1
4	Al_2O_3	%	60.12	60.52
5	Cr_2O_3	%	0.064	0.065
6	MgO	%	0.05	0.04
7	MnO	%	-0.01	-0.01
8	ZrO_2	%	0.28	0.32
9	P_2O_5	%	0.182	0.253
10	U	ppm	-10	-10
11	Th	ppm	-10	-10
12	V_2O_5	%	0.11	0.11
13	Nb_2O_5	%	-0.001	-0.001
14	CaO	%	0.32	0.38
15	SO_3	%	0.01	-0.01
16	K_2O	%	0.02	0.05
17	CeO_2	%	0.008	0.01
18	SnO_2	%	-0.002	-0.002

Table III. Sillimanite concentration at various stages of feed preparation and final process.

Sl. no.	Operation	Shell (%)	Quartz (%)	Sillimanite(%)	Other HM (%)
1	Feed	0.43	47.05	37.83	14.69
2	RER operation	0.44	53.06	41.61	4.89
3	Spiral operation	0.26	49.25	47.66	2.83
4.	Crossflow operation - 1	0.49	24.69	70.90	3.92
5	Wet table operation	0.31	3.31	75.99	20.39
6	Combined sillimanite	traces	45.68	50.38	3.94
7	Crossflow operation- 2	traces	21.14	73.70	5.16
8	ESPS and IRMS operation	traces	3.51	96.33	0.16
10	Flotation and IRMS operation	traces	1.62	96.00	2.38

DISCUSSION OF RESULTS

ProductQuality

It can be seen from the results that both the mineralogical purity as well as the chemical composition in terms of Al_2O_3 are better in the product by the new process.

Reduction in Calcium Contamination

The sillimanite produced in the new process contain less calcium mineral contamination.

Environmental Advantage

The new process totally avoids the usage of chemicals like oleic acid, sodium carbonate, and soda ash, resulting in no chemical effluent from the process. Also, the sillimanite product does not have the oleic acid coating normally found in the flotation product, which is an additional advantage.

CONFIRMATORY TEST WORK

The test work was repeated with a sample kept specifically for the purpose. The analysis of the product confirmed the findings of the original test work.

CONCLUSION

It evident that for production of sillimanite from the tailings of the heavy mineral placer deposits, the dry separation process is better in terms of product quality, recovery, and utility. An additional advantage is the avoidance of the environmental liabilities associated with the flotation process. Based on the study, a plant-scale facility to produce sillimanite by this process has been implemented and continuously operated for the last six months.

ACKNOWLEDGEMENTS

The authors express their sincere thanks to the management and directors of V.V. Mineral for their encouragement, provision of all facilities to carry out the test work, and for their permission to present this paper. The authors also thank all the executives and staff of VVM involved in carrying out the test work for their co-operation

Chellakkan Swamydas

Chief Adviser and Technical Expert
V.V. Mineral

Associated with the Beach Sand Minerals Industry for the last 46 years. Served with Indian Rare Earths Limited, for 34 years. Joined as an engineer and rose to the level of General Manager, heading one of the operational unit. Hands on experience in production, maintenance, commercial, project, R&D, corporate planning, regulatory affairs and safety and environment. Presently working as a consultant for setting up greenfield projects and being retained as chief adviser and technical expert by V.V. Mineral, a leading BSM mining company.

Measurement of the variability of mineral band position on gravity spiral concentrator with a view to process control

H.S. Baloyi, A.M. Russell, and D.N.C. Phillpotts

Mintek, Johannesburg, South Africa.

Spiral operation is rarely optimized due to the large scale of typical spiral circuits, which presents difficulties in manually adjusting spiral splitter positions. The implementation of online process control that enables spiral units to respond to changes in feed conditions would contribute significantly to the optimization of such units. Despite some previous industrial and academic interest, difficulties in measurement and actuation techniques have prevented large-scale control systems from being implemented on industrial spiral circuits. This paper proposes a new technique for the actuation of spiral units that is suitable for large-scale retrofitting on industrial spiral circuits. Test measurement data from an industrial mineral sands spiral plant in South Africa is presented in order to quantify the extent of the mineral band variation on the surface of the spiral. Measurements of the difference in the colour or intensity of mineral bands by an array of contrast sensors and a camera are used to infer the mineral band interface position. Water nozzles triggered by the readings of the mineral band position are used to move the mineral band to the correct position before the static splitter - with a view to protecting the concentrate grade and/or recovery for the concentrate port.

INTRODUCTION

Spiral separators are widely used for gravity separation in a variety of applications including coal, iron ore, industrial mineral sands, and chromite. A spiral separator consists of an open trough that is twisted helically downward about a central axis. Slurry is fed from the top and solids separate laterally as they move downwards. This often results in the formation of distinct mineral bands on the spiral surface with differing colours or intensities. The mineral band colours are dependent on the type of ore being processed. In the case of mineral sands, the heavy high-value mineral forms a dark band on the inner portion of the spiral, while a lighter band is formed by the low-value gangue material on the outer portion of the spiral (Matthews *et al.*, 1999).

Portions of the high-grade mineral band are collected down the length of the spiral in auxiliary splitter ports with manually adjustable splitter blades. The positioning of the splitters blades relative to the mineral band interface is crucial to maintaining the product composition within acceptable limits. Generally, industrial practice is for the splitter position to be manually adjusted. Large fluctuations in feed properties (such as feed grade, material density, solids percentage, and flow rate) result in the mineral band interface position varying, which requires the splitter to be moved in order to maintain the grade of the product or recovery of the valuable mineral. Spiral circuits often consist of vast numbers of spirals, which poses an immense challenge to the manual adjustment of the splitters in response to low-frequency variations in feedproperties. This impracticality generally results in operational neglect and subsequent loss of valuable minerals due to the splitters being conservatively positioned in order to maintain the product grade (Vermaak *et al.*, 2008).

This study focuses on the development of a measurement and basic control system for spiral concentrators that can be retrofitted into existing spirals, as opposed to one that would require redesigned process equipment. Low-cost equipment, easy signal processing, and robustness are key requirements of the system due to the number of spiral units on typical plants.

PREVIOUS ATTEMPTS AT MEASUREMENT AND CONTROL ON SPIRAL SEPARATORS

There have been several attempts by various investigators to use machine vision to measure the position of the mineral bands with a view to process control (Welsh and Deurbrouck, 1973; Vermaak *et al.*, 2008; Gold, (1991). Light-sensitive cells, such as photocells, have long been used to detect the colour of minerals moving on a surface, the signal being used to control the actuation of a linked sorting device. The patent by Sweet (1928) discloses an ore separation device that employs a stationary photocell and light for control based on the colour of light reflected by the materials passing the device. Gold (1991) developed a machine-vision-based control system for a spiral separator using a motor to move the final product splitter blade to match the detected band position.

Hendrickson (1962) describes the use of an automated splitter/collection port and a combined control strategy using pulp density as a measurement. Mohanty and Mahajan (2007), followed by Zhang *et al.* (2008), reported the development of a control system through the testing of several slurry density sensing methods. A microprocessor-based control system was developed by Zhang *et al.* (2008), who investigated several measurement systems, including ultrasonic, pressure mats, and mechanical systems. A mechanical strain gauge was used for density sensing and a microprocessor-based servo motor was used to adjust the splitter position.

Vermaak *et al.* (2008) developed a model of the mineral band as a function of feed properties (feed flow rate, grade, viscosity, and solids composition), using data obtained with optical band measurement techniques. The model may be used to adjust the mineral band as measured with machine vision to the desired width through the manipulation of feed conditions. However, the model requires measurement and control of many parameters, some of which are not practical or would require a large investment given that most spiral concentrators have little or none of the required instrumentation.

A European patent (Cooke, 2009) describes the use of inflatable bladders that contact the spiral flow and thereby divert the material towards the appropriate port.

Mohanty *et al.* (2014) discussed the adjustment of a coal cleaning spiral splitter position using an electrical conductivity measurement. The splitter position is mechanically manipulated based on the solid density gradient output signal. They found little other literature on spiral control. The limited academic investigations into spiral control have not yet found any feasible, long-term industrial application. It is speculated that this is due to the complexity and cost of a system that is able to control a large spiral plant. The design of a viable system will need to consider these factors.

COLOUR-BASED MEASUREMENT OF MINERAL BAND INTERFACE

In this work, the colour differences between the minerals across the spiral surface are measured by several laser contrast sensors, which produce a binary output that is used to infer the mineral band interface position. A camera system is used alongside the sensors as a means of verifying, through image analysis, the mineral band position inferred by the sensors.

Contrast sensors are commercially available devices that are commonly used in automation applications in the manufacturing, printing, and packaging industries. The sensors are small enough such that they can be placed between the spiral starts, close to the spiral surface. They use laser or LED light sources and therefore, measure a small point on the spiral surface and can be precisely positioned. The contrast sensors are calibrated to output a high or low voltage, depending on the reflected light intensity from the minerals on the spiral surface directly below the sensor. Experimentation at Mintek showed that red light (at 625 nm) provides the greatest contrast between typical mineral colours found on chromite and mineral sand spiral plants in South Africa.

The sensors offer cost advantages over a camera system and their discrete outputs are easy to process using simple micro-controllers. The actuation method proposed below potentially only requires an indication of the presence or absence of a material colour at a discrete point, making the use of a single simple sensor attractive.

The analysis of camera images used similar techniques to that of Gold (1991).

NOVEL METHOD OF ACTUATION USING NOZZLES

A novel actuation system (the subject of a patent) that makes use of jets of pressurized water, slurry, or air to manipulate the minerals on the surface of the spiral is proposed here. When the actuation medium impacts the spiral surface a 'blast zone' is created. This forms a barrier that effectively moves the edge of the mineral band towards the desired collection port (often demarcated by the splitters). The flow to the nozzles is actuated by solenoid valves. Several nozzles may be used together to allow for varying levels of material diversion.

This method offers the following advantages over other methods of automated spiral splitter adjustment:

- It does not require altering of spiral feed properties
- The required equipment can be retrofitted to existing spirals (including tightly spaced triple-start spirals) without modification to the spiral
- The device does not require motors or servos for actuation, which are expensive and difficult to fit in the constrained spaces around compact industrial spirals
- There are no moving parts
- Only relatively simple signal and control infrastructure is required
- The lack of moving parts (apart from the solenoid valve internals) and low electrical power requirement compared to mechanical actuation are safer for operators.

The nozzles are positioned such that they are inactive when the mineral band is at the required position, which will typically be at the edge of the splitter blade. It is envisaged that the splitter blade positions will be manually set and then left static under most conditions. Testing has shown that a water flow rate in the order of about 2 L/min through each nozzle is required for effective actuation. In comparison, approximately 58 L/min of water is present in the feed to a typical 75 cm spiral, thus the actuation water addition will have little impact on the slurry properties in most applications. If it is anticipated that the actuation water addition will have an adverse effect on the slurry properties, the feed water can be reduced. The water required for actuation can be tapped from the wash water pipelines, typically situated on spiral circuits, and can be recycled along with other water on the plant. Industrial tests making use of air or slurry for actuation have not yet been conducted. A sufficient supply of compressed air is generally not available on spiral plants and it is envisaged that the use of slurry for actuation may cause difficulties with the current experimental setup.

Actuation Configuration

Figure 1A shows the position of the mineral band approximated as a normal distribution. Figure 1B uses the cumulative frequency of the band position. This is equivalent to the frequencies of the states of discrete sensors that are triggered by colour (they remain on even if the mineral band position is far to the right-hand side). The highlighted area represents the amount of desired material entering the reject port when all the sensors to the right of x are triggered (*i.e.* the band interface position is to the right-hand side of x). Figure 1C shows all material (shaded region) that can be pushed over to the collection port (on the left-hand side) by a nozzle creating a circular blast zone having a diameter of at least $2x$ (centred at x). Desired material to the left of this splitting point still reports to the concentrate port. Using this nozzle without ensuring only desired material is in this region will result in the collected grade being impacted adversely by the inclusion of gangue material in the diverted flow. The union of the regions is shown in Figure 2D. This represents the desired material that can be moved from the reject port to the collection port when the mineral band triggers the sensor (at x) which is used to turn on the actuator. In this case, control is used to protect the recovery of the valuable mineral.

This area will change with the position of x and there will be a maximum area for some position of x. This value will inform the choice of an optimum nozzle placement and therefore an optimum nozzle blast zone and size.

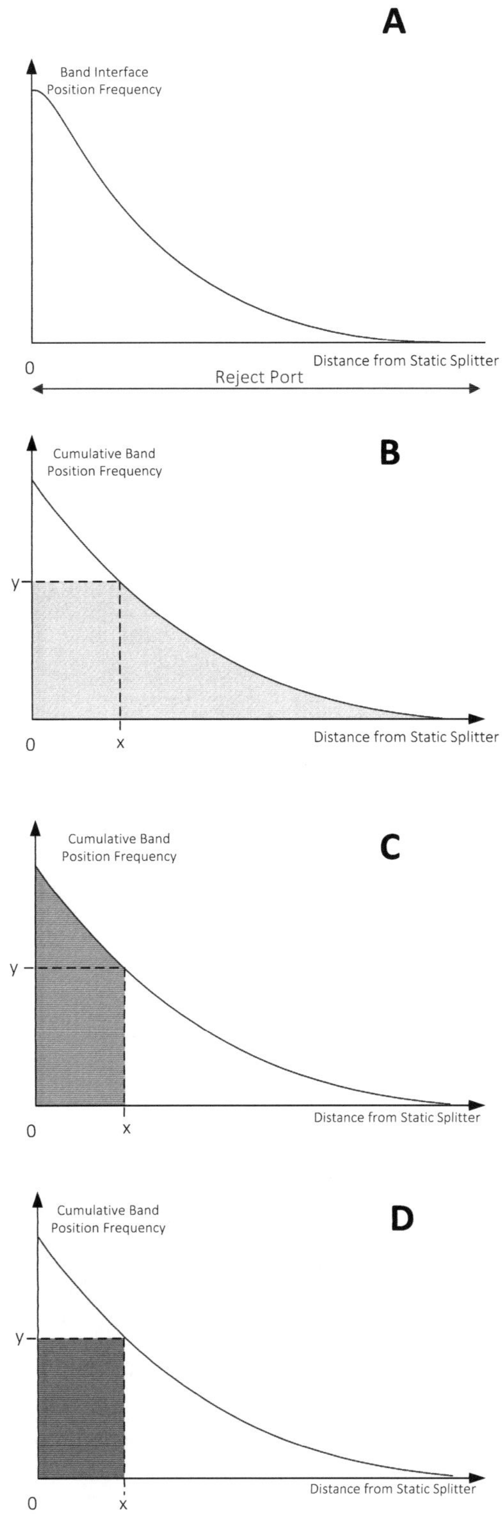

Figure 1. Mineral band interface position histograms representing desired material that can be pushed towards a collection port (A: mineral band interface position histogram; B: desired material that is lost when sensors up to x are triggered; C: material that can be pushed to collection port when nozzle at x is on; D: desired material that can be collected by a sensor-triggered nozzle).

The proposed configuration is shown in Figure 2. There are two distinct mineral bands containing heavy (valuable) and light (gangue) minerals respectively. The splitting device is placed so as to collect the heavy minerals into the concentrate/collection port. The position of the mineral band interface as detected by discrete sensors 1 and 2 measuring the colour of the mineral band is to the right of the static splitter. This means that some valuable material is lost to the gangue port. In this case, both sensor 1 and sensor 2 detect dark material. Nozzle 2 would then be turned on (nozzle 1 remains off). This would result in a 'blast zone' on the surface of the spiral that pushes all of the material to the left of the indicated apex of the blast zone (which would otherwise report to the gangue port) into the concentrate port. This would result in a reduction in the amount of valuable mineral reporting to the gangue port, while ensuring that no additional gangue material reports to the concentrate port.

Similarly, if the band is to the left of the static splitter, the material collected in the concentrate port would be contaminated. If neither sensors 1 or 2 detect dark material, nozzle 1 would become active, creating a blast zone that will reduce the amount of gangue material entering the concentrate port. This will result in an increase in the valuable mineral grade of the concentrate. A simpler configuration of only one sensor and one nozzle would allow for only grade or only recovery protection to be implemented separately.

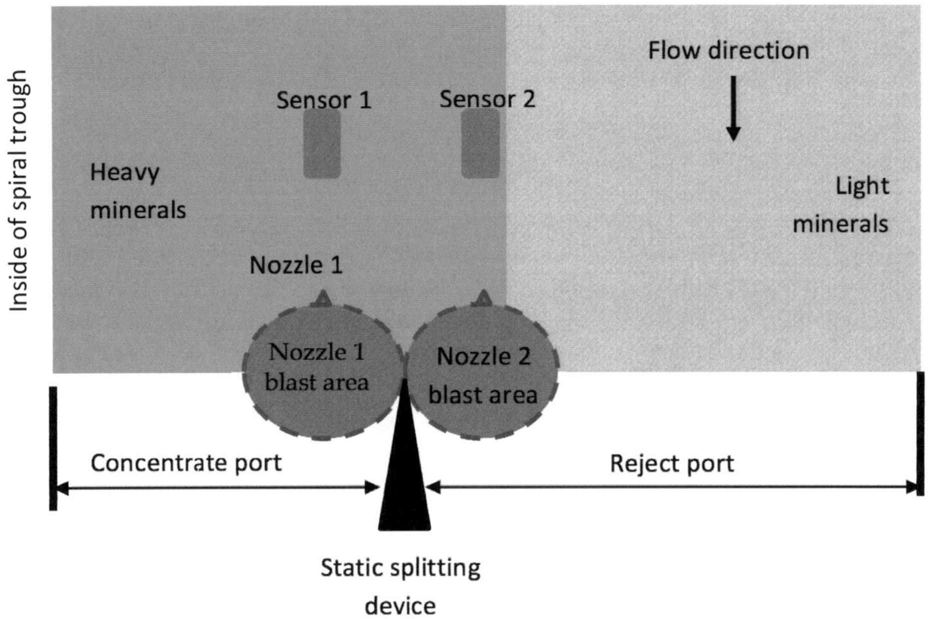

Figure 2. Proposed nozzle configuration.

The size of the nozzle's blast zone will influence the distance over which a band, targeted by an appropriately positioned splitter, can be manipulated in both the grade and recovery protection scenarios. The size and shape of the spray cone exiting the nozzle are mainly a function of the nozzle shape within the working pressure range of the nozzle. A pressure regulator should be used as wash water supply pressure may vary. The supply pressure affects the ability of the spray to cut through the solids and slurry on the spiral surface. This influences the size of the blast zone and the distance before the minerals converge again. The supply pressure must be adjusted to create the required minimum blast zone. Large and widely spaced blast zones can only fully target a large deviation and will be ineffective against much smaller variations, while a smaller nozzle will have a limited effect on large variations. Using a number of adjacent, progressively-triggered nozzles would allow for finer control and improved benefits.

Simulations of the effect of blast zone diameter on grade and recovery were performed to assess the potential of the scheme shown in Figure 2. The following assumptions were made during the simulation:

- There is a constant, steady-state mass flow rate through the spiral
- There is an evenly distributed flow on the spiral surface
- The overall feed composition (50% desired material) does not change (variations in the band are not due to variation in feed grade for the sake of the simulation)
- The variation in the mineral content around the interface follows a normal distribution with a mean at 5.0 cm and a standard deviation of 2.0 cm^2. This distribution and variance is a fair match for the camera-based mineral band position presented later
- The static splitter is positioned at the band mean interface (5.0 cm from the inside of the spiral)
- The split between the mineral bands is perfect
- The spray nozzle creates a circular blast zone on the spiral surface, such that the material split point is at the centre of the leading edge of the circle
- The blast zone is large enough to reach across to the static splitter
- The nozzles are close enough to the collection ports such that the displaced flow does not converge again before the static splitter.

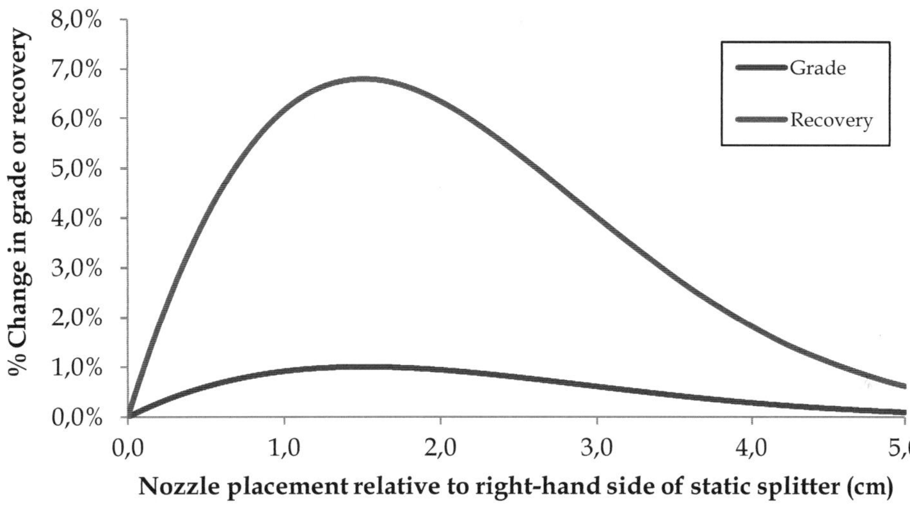

Figure 3. Effect of nozzle placement on grade and recovery.

Figure 3 shows that an additional 6.8% of the feed material would be recovered and the concentrate port grade would be 1.0% higher (as a result of the additional pure desired material that is recovered) in this recovery protection scenario. The greatest improvement occurs with the nozzle centred at 1.5 cm to the right of the static splitter. This would require a nozzle that can create a blast zone 3.0 cm in diameter if the spray is directed perpendicular to the spiral surface, creating a circular blast zone. If the spray is directed with a horizontal component, the required size of the blast zone may be smaller. The use of multiple progressively-triggered nozzles with smaller blast zones will allow the effective overall nozzle position and diameter to be manipulated, allowing finer control and improved benefit in cases where the mineral band variation may change. Observations during testing suggested that multiple smaller nozzles require less water than a single large nozzle, due to the improved penetration of the slurry surface by multiple nozzles.

Simulation of a similar grade protection scenario suggests a maximum increase of 6.1% in the grade of material collected.

The theoretical modelling of the mineral band actuation does not provide any guarantee that the economic benefits gained from the control system will offset its cost (construction, installation, and maintenance). The effects of recycle streams, multiple stages, and interstage flow changes make the calculation of mineral losses in a particular stage more complex, because lost valuables may be recovered later. The system would largely target the areas of the spiral circuit where losses are irrecoverable downstream. The economic return on the system in this study can be fully evaluated only by construction and on-site testing.

TEST WORK

Preliminary testing of the system hardware was done at a chrome spiral plant. The testing of the system was then carried out at the Tronox Namakwa Sands mine, located on the west coast of South Africa. The mine beneficiates heavy minerals to produce titanium dioxide, zircon, rutile, and high-purity iron products. Five contrast sensors were mounted on a single spiral and two static spray nozzles were mounted downstream from the sensor position. Plant wash water was fed to the nozzles, controlled by solenoid valves linked to a single-board computer (a Raspberry Pi). The single-board computer had remote connection capabilities over a cellular network, allowing monitoring of the system from Mintek's Johannesburg offices. Images from a camera focusing on the spiral surface in the region of the contrast sensor measurement points were used to verify the band position data from the contrast sensors through image analysis.

Placement of Control System on a Spiral

The double-start spiral on which tests were conducted consists of three auxiliary splitters (top, middle, and bottom) and one final product splitter (Figure 4). In practice, the auxiliary splitters are adjusted to separate the concentrate band from the middling stream. The upper two auxiliary splitters are usually conservatively positioned, in order to minimize gangue entrainment in the concentrate, and therefore do not require frequent adjustment. The final product splitter separates the concentrate, middling, and tails streams, and is usually adjusted to balance recirculating loads through the plant. Therefore, the lowermost auxiliary splitter is the primary target for optimization of grade and recovery.

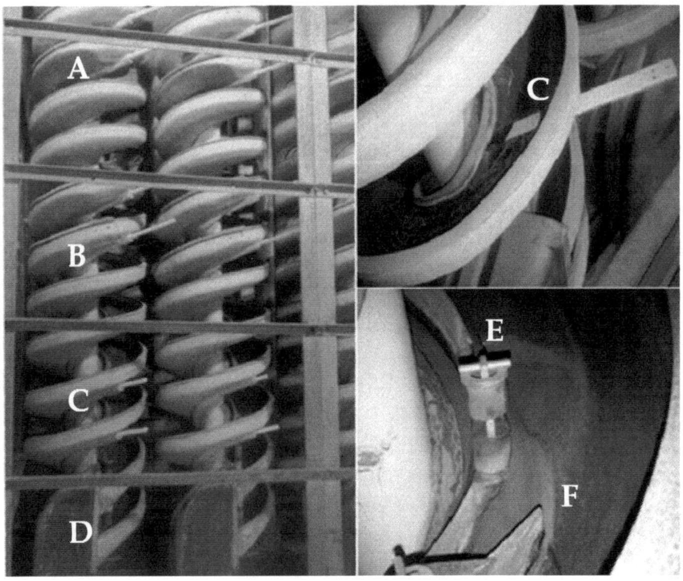

Figure 4. Potential spiral control positions (A, B, C: top, middle and bottom auxiliary splitter positions; D: product splitter; E: concentrate/middlings product splitter position; F: middlings/tails product splitter position) (Cloete, 2013).

Layout of the Control System

The system was mounted between the diverter and the auxiliary splitter of a cleaner spiral, as shown in Figure 5. The auxiliary splitter was then positioned at the median mineral band position. The spray nozzles were mounted downstream from the contrast sensors, on either side of the auxiliary splitter blade. A 3D printed bracket (shown in Figure 6) was used to cant the nozzles and sensors so that they were appropriately positioned in the tight confines between the spiral starts. This was mounted above the spiral surface using a bracket from the external steel framework around the spirals. The sensor measurement points (visible as red dots in Figure 5) were spaced 1.0 cm apart. This is close to the theoretical optimum suggested by the simulation exercise. Sensor packaging issues prevent closer spacing at this time. A camera was placed above this mounting to record images for verification purposes. The recorded images were processed automatically and some were manually inspected in order to check the discrete measurements from the sensors.

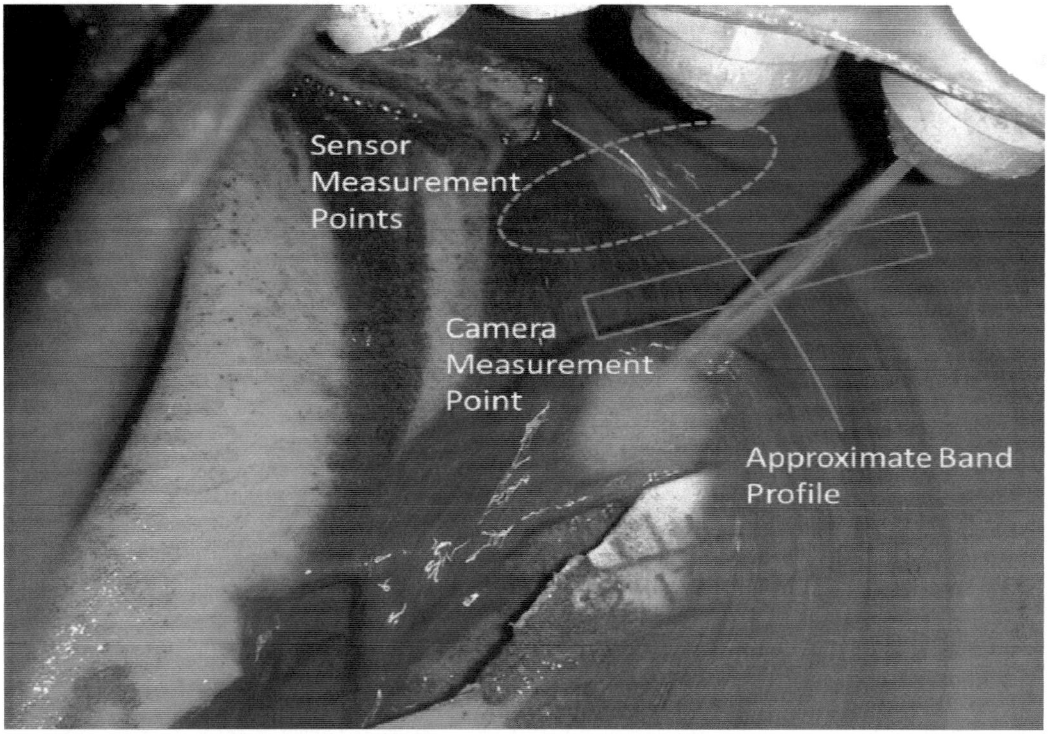

Figure 5. Layout of the measurement sensors and the actuating nozzles on a spiral.

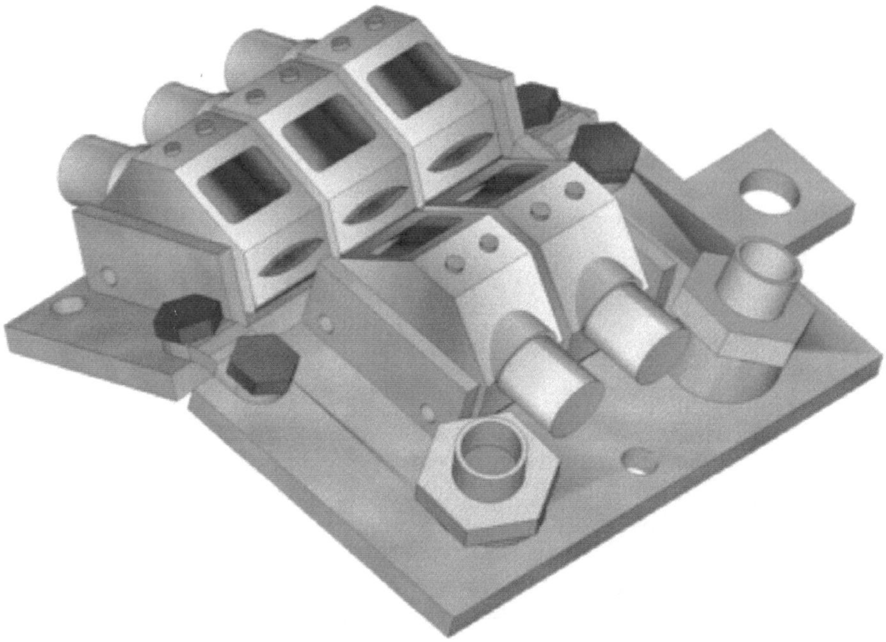

Figure 6. Sensor and nozzle bracket.

RESULTS AND DISCUSSION

Mineral Band Interface Variability

Image analysis was used to determine the position of the mineral band interface. The accuracy of the position obtained using this technique was verified by inspection and used as a basis for comparison of the sensor measurements.

Figure 7 shows time series trends of the mineral band interface position as measured by both the camera and by the contrast sensors. Rapid short-term deviations as well as some sustained offsets are observed. It is expected that sustained offsets might be experienced by all the spirals in a process section, while shorter term deviations may be unique to a particular spiral start.

It can be seen that in some cases, the contrast sensors track outward movements (increases) in the mineral band interface position while inward movements in the mineral band position are tracked relatively poorly. The poor tracking of inward movements is due to the nonlinearity of the band position. Outward band position movements at the splitter port where the band position is measured by the camera correspond to proportionally large outward band position movements immediately following the diverter, allowing the sensors to track the movement. Conversely, inward band positon movements at the splitter port do not result in proportional inward movements at the diverter position and hence the sensors do not register a proportional inward movement at the contrast sensor measurement point, which is closer to the diverter than the camera position measurement point. It is anticipated that moving the contrast sensor measurement point further downstream from the diverter will improve the tracking of inward mineral band movements.

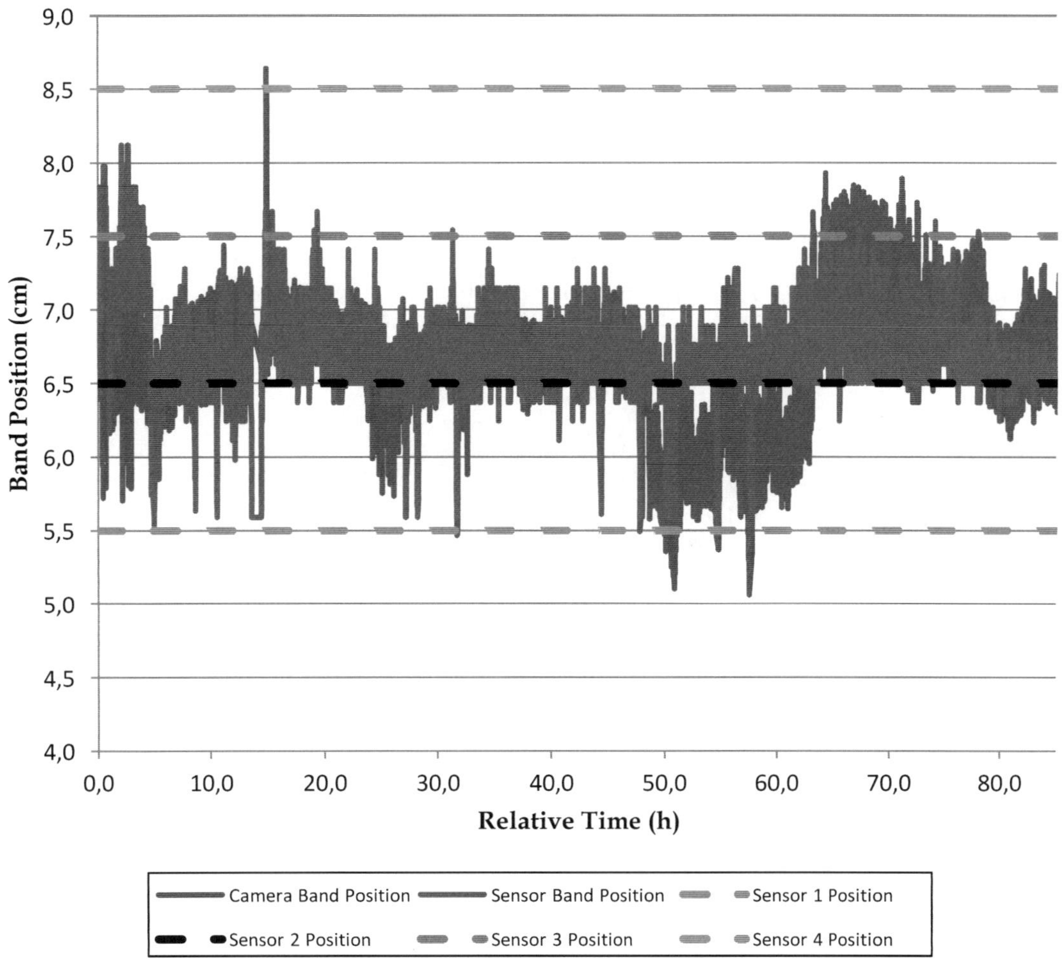

Figure 7. Band interface position as measured with a camera and sensors on a cleaner spiral.

A bagplot constructed according to the method of Rousseeuw, Ruts, and Tukey (1999) is used to further visualize the correlation between a random sub-sample from the camera and contrast sensor data pairs (shown in Figure 8). The means of both data-sets are plotted as a single point. The centre mean lies on the identity line, indicating that the means of the two data-sets are approximately equal. A shaded bag is drawn, which encompasses the closest 50% of the data relative to the centre mean. The bag area is inflated by a factor of three, and all points that lie outside of this region are deemed outliers. A shaded fence is then drawn around all of the data points that lie between the bag edge and the outliers. The orientation of the bag is skewed towards the horizontal axis, with a large portion of the bag area lying above the identity line (where the measurements would be equal).

Therefore, in the majority of cases, the contrast sensors over-predict the mineral band interface position. The band interface also generally moves through a small range, while the sensors only produce discrete measurements at 1.0 cm intervals. This means a very strong correlation cannot be expected and the spacing of the sensors must be reduced to capture the variability that the camera-based measurement is capable of measuring.

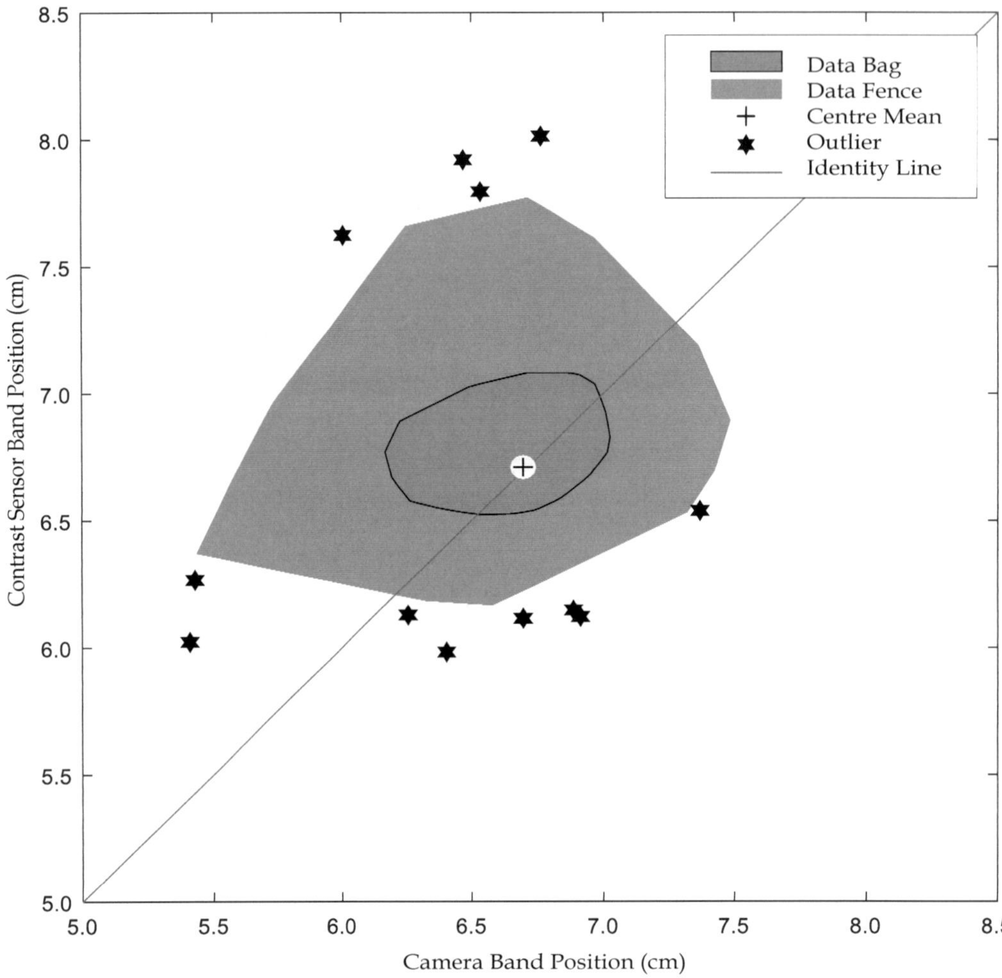

Figure 8. Bagplot of sensor and camera data.

Due to the small area of the contrast sensor measurement points, fouling, even by small particles, can adversely impact the performance of the sensors through dispersion of the transmitted or reflected laser light. Blinding of the sensors through this mechanism led to instances where the mineral band position measurement was observed to remain constant towards the latter part of testing. These portions of data were identified and excluded from the analysis in Figure 7 and Figure 8.

Due to the issues with the resolution and reliability of the sensors, the remainder of this analysis relies on the camera measurements.

Control Potential

The actuation system hardware is considered to be robust, with only a few temporary water line blockages being encountered due to salinity and silt in the water. This can be prevented by making use of larger diameter distribution lines and periodic automated flushing.

Figure 9 shows the frequency of mineral band variations as a function of the distance from the auxiliary splitter edge. This distribution was constructed using the band position data obtained from the monitoring camera. The ideal splitter position is chosen to be the modal band position, with negative relative band position values being those on the inner part of the spiral relative to the splitter position and positive relative band position values being those on the outer part of the spiral relative to the splitter position. The mineral band interface is observed to vary within a small range (2.0 cm), which is expected since the tests were carried out on a cleaner spiral. It was observed that 83.2% of the mineral band interface deviation occurred within a 1.0 cm range of the modal value. Deviations from the auxiliary splitter position are evenly split around the ideal splitter position. This indicates that, in this process section, opportunities for grade protection and recovery optimization are equal. This means that it may be beneficial to implement a control system capable of both control scenarios.

An ideal control system would be one that is capable of correcting all deviations of the mineral band from the auxiliary splitter position. In essence, an ideal controller would ensure that the desired grade is maintained with the maximum achievable recovery at all times. If it is assumed that there is a direct correlation between the mineral band interface position and the valuable mineral grade of the material on the spiral surface, an ideal control system would maintain the mineral interface position in line with the edge of the auxiliary splitter blade at all times. In order to simplify this analysis, it is also assumed that the respective mineral bands contain only pure materials. This results in a simulated 95.1% recovery and collected grade of 84.3 %.

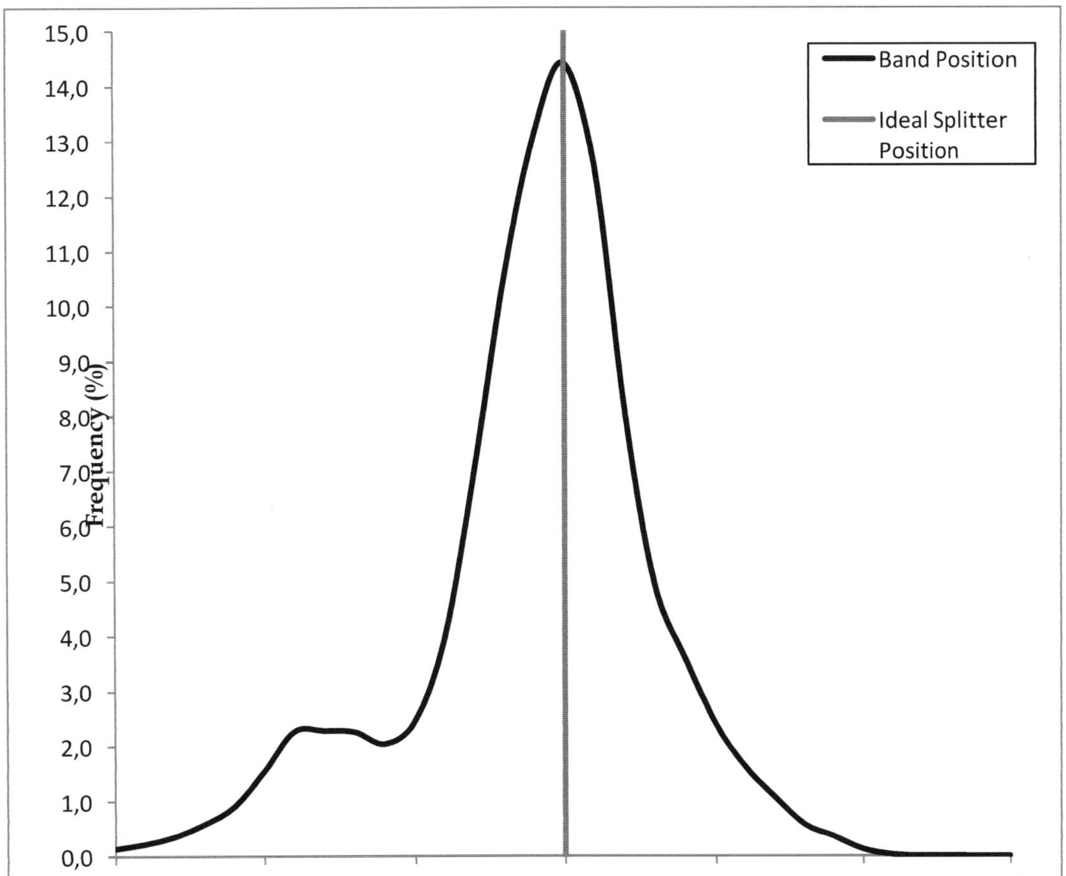

Figure 9. Mineral band interface position variation for a mineral sands cleaner spiral.

An estimate of the potential grade and recovery improvements that can be achieved through the implementation of an ideal controller can be obtained from the mineral band variation histogram shown in Figure 9. Grade protection can be achieved in the region where the relative band position is less than -0.5 cm, and recovery protection can be achieved in the region where the relative band position is greater than 0.5 cm. These limits are selected based on the current physical size constraints imposed by the sensor packaging.

Analysis of these areas suggests that controlling the typical variation measured on an industrial spiral would result in an additional 4.2% recovery using a simple single nozzle and single discrete sensor positioned on the reject port side of a static splitter (at a positive relative band position). The concentrate port grade would be 0.6 % higher as a result of the improved recovery of desired material. Grade protection with a single nozzle and sensor on the concentrate port side of the splitter (at a negative relative band position) could improve the concentrate port grade by a further 4.2% (without adversely affecting the recovery) using the tested measurement configuration. A combined scheme could theoretically result in a 4.2% increase in recovery and a 4.8% increase in concentrate grade. The required sensor spacing is only marginally within the present minimum possible sensor spacing. It must be stressed that these improvements will be attained only under perfect control conditions and rely on several assumptions as stated previously.

CONCLUSIONS

A water-nozzle-based control system can be used to effect grade and/or recovery protection on industrial spiral circuits. The system is intended to be retrofitted to existing spirals on large plants and consists of low-cost hardware, processing, and signal handling equipment.

Laser contrast sensors were tested as a simple means of identifying the mineral band position on a spiral. Discrete measurements from these sensors (if accurate) are well suited to this spiral actuation concept.

The quantification of the mineral band variance is important in order to determine if sections of spirals in the plant will benefit from control, as well as to determine the required placement of the actuating nozzles. Industrial testing has revealed that mineral band variations consist of both short-term movement and long-term offsets. The full extent of the benefits can be realized only by targeting both of these types of variation. It was shown that the mineral band position measured by the contrast sensors correlates quite poorly with the band position measured by a monitoring camera.

An industrially successful spiral plant control system would require a robust measurement and/or actuation system that requires little maintenance or human intervention. Difficulties were experienced with blinding of the sensors by slurry droplets, resulting in the degradation of the measurements. Sensor spacing is a current limitation imposed by the physical design of the sensor housing, preventing increased measurement resolution from being achieved. An improved measurement system is being investigated.

Simulated results using the actual measured band interface variation using the camera images suggest that a 4.2% increase in overall recovery, and an added increase in grade of 4.8% can be achieved (without adversely affecting the recovery) using the tested measurement configuration. This requires perfect targeting and sensing of the mineral band interface at set discrete positions. Metallurgical testing will reveal the extent to which this is possible.

Metallurgical testing is ongoing, with opportunities to use nozzle actuation with grade, manual, or modelled inputs also being explored.

ACKNOWLEDGEMENTS

The authors gratefully acknowledge the site personnel and management of Tronox Namakwa Sands for the assistance provided during the test period. Mintek also acknowledges the assistance, information, and samples from other industrial sites and experts.

REFERENCES

CLOETE, D. (2013). Wet gravity optimization. Unpublished presentation, Tronox Namakwa Sands, Vredendal.

COOKE, G. (2008). An adjustable spiral concentrator. European patent. EP20090802279.

GOLD, D. (1991). The investigation and design of a machine vision system for the detection and control of the separation in a spiral ore concentrator. MSc thesis University of Cape Town, South Africa.

HENDRICKSON, L.G. (1962). Method and apparatus for controlling spiral concentrators - product separator. US patent 3235079 A.

MATTHEWS, B.W., FLETCHER, C.A.J., and PARTRIDGE, T.C. (1999). Particle flow modelling on spiral concentrators. *Conference on CFD in the Minerals and Process Industries*, Melbourne, Australia, 6-8 December 1999. CSIRO Minerals, Melbourne. pp. 211-214.

MOHANTY, M.K., MAHAJAN, A.M., WANG, H., and ZHANG, B.J. (2014). Automated system for coal spiral. US patent 20140238906 A1.

MOHANTY, M.K., MAHAJAN, A.M., and ZHANG, B.J. (2007). Automation and performance improvement of spiral concentrator. Final Technical Report, Project no. 06-1/2.1A-6. Department of Commerce and Economic Opportunities and Illinois Clean Coal Institute.

ROUSSEEUW, P. J., RUTS, I., and TUKEY, J.W. (1999). The bagplot: a bivariate boxplot. *American Statistician*, 53, 382-388.

SWEET, A.T. (1928). Metallurgical separator, US patent 1678884 A.

VERMAAK, M.K.G., VISSER, H.J, BOSMAN, J.B., and KREBS, G. (2008). A simple process control model for spiral concentrators. *Journal of the Southern African Institute of Mining and Metallurgy*, 108 (3), 47-154.

WELSH, R.A. and DEURBROUK, A.W. (1973). Product separator. US patent 3933249.

ZHANG, B., YANG, F., SAHOO, P.K., AKBARI, H., WANG, H., MAHAJAN, A.M., and MOHANTY, M.K. 2012. Development of an automated control system for a spiral concentrator - Phase 1. *Separation Technologies for Minerals, Coal and Earth Resources*. Young, C. and Luttrell, G.H. (eds). Society for Mining, Metallurgy & Exploration, Littleton, CO. pp. 129-137.

Hlayisi Baloyi

Process Control Engineer
Mintek

Hlayisi Baloyi graduated in 2012 with BSc in Chemical Engineering at UCT. Hlayisi started working for Mintek in 2013 in the Measurement and Control division. He has been involved in research and development of control in spirals as well as the commissioning of Mintek's advanced control system for flotation circuits.

Adrian Russell

Process Control Engineer
Mintek

Adrian Russell has been involved in the development of new control systems, as well as the commissioning and servicing of Mintek's advanced flotation control systems. Adrian is currently completing a postgraduate qualification in control engineering part-time.

Prereduction and DC open-arc smelting of carbon-based ilmenite pellets

K. Bisaka, I. Thobadi, and X. Goso

Pyrometallurgy Division, Mintek

The smelting of ilmenite to produce valuable titania slag and pig iron is an energy-intensive process. At process temperatures of about 1700°C, roughly 1 MWh of electricity per ton of ilmenite feed is consumed. Following successful laboratory-scale test work, a pilot-scale operation was undertaken to demonstrate prereduction and subsequent smelting of carbon-based ilmenite pellets. Five metric tons of carbon-based micropellets of 2 to 5 mm in size, consisting of ilmenite, anthracite, and an organic binder, were prepared for prereduction in batches of 250 kg. Prereduction was conducted in a fixed-bed furnace under a controlled atmosphere using CO gas at temperatures between 900 and 1100°C. The resulting prereduced pellets were successfully smelted in a 200 kW DC arc pilot furnace. The pilot test work and the prior laboratory test work demonstrated that consistent prereduction could be achieved due to improved heat and mass transfer within the carbon-based pellets, as well as diffusion of CO gas reactant to the reaction sites. Smelting of prereduced ilmenite appeared technically feasible and efficient; the furnace operation during the smelting process was stable with no visible sign of foaming. On average, furnace electricity consumption was measured at 0.6 MWh per ton of prereduced ilmenite pellets (extent of prereduction between 74–80%, with 65–73% metallization). Production of titania slag containing as little as 1.3% FeO and as high as 95% TiO_2 is achievable by this process. The test work demonstrated the technical feasibility, viability, and robustness of the smelting of prereduced carbon-based ilmenite pellets as one of the many alternatives to the current ilmenite smelting process.

INTRODUCTION

Electric smelting of 48–52% TiO_2-grade ilmenite is carried out for the production of 85–90% TiO_2 titania slag. This energy-intensive process operates at high temperatures of approximately 1700°C, and consumes around. 1.0–1.3 MWh per ton ilmenite. An amount of 4.5 Mt of titania slag is produced globally per annum, of which South Africa contributes 1.5 Mt (Gous, 2006; Kotze, Bessinger, and Beukes, 2006; Williams and Steenkamp, 2006). In South Africa, production is carried out in DC arc furnaces at Tronox Heavy Mineral Sands (previously Namakwa Sands) and Exxaro KZN Sands, while AC furnaces are used at Richards Bay Minerals (RBM). Altogether these furnaces have an installed capacity of 435 MW, which currently represents close to 1% of the national grid (Gous, 2006; Kotze, Bessinger, and Beukes, 2006; Williams and Steenkamp, 2006; Zietsman and Pistorius, 2006). The DC open-arc smelting process was jointly developed in South Africa by Mintek and Anglo American, while the QIT process used by RBM was developed by Quebec Iron and Titanium (QIT), Canada (Gous, 2006; Kotze, Bessinger, and Beukes, 2006; Williams and Steenkamp, 2006).

The availability and cost of energy (for example, the availability and price of electricity, coal, and natural gas) play a significant role in the development, implementation, and operation of energy-intensive processes. At national and global levels, this ultimately dictates the size of the industry as well as its capacity to create jobs and generate income. The gradual decrease in the global competitiveness of the South African ferroalloys industry can be cited as one example of the effect of rising costs and tight supply of electricity on the local industry. A gradual decrease in alloy production and increasing exports of raw ores are being observed. This business model decreases the prospect for production capacity expansion and job creation.

From a number of options evaluated to stimulate the ilmenite smelting industry, preheating and prereduction of ilmenite prior to smelting are able to mitigate, to some extent, the impact of electricity price and tight supply on this industry. There is no existing plant where ilmenite is preheated prior to smelting in the furnace, as preheating may create more operational challenges such as severe slag frothing (Kotze, Bessinger, and Beukes, 2006). Smelting of cold prereduced ilmenite pellets has been practised by Tinfos Titan KS in Norway (Murty, Upadhyay, and Asokan, 2007). Although it is a successful process, the Tinfos process uses an elaborate feeding system in which bentonite-bound pellets are reduced in a rotary kiln using coal in excess of the stoichiometric requirements. The lumpy pellets are then magnetically separated from fines and unreacted coal, and are smelted in electric furnaces. The Tinfos process produces a low-grade slag of about 77% TiO_2, with the main impurities being MgO and SiO_2 (Nell, 2007).

Through extensive investigation of the solid-state reduction of ilmenite, the smelting of prereduced ilmenite in a DC arc furnace was developed at Mintek. In this process, provisionally patented as the 'Impril process' (SA Patent 2015/08501), carbon-based ilmenite micro-pellets 2–5 mm in size consisting of ilmenite, anthracite, and organic binder are heated in a fixed-bed configuration, at high temperatures in the range of 900–1100°C in the presence of CO gas. CO could potentially be replaced by other reducing gases such as syngas, hydrogen, and natural gas. This process would be able to produce consistent prereduced ilmenite pellets for direct smelting in the furnace.

Following successful laboratory-scale tests, the Impril process was demonstrated at pilot scale. The results are summarized in this paper. The aim of the test work was to demonstrate the technical feasibility of the smelting of prereduced carbon-based ilmenite pellets in a DC open-arc furnace and the benefits that this could generate. Preparation of carbon-based ilmenite pellets and their pre-reduction in a fixed-bed configuration, as well as the smelting of the resulting prereduced carbon-based ilmenite pellets in the DC open-arc furnace, were evaluated individually.

TEST MATERIALS

Five tons of ilmenite concentrate and one ton of anthracite required for the test work were sourced locally. The as-received ilmenite was sampled using a riffle splitter and a representative sample was used for characterization - the methodologies are described in the next section. The ilmenite had a particle size in the 38–150 μm range. The chemical analysis of the concentrate is summarized in Table I. The concentrate is composed mainly of iron oxides and TiO_2. Iron is present as Fe^{2+} and Fe^{3+}. The chemical analyses of the anthracite used is summarized in Tables II and III. The anthracite was milled to 85% passing 106 μm to facilitate its incorporation into the ilmenite pellet recipe. The brand name and composition of the organic binder used for the pellets are not disclosed since it is proprietary.

Table I. Bulk chemical composition of the raw ilmenite (mass %).

MgO	Al_2O_3	SiO_2	CaO	TiO_2	V_2O_5	Cr_2O_3	MnO	FeO	Fe^{3+}	Fe^{2+}	Fe^0	Fe/Ti
0.28	0.44	0.57	0.05	46.6	0.25	0.09	1.08	47.87	23.71	13.50	<0.05	1.33

<0.05%: The analyte concentration could not be accurately quantified as it is below the limit of detection (LOD)
Total Fe in the sample is expressed as %FeO

Table II. Summary of the bulk chemical composition of the anthracite (mass %).

Moisture	Ash	Volatiles	Fixed carbon	Total carbon	Total sulphur
3.57	4.74	6.19	89.1	90.7	0.56

Table III. Bulk chemical composition of the anthracite ash (mass %).

MgO	Al_2O_3	SiO_2	CaO	TiO_2	V_2O_5	Cr_2O_3	MnO	FeO
1.43	6.42	18.78	2.74	1.43	1.61	1.25	1.07	65.27

Total Fe in the sample is expressed as %FeO

Analytical and characterization methods

The bulk ilmenite material was sampled using a riffle splitter. A subsample was characterized by particle size distribution (PSD), X-ray diffraction (XRD) analysis, scanning electron microscopy (SEM), and bulk chemical analysis. The raw materials and products were chemically analysed by appropriate accredited methods using inductively coupled plasma optical emission spectroscopy (ICP-OES) and X-ray fluorescence (XRF). Phosphorus content was determined by spectrophotometry, and carbon and sulphur by combustion methods (LECO). Wet chemistry for speciation was carried out to determine the content of the respective iron species by selective leaching. SEM analysis was carried out for the determination of the textural relationships between phases and the evaluation of the morphology of the grains and pores, as well as any features that may affect the gas-solid interactions.

PELLET PRODUCTION AND TESTING

Pellet preparation

Pellets were prepared in a pilot-scale pelletizing unit. This unit is comprised of an inclined rotating pan of 985 mm diameter and 170 mm depth, a feeding system, and a water spraying system. It is equipped with a variable-angle mechanical system and a variable-speed electric motor which is used by operators to set the angle and speed of the rotating pan. A blend of solid materials consisting of raw ilmenite, milled anthracite, and binder was fed to the rotating pan. Each blended batch was prepared by mixing 200 kg of ilmenite, a stoichiometric amount of milled anthracite, and 1% of the organic binder in a Jones mixer for about 30 minutes. The stoichiometric amount in this case is defined as the amount of pure carbon required to reduce all the iron oxides to metallic iron without any co-reduction of titanium oxides. Water was sprayed onto the blend in the rotating pan, thus binding and rolling the grains together into pellets. The feed rates of the blend of solid materials and process water were manually regulated during the pelletizing process in order to produce micro-pellets of 2–5 mm size. The pellets were intermittently discharged from the rotating pan using a steel scoop once they were of required size and physical quality.

In total, about 4.5 t of carbon-based ilmenite pellets were produced. The green pellets were indurated by drying in an ambient environment for 24 hours, stockpiled for 4 days and packed into 1 m^3 bags. The mechanical strength of the pellets was measured using a basic compression test machine. The procedure consisted of randomly selecting 10 to 15 pellets and subjecting these pellets to the standard compression test. The weight, mean diameter, and the force to crack each pellet were recorded. Based on the force and mean diameter, the compressive stress/strength of each pellet was calculated. The average compression strength of the pellets was about 1.50 MPa. This value may be considered as a reference value for future work.

Prereduction of ilmenite pellets

The pilot prereduction unit is presented schematically in Figure 1. This unit is a top-hat furnace with an internal muffle. The pellets were loaded at a height of 100 mm (250 kg pellets) in a single tray of dimensions 1700 mm × 900 mm. The loading area was a mesh screen which acted as a distribution plenum for the reducing gas and argon gas that were blown through the muffle load throughout the operation. Firing commenced once the loaded tray was placed inside the furnace. Two K-type thermocouples were used to monitor the process: one was used to measure and control the muffle temperature and the second measured the internal temperature at the coldest part of the load, which was located in the centre of the tray at 450 mm from the edge, 50 mm deep into the product. The burden thermocouple was used to follow the progress of the heating.

CO gas, supplied from cylinders, was fed through a pipe connected at the bottom of the furnace tray. Due to operational challenges related to the logistics and supply of gaseous reductant to the process, CO was only used intermittently to flush the furnace atmosphere every 12 minutes. An amount of 5 kg CO was used per firing cycle, or about 18 g/min/m^2 if CO is spread over the 3-hour blowing cycle. Each firing cycle lasted for about 3–5 hours. The process gases from the furnace were released through an exhaust pipe. CO was replaced by Ar once the prereduction period was completed. The furnace was then switched off and the prereduced material was left to cool inside the furnace in the argon gas environment.

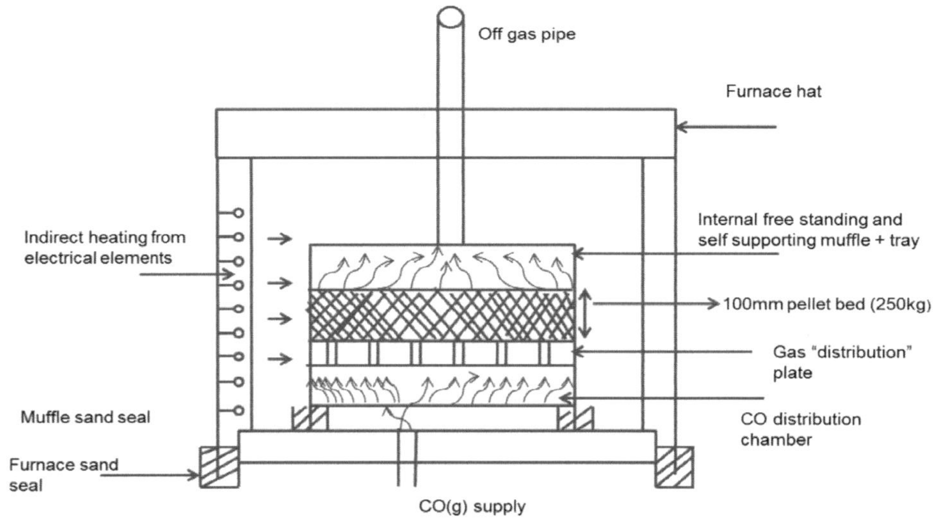

Figure 1. Schematic representation of the top-hat furnace used for prereduction.

Batches of indurated pellets of 250 kg each were heated in the muffle furnace. A typical profile of the furnace temperature (TOP °C) and the burden temperature (BOTTOM °C) is presented in Figure 2. The furnace temperature was gradually raised over a period of about 2 hours to 1100°C; this was subsequently followed by a 2-hour stable operation period at 1100°C. The muffle furnace operation was consistent and stable. With these operating conditions, the burden temperature was measured at between 800 and 1050°C. A drawback of the test conditions is that CO was blown through the burden only intermittently, as explained earlier. Overall, the furnace operation was consistent, resulting in fairly uniform prereduction conditions for the carbon-based ilmenite pellets.

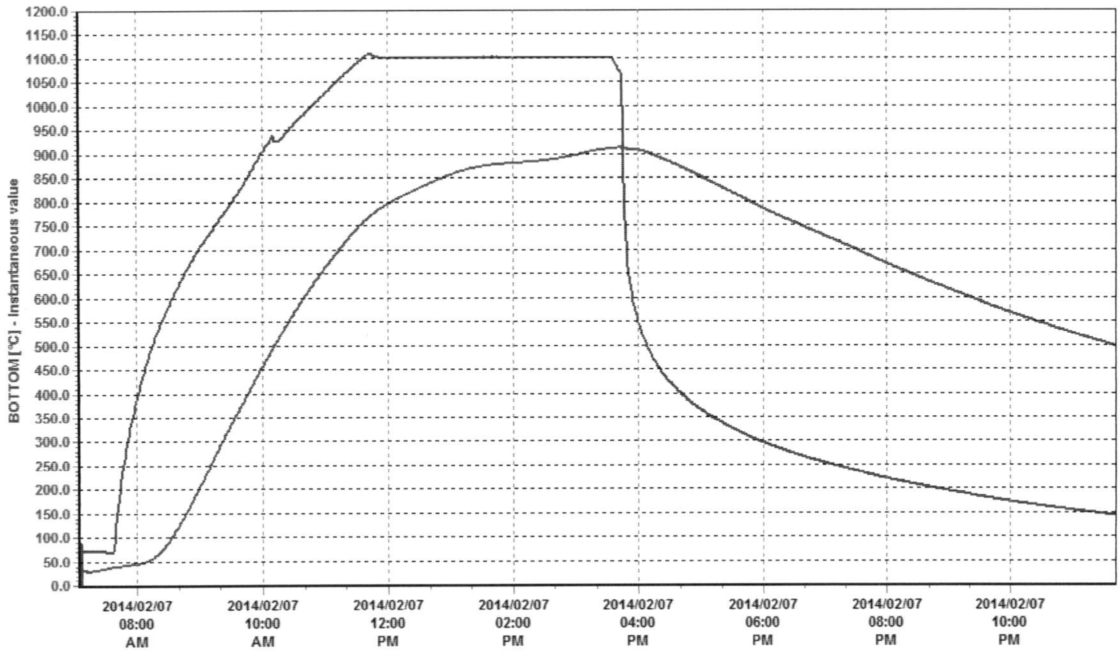

Figure 2. Typical temperature profile during the prereduction process (red=muffle temp; blue=burden temp).

Figure 3 shows a top view of the cooled tray of prereduced carbon-based ilmenite pellets. There was no visible disintegration or sintering of the pellets, and burden permeability was therefore preserved throughout the prereduction test. This also demonstrates that the carbon-based pellets are able to withstand high temperatures of up to 1100°C, suggesting that prereduction in a fixed-bed configuration is possible for these particular carbon-based pellets.

Figure 3. Prereduced pellets of 3–5 mm size in the furnace tray.

In total, about 3.6 t of prereduced pellets were produced. The pellets were bagged in 1 m^3 bags, from which five composite samples were collected using a pipe sampling technique. The samples were characterized by both chemical and mineralogical analyses. The chemical analyses of the five composite samples are given in Table IV.

Both the raw and prereduced ilmenite materials were chemically analysed; specifically, the analysis of the oxidation state of iron (Fe^{3+}, Fe^{2+}, and Fe^{0}) is used to calculate the prereduction and metallization yields for the reduced pellets.

Table IV. Chemical compositions of the prereduced pellets.

	MgO	Al_2O_3	SiO_2	CaO	TiO_2	V_2O_5	Cr_2O_3	MnO	Total Fe	Fe^0	Fe^{2+}	C	Ti^{3+}	Fe/Ti ratio
Bag 1	0.53	0.33	0.31	0.10	44.4	0.36	0.07	1.05	34.92	25.55	9.37	7.23	6.55	1.32
Bag 2	0.55	0.30	0.26	0.07	44.5	0.36	0.08	1.06	35.23	25.44	9.79	7.72	6.76	1.32
Bag 3	0.50	0.30	0.24	0.15	43.5	0.33	0.07	1.03	33.60	21.90	11.70	7.97	5.45	1.29
Bag 4	0.48	0.32	0.39	0.11	42.4	0.33	0.07	1.05	34.46	22.66	11.80	8.01	4.78	1.35
Bag 5	0.36	0.45	0.64	0.18	41.9	0.33	0.07	1.08	36.01	24.21	11.80	7.20	4.66	1.43
Average	0.48	0.34	0.37	0.12	43.3	0.34	0.07	1.05	34.84	23.95	10.89	7.63	5.64	1.32
St dev	0.07	0.06	0.16	0.04	1.17	0.02	0.004	0.02	0.90	1.64	1.21	0.39	0.98	0.02

Negligible reduction of titanium oxides is assumed throughout and the prereduction yield was therefore calculated based on the mass balance of oxygen associated with each gram of iron before and after prereduction. Equations [1] and [2] were used for the calculation of the prereduction and metallization yields, respectively.

$$Prereduction\ yield = 100 \cdot \frac{(Oxygen\ removed\ by\ the\ prereduction\ process)}{(Oxygen\ associated\ with\ each\ gram\ of\ iron\ in\ ilmenite\ sample)} \quad [1]$$

$$Metallization\ yield = 100 \cdot \frac{Fe^0}{Fe_{total}} \quad [2]$$

The calculated degrees of prereduction and metallization for the five composite samples are presented in Table V.

Table V.: Pre-reduction and metallisation degrees of ilmenite pellets.

Composite sample no.	Prereduction, %	Metallization, %
Bag 1	79.7	73.2
Bag 2	78.9	72.2
Bag 3	73.6	65.2
Bag 4	74.1	65.8
Bag 5	75.1	67.2
Average	76.3	68.7
St dev	2.8	3.7

Tables IV and V show that pellets prereduced to a consistent extent were produced as a result of the uniform furnace operating conditions. This would also suggest that within the range of operating conditions and pellet recipes investigated, the fixed-bed prereduction configuration would be appropriate for the prereduction of the prepared pellets. Design and operation of a controllable smelting process for prereduced pellets is possible in these conditions.

As per the prior laboratory test work results, the actual yields can be improved. In the laboratory test work, reduction yields of over 95% were achieved when a continuous CO flow was maintained and prereduction was carried out at a temperature of 1000 °C. The lower prereduction yields achieved in the pilot-scale work were attributed to lower firing temperatures recorded in the burden, as well as the significantly lower and intermittent flow of CO blown through the burden.

DC SMELTING OF PRE-REDUCED ILMENITE PELLETS

Pilot smelting set-up

Mintek's 200 kW DC arc furnace facility, used for the smelting phase of the work, is schematically presented in Figure 4. It consists of a cylindrical 1 m outer diameter steel shell which is water-spray cooled and lined with a single layer and three rows of magnesite-chrome bricks, and a hearth consisting of a rammable magnesia refractory. The refractory design resulted in a furnace crucible internal diameter (ID) of 0.656 m. The furnace is equipped with an alumina-lined conical roof, and the shell is bolted onto a domed base. A single tap-hole was used for this test, with molten slag and metal tapped together and allowed to separate in the ladle. The furnace is equipped with a single, centrally-located graphite electrode of 40 mm diameter operating as a cathode, and an anode consisting of steel pins embedded in the hearth refractory. The feed system comprises individual hoppers used to feed raw ilmenite, anthracite, and prereduced ilmenite pellets through a feed pipe connected to a feed port located in the furnace roof. The furnace is linked with an off-gas system for the cleaning of process gases prior to their release to the atmosphere.

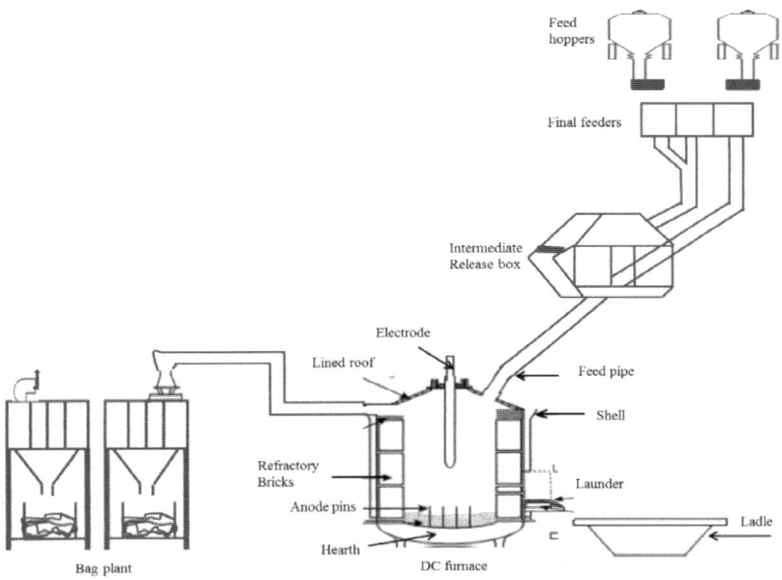

Figure 4. Schematic diagram of the pilot facility.

Smelting

Continuous smelting of partially reduced ilmenite pellets (approx. 70% prereduction yield) was carried out in a 200 kW DC open-arc furnace. The aim of the pilot smelting test was to demonstrate stable furnace operation as well as production of a consistent slag quality, in particular, a slag TiO_2 grade above 85%. The test work also had the objective of confirming the process specific energy requirement.

In total, 28 furnace batches were completed, producing 2.4 tons of titania slag and 1.1 tons of pig iron.

Two conditions were particularly interesting in this process: direct melting, and smelting of the as-received pre-reduced pellets. The results of these conditions were compared to those of the carbothermic smelting of raw ilmenite.

Smelting Conditions

As can be seen in Figure 5, the residual slag FeO content varied with the reducing conditions. The smelting process was controlled using the tapping temperature and the residual FeO content in the tapped slag.

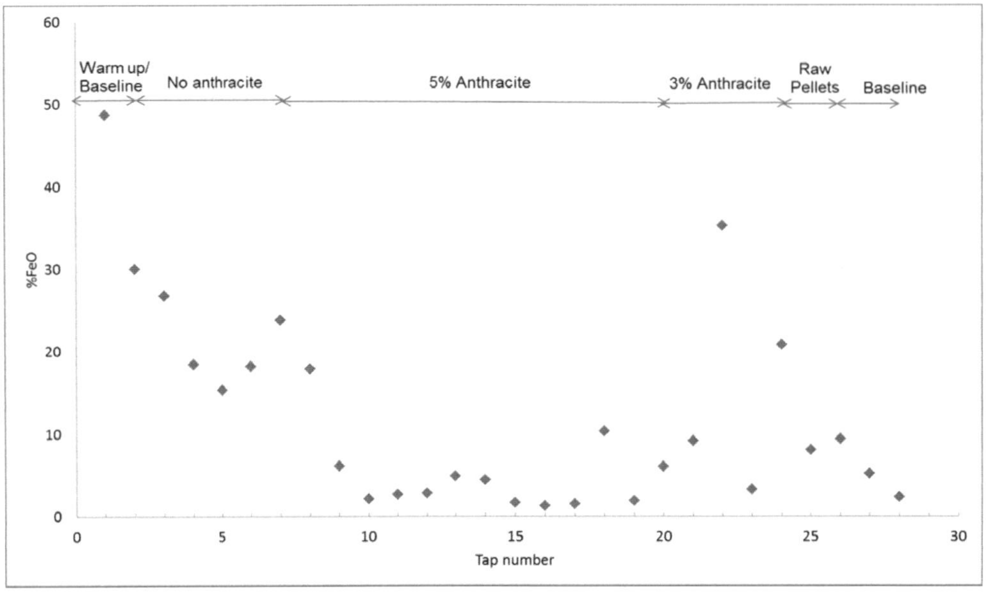

Figure 5. Evolution of FeO content in the titania slag throughout the campaign.

Titania Slag Analyses

Operation of smaller pilot furnaces is generally challenging. Eroded refractory contaminated the final slag and metal with chromium and magnesium species (oxide and metal). The standard and weighted averages of the chemical compositions of the slag produced per condition were calculated and are summarized in Table VI.

Slag containing a weighted average of 22% FeO was produced in test conditions during which the pellets were processed as-is (melting of prereduced pellets). However, this test condition did not achieve steady state due to acute erosion of the furnace crucible. Pyrosim, Mintek's in-house thermochemical simulation software, was used to calculate the degree of metallization that would result in production of a slag containing about 20% FeO. This was found to be about 70% metallization (Jones, 1992), indicating that the pilot prereduction results were acceptable.

As the efficiency of the furnace shell water-spray cooling was fairly low, increasing the reduction of FeO and thus increasing the slag liquidus temperature was expected to aid building of a freeze lining, which would also limit refractory erosion.

Subsequent smelting tests were carried out with the addition of 5% anthracite to the smelting recipe, which resulted in the expected lowering of FeO in the slag to below 10%. The composition of the slag produced in this period is given in Table VII. Slag FeO contents as low as 1.3–4.9% were achieved without visible signs of slag foaming. This condition was maintained for a longer period (Batch 10 to Batch 17) during which stable furnace operation was demonstrated and slags of consistent FeO content were produced.

The concentration of TiO_2 in the slag was affected by a combination of factors, including the contamination of the slag with foreign species from the refractory, ilmenite ore grade, and the extent of reduction in the smelting process (or residual slag FeO content). However, considering a specific smelting operation the TiO_2 content is a function of the level of residual FeO in the slag. The FeO level should be such that it ensures both optimal slag chemistry and optimal furnace operation (such as developing a freeze lining and preventing foaming).

Results for this particular test work suggest that smelting of partially reduced ilmenite and operating the furnace with lower FeO content in the slag is technically possible.

Table VI. Quality of slag product per condition.

Batch		TiO_2	FeO
2–7 Remelting	Weighted av.	60.02	22.27
	Av.	60.96	22.08
	St dev	4.81	5.69
8–20 5% anthracite	Weighted av.	85.86	4.81
	Avg	87.25	4.84
	St dev	11.09	4.69
10–17 5% anthracite	Weighted av.	87.12	2.67
	Av.	88.73	2.62
	St dev	8.76	1.35

In conventional DC ilmenite smelting, FeO content in the final slag is 8–12% (Zietsman, 2004). The content of both TiO_2 and FeO is about 96% in typical South African commercial ilmenite smelting operations, the balance being the impurities. A slag with greater than 90% TiO_2 would be produced by lowering the FeO below 6%, while a level of 86% TiO_2 in the slag may be achieved with an FeO content of about 10%. Production of a higher TiO_2 slag leads to increased reductant and energy requirements, increased refractory wear, tapping issues, and foaming (Maharajh, 2015).

Smelting of prereduced ilmenite can avoid these challenges, at least partially. Slag foaming under current smelting conditions is a result of two main factors: the amount of reaction gas generated, and slag properties such as surface tension, viscosity, and density. Through smelting of prereduced ilmenite (approx. 70% prereduction yield), the amount of reaction gas is decreased by as much as 70% compared to conventional ilmenite smelting processes. Decreasing the reaction gas volume to this degree is expected to have a significant effect on the likelihood and intensity of foaming incidents during abnormal operating conditions in the furnace. The physical properties of slag were investigated and were found not to be significantly different within the range of composition investigated. Modelling of a binary slag composed of TiO_2 and FeO was carried out to illustrate that the surface tension is relatively insensitive to the slag composition. Slag foaming or slag foaming intensity is therefore expected to depend primarily on the amount of reaction gas released by the process.

Conditions used in conventional DC ilmenite smelting in South Africa can be applied to the present process test work to remove the effects of refractory erosion and ilmenite ore quality on the slag TiO_2 grade. By targeting a residual slag FeO content lower than 6%, the Impril process should be able to produce a slag TiO_2 grade of at least 90%. Due to the benefit of prereduction, this process may be able to operate stably with slag containing less than 6% FeO. As can be seen in Table VII, this was largely achieved.

Table VII. Analysis of slag (mass %) from the stable smelting operation – Batch 9 to 17.

Batch	*TiO_2	Cr_2O_3	MnO	FeO	V_2O_5	SiO_2	CaO	Al_2O_3	MgO
10	84.74	1.34	1.37	2.06	0.30	1.31	0.34	1.96	6.58
11	87.84	0.60	1.18	2.59	0.21	1.16	0.28	2.49	3.65
12	83.15	0.31	1.68	2.73	0.29	1.20	0.38	2.76	7.50
13	88.86	0.45	0.68	4.89	0.17	0.49	0.31	1.45	2.70
14	88.19	0.45	0.99	4.37	0.16	0.43	0.35	1.53	3.53
15	91.27	0.31	1.47	1.58	0.15	0.28	0.32	1.47	3.15
16	94.28	0.08	1.32	1.25	0.10	0.21	0.17	0.98	1.61
17	93.32	0.09	1.21	1.48	0.10	0.16	0.14	1.10	2.40
Average	88.96	0.45	1.24	2.62	0.19	0.66	0.29	1.72	3.89
St dev	3.89	0.40	0.30	1.35	0.08	0.48	0.09	0.64	2.06

* By difference

However the ultimate goal of the Impril process is to process fully prereduced ilmenite pellets with at least a 90% prereduction yield to reduce the electrical energy requirement. It is postulated that a favourable effect on the quality of the final products will also be achievable.

Pig Iron Analyses

Iron alloy was successfully produced during both the direct melting and smelting of prereduced carbon-based ilmenite. The Fe content varied between 91% and 97% due to contamination by Ti, Cr, Si, and Mn. Apart from Cr, which comes mainly from the furnace refractory lining, the other elements present in the alloy come from the ore. These contaminants can, however, be removed through refining of the raw pig iron. The content of the minor elements C, P, and S met the requirements of commercial pig iron. C was below 2% on average, the requirement being 3.5–4.5%. S and P were both low and within the requirements for pig iron, namely below 0.05% and 0.12%, respectively. The contents of S and P are related to their concentrations in the raw ilmenite.

Furnace Electrical Consumption and Operation

The 200 kW DC open-arc furnace was operated at a power level in the range of 115–140 kW and corresponding voltage of 100–115 V. Consistent furnace heat losses in the range between 60–90 kW were measured, and the specific energy requirement (feed rate ratio at constant power) was varied to produce molten material at a temperature of about 1750°C. Average tapping temperatures were scattered within a range between 1670 and 1780°C.

The specific energy requirement (SER) for the direct m
elting and subsequent smelting of prereduced carbon-based ilmenite pellets was measured between 0.6 and 0.7 kWh/kg prereduced ilmenite. The SER of the direct melting was closer to the lower value, while smelting required more energy with the SER closer to the higher limit as expected. The energy requirement for smelting of prereduced ilmenite is dependent on the degree of prereduction and the extent of the reduction in the smelting furnace at a given temperature.

Table VIII gives a summary of the furnace operation for the various smelting conditions investigated. There are minor variations of power and voltage between the conditions. Stable furnace operation was achieved irrespective of the condition investigated, and with no visible sign of slag foaming. This continued to be the case even after increasing the FeO reduction.

As expected, the degree of reduction achieved will impact the electricity consumption during smelting. A 30–40% reduction in furnace electricity required relative to the conventional smelting process can be achieved in the Impril process, assuming that a prereduction yield of at least 70% can be achieved.

Table VIII. Average electric parameters per condition.

Condition		Set power, kw	Actual power, kW	Actual voltage	Set HL, kW	Actual HL, kW	Power input kWh	Power output kWh	Set SER kWh/kg	Actual SER, kWh/kg	Tapping temp. °C
Melting only	Average	115	117	101	85	79	2434	1935	0.59	0.59	1672
2 - 7	Std dev	8	28	16	8	7			0.04	0.07	72
5% anthracite	Average	125	113	114	87	62	5552	4312	0.69	0.64	1763
8 - 20	Std dev	15	15	17	8	11			0.02	0.14	76
5% anthracite	Average	125	120	106	90	63	3464	2624	0.69	0.67	1777
10 - 17	Std dev	9	8	8	0	9			0.02	0.04	90
Ilmenite	Average	160	149	78	90	81	1202	727	1.30	1.33	1750
Baseline (26 - 28	Std dev	0	3	31	0	10			0.00	0.10	68

Achieving a specific energy requirement of 0.45–0.50 kWh/kg prereduced ilmenite may be possible under optimized conditions. Additional large-scale test work would enable confirmation of these estimates.

Compared to conventional ilmenite smelting, different furnace behaviour may be expected when smelting or melting prereduced ilmenite since titania slag is highly conductive. These differences may include arc length, power, and stability, and electrode movement effects not observed during the small-scale pilot test. Arc resistivities were measured for the various conditions investigated (melting and smelting) in order to predict the furnace arc stability. Arc resistivity was found to be in the range of 0.0168 and 0.0240 Ω.cm, close to 0.0175 Ω.cm, a typical value for arc resistivity in smelting processes with CO-rich atmospheres.

CONCLUSION

Through extensive investigations, Mintek has developed the 'Impril' process, an improved version of the original Mintek-patented process for DC smelting of ilmenite. The new process consists of DC arc smelting of prereduced carbon-based ilmenite pellets. The objectives of Impril are to decrease furnace electricity consumption, improve product quality, and improve furnace operability.

Carbon-based ilmenite pellets were successfully prepared. They were composed of the as-received ilmenite, milled anthracite, and an organic binder. A mechanical strength of up to 1.5 MPa was achieved for indurated pellets under ambient environmental conditions.

Hybrid prereduction of carbon-based ilmenite pellets was successfully carried out in a fixed-bed configuration at temperatures of about 1000°C. The tests used a CO reducing gas environment and a residence time of 1–4 hours. This prereduction process was successfully demonstrated at laboratory and pilot scale. Process performances appeared to be strongly related to the operating temperature, CO flow rate, and residence time. Consistent prereduction and metallization yields were achieved, which are important parameters for designing a controllable smelting process. Prereduction yields greater than 90% were achieved as a result of the combined actions of solid carbonaceous reductant, carbon-based binder, and CO.

Smelting of cold prereduced pellets with a prereduction yield ranging from 74 to 80 % in a 200 kW DC arc furnace was successfully completed. The test results successfully demonstrated the Impril process concept, and the indicative results supported the theoretical calculations and laboratory results. A residual slag FeO content in the range of 1.3–4.9% was achieved with a 5% additional anthracite addition to the recipe. Direct melting of the prereduced pellets was also successfully carried out with the slag composition in line with the expected outcome, based on the degree of prereduction achieved during the pretreatment step. The furnace operated consistently with no major challenges such as slag foaming or furnace instability. The measured arc characteristics showed that the electrical behaviour in smelting of prereduced ilmenite is expected to be similar to the conventional smelting processes. A specific energy requirement (SER) of about 0.6–0.7 MWh/ton prereduced ilmenite was measured, which represents about 30–40% electricity savings compared with conventional smelting.

The Impril process may also enable production of higher TiO_2 grade slag (above 90%) due to the reduction of FeO in slag, and it was demonstrated that a saleable pig iron quality could be produced. Relative to other existing ilmenite smelting processes, the Impril process may also contribute towards a reduction in the occurrence of slag foaming as a result of the significantly lower quantity of gas generated. Producing high TiO_2 grade slags will have a positive impact on the downstream processes for the production of TiO_2 pigments and Ti metal and alloys.

The pilot-scale test work demonstrated the technical feasibility, viability, and robustness of the smelting of prereduced carbon-based ilmenite pellets as one of the many enhancements to the current ilmenite smelting process.

REFERENCES

Dewan,, M.A.R., Zhang, G., and Ostrovski, O. (2010). Carbothermal reduction of a primary ilmenite concentrate in different gas atmospheres. *Metallurgical and Materials Transactions B*, 41B, 182-192.

Grey, I.E. and Reid, A.F. (1974). Reaction sequences in reduction of ilmenite: 3 - Reduction in a commercial rotary kiln; an X-ray diffraction study. *Transactions of the Institution of Mining and Metallurgy Section C*, 84, C39-C46.

Grey, I.E., Reid, A.F., and Jones, D.G. (1974). Reaction sequences in the reduction of ilmenite: 4 - Interpretation in terms of the Fe-Ti-O and Fe-Mn-Ti-O phase diagrams. *Transactions of the Institution of Mining and Metallurgy Section C*, 83, C105-C111.

Gous, M. (2006). An overview of the Namakwa Sands ilmenite smelting operations. *Journal of the South African Institute of Mining and Metallurgy*, 106, 379-384.

Jones, D.G. (1974). Reaction sequences in the reduction of ilmenite: 2-gaseous reduction by carbon monoxide. *Transactions of the Institution of Mining and Metallurgy Section C*, 186-192.

Jones, R.T. (1992). Pyrosim, Steady-state simulation of pyrometallurgical processes, http://www.mintek.co.za/Pyromet/Pyrosim/Pyrosim2.htm

Kotze, H., Bessinger, D., and Beukes, J. (2006). Ilmenite smelting at Ticor SA. *Journal of the South African Institute of Mining and Metallurgy*, 106, pp 165-170.

Maharajh, S. (2015). Value-in-use model for chlorination of titania feedstocks. *Journal of the Southern African Institute of Mining and Metallurgy*, 115, 385-394.

Merk, R. and Pickles, C.A. (1988). Reduction of ilmenite by carbon monoxide. *Canadian Metallurgical Quarterly*, 27 (3), 179-185.

Murty, C.V.G.K., Upadhyay, R., and Asokan, S. (2007). Electro smelting of ilmenite for production of TiO_2 slag - potential of India as a global player. *Proceedings of the 11th International Ferroalloys Congress (INFACON XI)*, New Delhi, India. pp. 823-836.

South African Provisional Patent Application no 2015 / 08501. Improved ilmenite smelting process (Impril)

Williams, G.E. and Steenkamp, J.D. (2006). Heavy mineral processing at Richards Bay Minerals. *Journal of the South African Institute of Mining and Metallurgy*, 106, 181-188.

Zietsman, J.H. and Pistorius P.C. (2006). Modelling of an ilmenite-smelting DC arc furnace process. *Minerals Engineering*, 19, 262-279.

Zeitsman, J.H. (2004). Interactions between freeze lining and slag bath in ilmenite smelting. PhD thesis, University of Pretoria. pp 286-287.

Kabwika Bisaka

Head - New Technology, Pyrometallurgy Division
Mintek

Kabwika has vast experience of more than 25 years in the field of Pyrometallurgy covering the operation and management of commercial plants, as well as research and development for the production of the following commodities: copper, cobalt, ferroalloys, ilmenite, rare earths, and coal. He joined Mintek in 1997 and since then has worked at various capacities in the Pyrometallurgy Division where he heads the New Technology group. His main research activities are directed towards but not limited to energy saving, minimisation of CO_2 emissions, beneficiation of complex ores as well as the minimisation of wastes.

Online elemental analysis of rutile and zircon concentrates from mineral sands

L. Koresaar

Outotec (Finland)

Blending and separation methods are generally used to separate the components of heavy mineral sands, mainly the titanium minerals rutile and ilmenite (TiO_2), and the zirconium-bearing mineral zircon ($ZrSiO_4$). A new online elemental analyser, Courier® Dx, based on XRF spectrometry principles, has been developed for the dry sands process. The elemental analysis is used to minimize the out-of-specification rejects of the process, thus improving the recovery.

INTRODUCTION

Heavy mineral sands consist of minerals that have specific gravities in excess of 3. Titanium- and zirconium-bearing minerals, and to some extent rare earth element-bearing minerals, are the most economically important. The majority of TiO_2 feedstocks are utilized in white pigment manufacturing around the globe. The TiO_2 pigment is used as an additive in paints, plastics paper, and ink. Zircon is generally a by-product or a co-product of the mineral sands separation process, and its main use is in the ceramics industry. Ilmenite is the most abundant titanium-bearing mineral, but has a lower TiO_2 content than the more valuable leucoxene and rutile.

Titanium and zircon are extracted from mineral sands deposits using concentration and separation processes. In general, extraction is by wet gravity separation, followed by dry magnetic and electrostatic separation to separate the individual minerals. As these concentration and separation methods have restrictions in the throughput, the feed has to be distributed to parallel multistage circuits. The inherent complexity of the separation systems has made the solid-solid separation less amenable to traditional automated process control (Luttrell and Mankosa, 2002).

For dry processes there has been a restricted selection of reliable and cost-efficient online analysers. Optical reflectance spectroscopy analysers have been used for online analysis, but they require frequent calibration due to variance in properties not related to elemental composition (Haavisto *et al.*, 2008). Prompt gamma neutron activation analysis (PGNAA) may be used for this purpose, but these instruments require a radioisotope source, which gives rise to problems with safety, transportation, and disposal of the radioisotopes.

X-ray fluorescence (XRF) analysers have been used for several decades as the preferred on-stream analysis method for wet mineral concentrators (Twidle and Engelbrecht, 1984). XRF analysers employ high-energy electromagnetic radiation to excite electrons in the inner shells of atoms (Figure 1). The excited electron is ejected from the atom and an outer shell electron with higher energy fills this space. The higher energy electron emits a photon with a wavelength corresponding to the energy difference between the energy shells. As each element has unique energy levels, the photons emit radiation at specific wavelengths. By measuring the radiation at the characteristic wavelengths, the concentration of the element in the sample can be calculated.

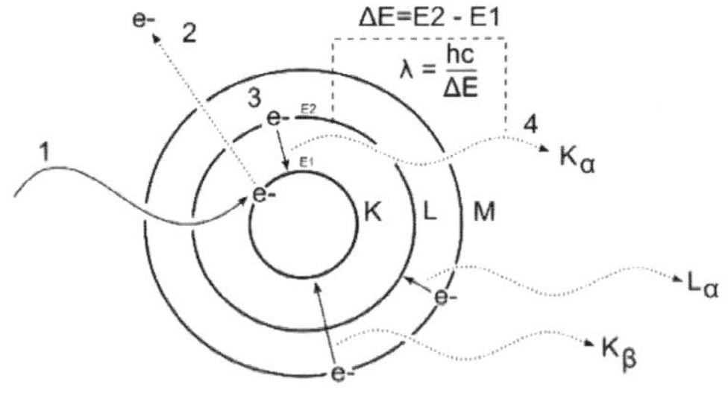

Figure 1. X-Ray fluorescence principle. Incoming X-ray radiation (1) has correct energy to eject an electron from the inner shell of the atom. (2) Electron in a higher energy state occupies the vacancy (3) by emitting a photon with specific energy/wavelength, corresponding to the energy difference of the shells (4).

ANALYSER SYSTEM

Online slurry analysers and their benefits for flotation control have been reviewed many times (Kongas and Saloheimo, 2001; Remes *et al.*, 2007). The Courier® Dx dry mineral sands analyser is based on the slurry analyser, with a modified sample introduction system. A schematic diagram pf the analyser probe is presented in Figure 2, and the secondary sampling system in Figure 3. A third-party primary sampler takes a smaller sample from the primary sample flow and feeds it to the analyser by gravity. The measurement chamber contains a motor which controls the sample flow for the measurement. Motorized sample presentation to the measurement cell allows a controlled and representative sample flow and good precision of the measurement. The sample introduction system also allows for manual sample measurement.

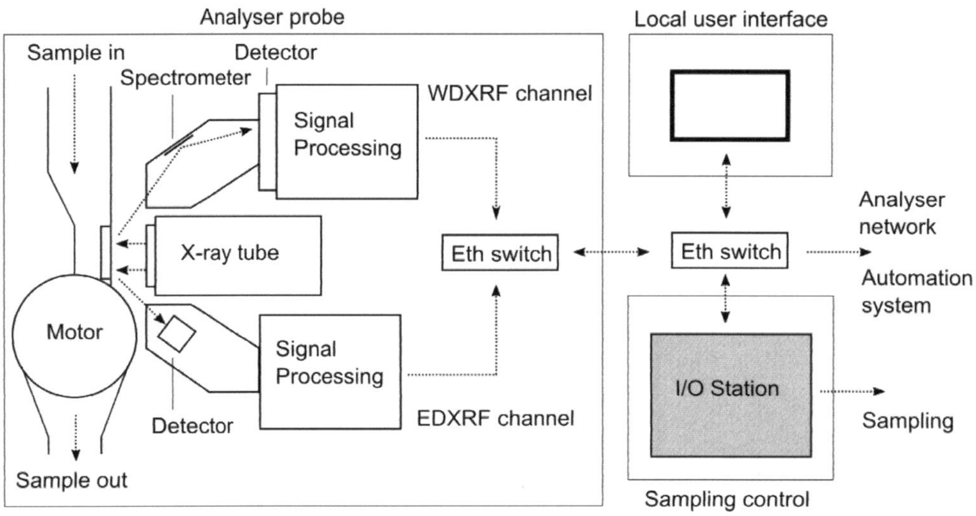

Figure 2. Schematic diagram of the analyser probe. The dry sample chamber is operated by motor to enable passing of enough sample material. The X-ray tube excites the electrons in elements in the sample. The characteristic fluorescence is isolated by the spectrometer and the photons are counted by the detector. The results are then sent to the local user interface and analyser network.

The repeatability of the measurement was tested with a rutile sample using an analyser equipped with the motorized sample presentation system. The operating frequency of the motor was varied between 0.5 and 4 Hz, and one 5 kg sample was measured 10 consecutive times to estimate the repeatability (Rousseau, 2001). A measurement cell motor speed of 1 Hz corresponds to 1.7 kg min^{-1} sample flow, and 2 Hz to 3.4 kg min^{-1} *etc.* The analyser was equipped with a Rh anode X-ray tube operating at 35 W power. The energy-dispersive channel was used to measure Ti and Fe. Wavelength-dispersive spectrometers were used to measure Zr, Th, Nb, and Pb. Each sample was measured for two minutes. The relative standard deviations of the measurements are presented in Table I. The best stability and repeatability was achieved when the motor was operated at 4 Hz frequency.

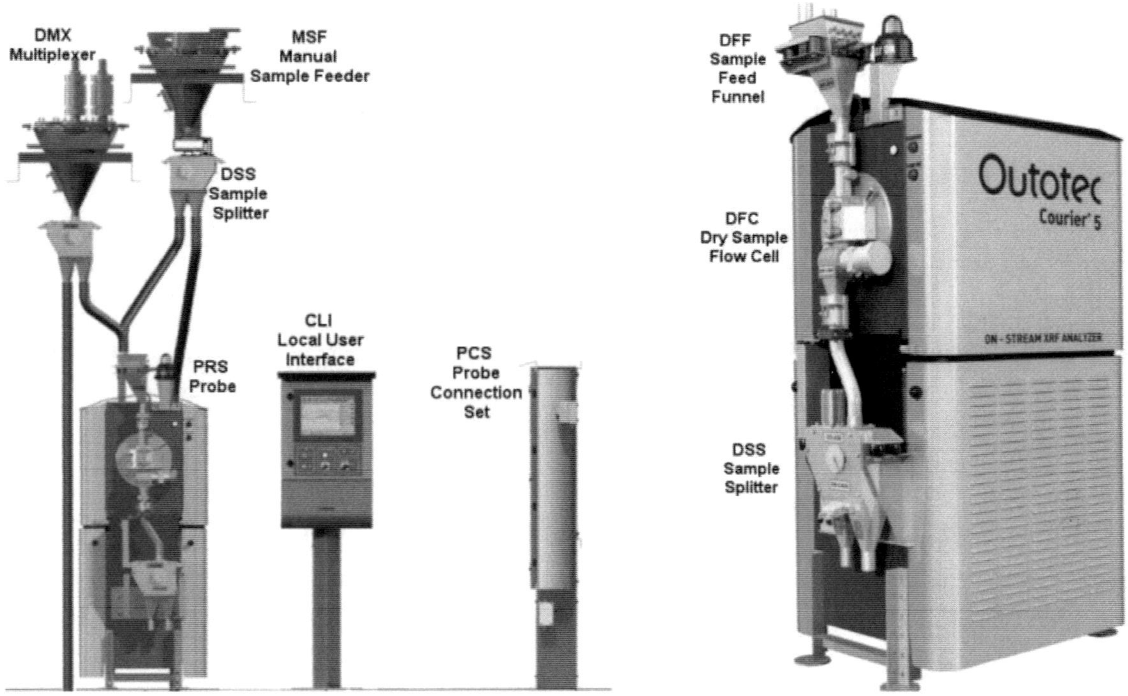

Figure 3. Two sample line analyser modules and a DMX multiplexer secondary sample presentation system. The analyser is also equipped with a manual sample feeder to allow manual sample measurement.

Table I. Sample information and repeatability of measurement with varying sample speed. The count rate is the average count rate of the ten repeats.

	TiO_2	ZrO	Fe	Nb	Pb	Th
Concentration (%)	95.4	0.97	0.92	0.33	0.025	0.008
Count rate (cps)	2090	469	155	452	128	49
Motor frequency (Hz)						
0.5 Hz - cps RSD (%)	0.57	3.45	3.10	2.40	0.77	1.07
1 Hz - cps RSD (%)	0.27	2.32	3.73	1.28	0.59	2.17
2 Hz - cps RSD (%)	0.24	1.26	1.68	0.64	1.25	1.60
4 Hz - cps RSD (%)	0.13	1.07	0.92	0.55	0.99	0.55

Any drifting of the X-ray tube, proportional counters, or multichannel analysers is compensated by internal reference measurement. A high-precision mechanism brings an internal standard sample in front of the X-ray tube between each measurement. The reading of the internal reference sample is used to compensate any drifts in the analyser. The analyser can be equipped with an air-cooled 35 W X-ray tube, or a water-cooled 200W X-ray tube when lower concentrations are analysed. An analyser with the higher power X-ray tube is also equipped with a water cooling system. The analyser requirements for installation are power supply and instrument air.

ELECTROSTATIC SEPARATION CIRCUIT CONTROL

Rutile and zircon are nonmagnetic and are separated from magnetic sands before the electrostatic separation circuit. Electrostatic separation is the final beneficiation step in the upgrading of rutile and zirconium sands. Both minerals have relatively high resistivity (Kelly and Spottiswood. 1989), but there is sufficient difference between them to allow for a good separation. Rutile is considered the conducting mineral, having the lower resistivity of the two minerals. Several stages of electrostatic separators are used to produce rutile concentrate, zircon concentrate, and out-of-specification concentrate, which is a combination of non-conductors from rutile cleaners and conductors from zircon cleaners. A simplified flow diagram is presented in Figure 4. The out-of-specification product is not of economic interest without further processing.

The Courier® Dx online analyser is installed to measure the TiO_2 content of the rutile concentrate. The result from the analyser can be used to control the rate of blending of the out-of-specification concentrate back to the feed. The control strategy allows stable rutile concentrate production. Variability of the feed quality is normalized by the volume of the blending pile. The control circuit could increase the throughput tonnage, without decreasing the grade, by several kilotons per annum (Luttrell and Mankosa, 2002).

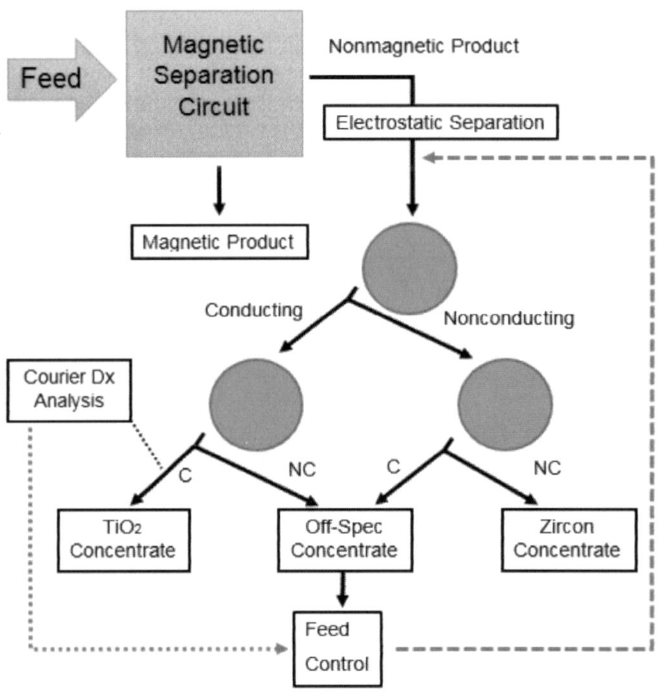

Figure 4. Simplified flow diagram of magnetic separation followed by electrostatic separation circuit, and the control strategy. Courier® Dx analysis is used to control the blending of off-specification concentrate to the circuit.

The Zr content in the rutile concentrate can be used to monitor for process disturbances. An increasing Zr content indicates out-of-range situations in the process. Online analysis of the concentrate allows quick operational response to these kinds of events.

RESULTS

One dry sands analyser was installed at a rutile concentrator plant. Empirical calibration of the analyser was used to account for real process situations, *e.g.* changes in particle size or ore type. Small portions of the analysed samples were collected for a comparative laboratory analysis and the intensities of the element channels were recorded. 74 samples were collected for the calibration of the analyser. Multivariate linear regression (Equation [1]) was used to find equations for each analysed element.

$$C\ (\%) = R_0 + R_1 f_1 + \ldots + R_n f_n \qquad [1]$$

where C is the analyser-predicted value of concentration, R values are empirical calibration constants, and f values are the variables used for the calibration model, *i.e.* element channels. Some calibration equations required more than one variable due to the matrix effects in the sample, where one element may increase or decrease the intensity of another element due to their relative emission or absorption energies. The concentrations of Ti were divided in high (>90%) and low (<90%) contents, and two calibration models were constructed for TiO_2 (Figures 6a and 6b) to improve the accuracy of the analysis. A universal model was used for Fe (Figure 5) and for Zr (Figure 7).

Statistical information on the calibration models is presented in Table II. The R^2 is the coefficient of correlation of the models. The absolute and relative σ of the models was calculated using Equations [2] and [3], respectively.

$$\sigma = \sqrt{\frac{\sum_i (y_i - \hat{y}_i)^2}{n-1}} \qquad [2]$$

$$Relative\ \sigma = 100\frac{\sigma}{\bar{x}} \qquad [3]$$

where y_i is the laboratory assay of the ith measurement, $\hat{y}_i$ is the analyser predicted value of the ith measurement, n is the number of samples, and $\bar{x}$ is the average of laboratory assays.

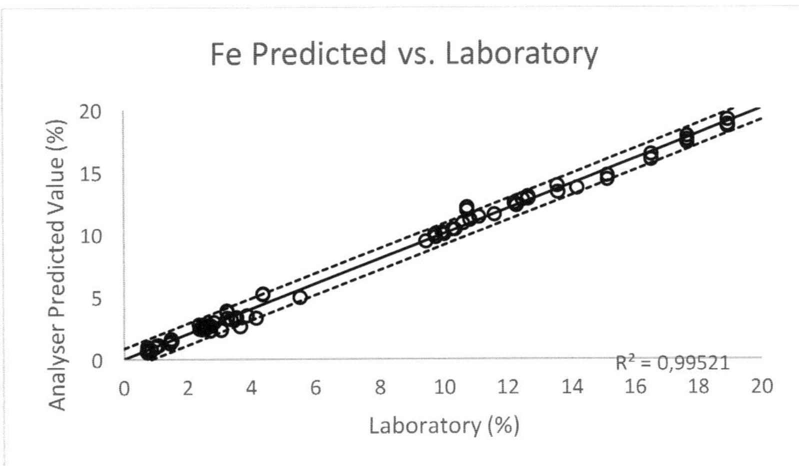

Figure 5. Calibration model for Fe. The x-axis represents the laboratory assays and the y-axis represents the analyser predicted values. The dashed lines represent the 95% confidence interval.

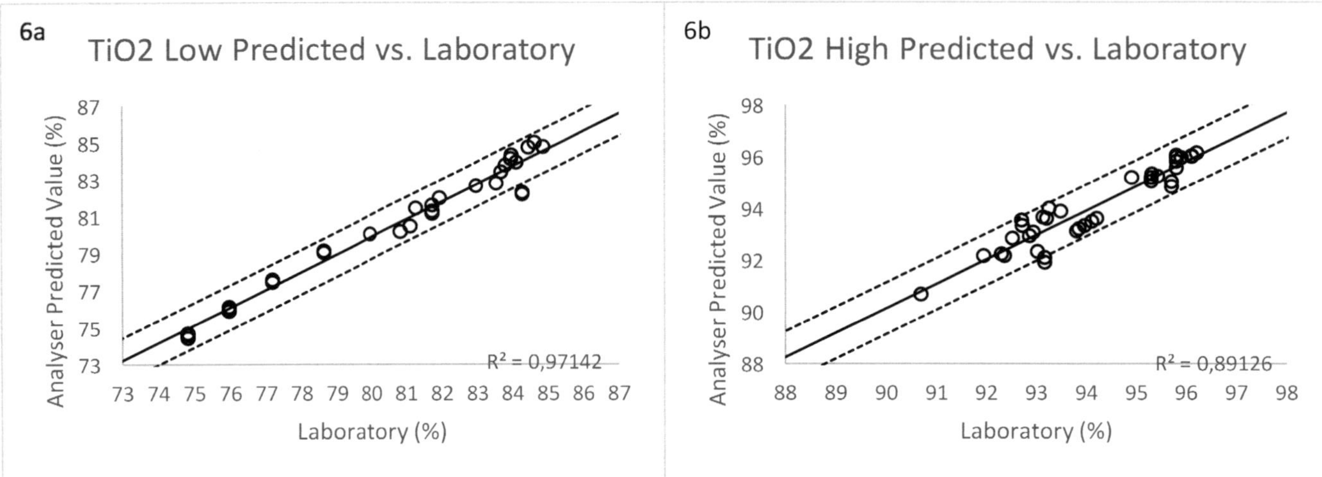

Figure 6. Calibration models for low TiO_2 (a) and high TiO_2 (b). The x-axis represents the laboratory assays and the y-axis represents the analyser predicted values. The dashed lines represent the 95% confidence interval.

Table II. Statistical information of the analyser calibration.

Assay	R^2	σ	Relative σ
Fe	0.9952	0.41	5.77
TiO2 Low	0.9714	0.61	0.75
TiO2 High	0.8913	0.50	0.53
Zr	0.9902	0.02	6.17

CONCLUSIONS

Online analysis of mineral sands is possible with good accuracy using an XRF analyser. The process of rutile concentrators can be optimized by analysing the final product. The analyser results can be used to detect abnormal process situations or to blend the out-of-specification concentrate back to the process. The results from the pilot analyser installation prove the analyser results are good enough for process control.

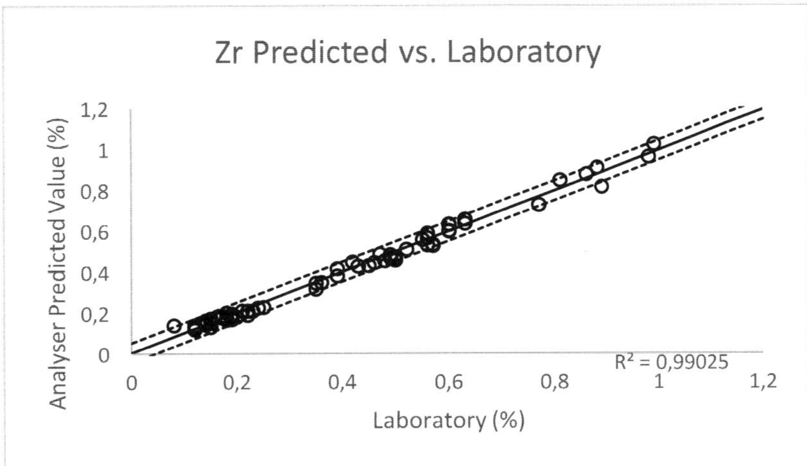

Figure 7. Calibration model for Zr. The x-axis represents the laboratory assays and the y-axis represents the analyser predicted values. The dashed lines represent the 95% confidence interval.

REFERENCES

Haavisto, O., Kaartinen, J., and Hyotyniemi, H. (2008). Optical spectrum based measurement of flotation slurry contents. *International Journal of Minerals Processing,* 88 (3-4), 80-88.

Kelly, E.G. and Spottiswood, D.J. (1989). The theory of electrostatic separation: A review. Part I. Fundamentals. *Minerals Engineering,* 2 (1), 33-46.

Kongas, M. and Saloheimo, K. (2009). New innovations in on-stream analysis for flotation circuit management and control, *IOCG Resources, Conference in Minerals Engineering 2009,* Luleå Sweden, February 2009.

Luttrell, G.H. and Mankosa, M.J. (2002). Strategies for the instrumentation and control of solid-solid separation processes. *Proceedings of the Mineral Processing Plant Design, Practice and Control, Vol 2.* Society for Mining, Metallurgy and Exploration Inc. Vancouver BC. pp. 2152-2163.

Remes, A., Saloheimo, K., and Jämsä-Jounela, S.-L. (2007). Effect of speed and accuracy of on-line elemental analysis on flotation control performance. *Minerals Engineering* 20, 1055-1066.

Rousseau, R.M. (2001). Detection limit and estimate of uncertainty of analytical XRF results. *Rikagu Journal,* 18 (2), 33-47.

Twidle, T.R. and Engelbrecht, P.C. (1984). Developments in the flotation of copper at Black Mountain. *Journal of the South African Institute of Mining and Metallurgy,* 84 (6), 164-178.

Lauri Juhani Köresaar

Analyzer Application Specialist
Outotec (Finland) Oy

Graduated from the University Of Edinburgh with BSc (Hons) Chemistry degree. After graduating has been working for Outotec. First as a research technician where responsibilities were in the field of product/concept development of online analysers. Has been working as an application specialist for online analysers since 2014, where responsibilities include developing new applications for analysers, performance assessments and technical support of analysers.

A case study of geotechnical characterization and 3D modelling of the biologically cemented Mandena heavy mineral sand deposit

L.I. Boshoff,[1] A.H. Bracken,[1] and F. Hees[2]

[1]SRK Consulting, Johannesburg, South Africa
[2]Rio Tinto Iron and Titanium-RTIT, Johannesburg, South Africa

Biological cementation of the Mandena heavy mineral deposit in Madagascar was causing difficulty in dredge mining and hence in planning and control of production rates. An investigation was initiated to better evaluate the geotechnical properties of the orebody. The investigation was aimed at defining the distribution and geotechnical properties of the cemented strata, in order to generate a site-specific geotechnical model for the Mandena deposit. The geotechnical characterization was developed by an integrated approach using several ground investigation techniques such as piezocone (also known as CPTu) testing, geotechnical drilling, standard penetration testing (SPT), as well as laboratory testing. The final outcomes of the investigation included a 3D geotechnical model showing the distribution and extent of the cemented areas as well as the typical geotechnical parameter ranges associated with each of the defined geotechnical units. The interpretations developed were incorporated into the mine and production planning to ultimately optimize operational planning and control.

INTRODUCTION

QIT Madagascar Minerals (QMM) is a heavy mineral sand mining operation at Mandena, near Fort Dauphin in Madagascar. During the final planning and commissioning stages of the project, the need to fully characterize the unusual geotechnical conditions was identified and a geotechnical investigation was initiated.

NATURE OF THE PROBLEM

During the commissioning phase it was found that mining production was being impacted by biologically cemented/indurated layers encountered within the sands. Upon closer inspection of the indurated horizons in the dredging face, it was discovered that these layers contain a high loading of both bacteria and fungi. It appears that the bacteria produce an extracellular (outside the cell) polysaccharide (*i.e.* sugar and carbon salts which form a gelling biofilm to trap nutrients). There are a number of different bacteria. The fungi, on the other hand, produce a network of fibrous strands that bind the sand grains together. Thus both bacteria and fungi probably contribute to the 'apparent cohesive' strength or binding of the indurated sand (Lynn, 2008).

The main challenge of this project was posed by the many unknowns surrounding the nature and extent of this biological cementation that was presenting a geotechnical problem for the mining process. Typically, the following questions arose:

- Is this a biological, geological, or geotechnical problem, or all of the above?
- How does one go about modelling a phenomenon that is not necessarily governed by apparent geological processes or markers?
- What tools and techniques would be suited to cope with the variable conditions being investigated? How effective will they be?
- How does one ensure the tools and techniques used can provide useful data that can be verified?
- How does one combine all the available data collected to generate a practical geotechnical model that will assist in a mining production environment?
- How does one approach the modelling of geotechnics and relate this to the geological resource model and mine plan?

DATA GATHERING AND EVALUATION

General

The scoping of the investigation was completed during various brainstorming sessions that considered existing data, the most crucial engineering parameters required for mine and dredge design, and also which geotechnical investigation techniques available could potentially obtain the required data effectively. The scoping study also included trials using different rotary and vibracore drilling techniques to confirm which would be most appropriate to use.

This process culminated in an investigation methodology involving two phases. Phase 1 focused on high-density geotechnical rotary core drilling, logging, sampling, and standard penetration testing (SPT) as well as piezocone with pore pressure measurement (otherwise known as CPTu) testing within a limited area. Soil laboratory testing and rock testing were also completed on samples retrieved from the boreholes. It should be noted that following the trials, it was decided not to use a vibracore drill as it was found that the vibration of the drill broke the intergranular bonds in the indurated layers, resulting in the indurated materials being recovered as soils.

Phase 2 focused on a lower density of testing on the remainder of the orebody. During Phase 2 all available historical information was used, including existing exploration borehole log data, where appropriate, to generate a 3D geotechnical model and typical geotechnical parameter ranges for the most important geotechnical properties.

In-situ measurements using a piezocone (CPTu) rig provided direct measurement of cone resistance (q_c), sleeve friction (fs) and pore water pressure (u), which can be used to derive various geotechnical parameters. Due to the expected variable conditions, refusal of the CPTu equipment was expected on the hardest indurated layers. A standard drilling rig or hydro vibracore rig (sonic method) was therefore provided to advance through the hardest layers when encountered.

Selected CPTu locations were paired with high-quality rotary core drilling, using a triple tube barrel and a mud flush intended to obtain cores of indurated sands and more cohesive/dense soil layers. Limited SPT was also completed in the holes and used to provide a comparison with the CPTu data. Detailed geotechnical logging of the core was completed. Laboratory testing of sampled core assisted in confirming the parameters derived from the CPTu interpretations and logging and to determine parameters that could not be derived from the CPTu testing due to refusal on the very hardest indurated sand layers.

The combined Phase 1 and Phase 2 geotechnical database consisted of 147 CPTu locations, and 48 borehole and SPT locations. More than 200 disturbed samples and approximately 80 core samples were tested in the laboratory. Figure 1 provides a general location map and test location layout of the area investigated.

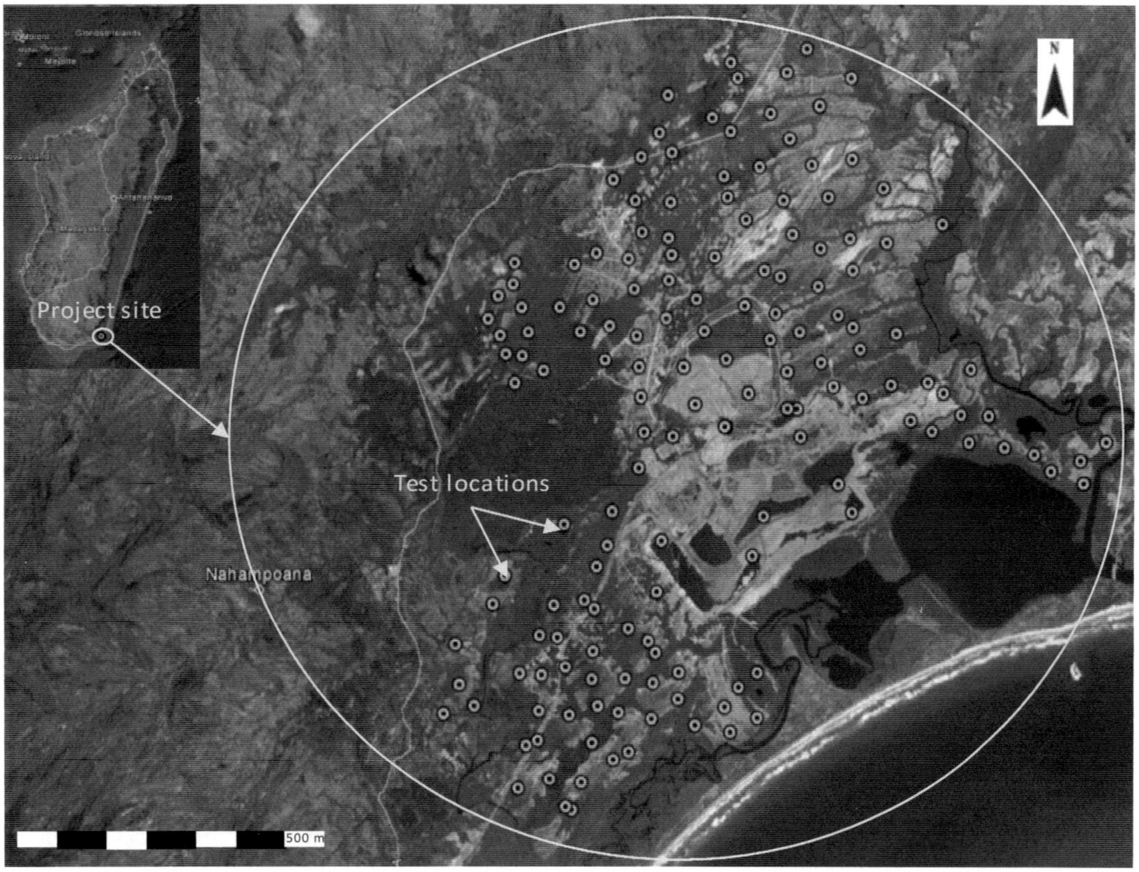

Figure 1. Schematic of project site locality and planned drilling and test location layout.

Field Logging of Rotary Drilling Samples

Poor recoveries obtained during drilling provided little material within the un-indurated zones and this posed various challenges during field logging. To ensure consistent capturing of information that could later be calibrated and refined when combined with the other data gathered (such as the CPTu and laboratory information), a site-specific approach to logging was adopted to ensure a consistent account of the indurated materials was captured. Generally, zones registering SPT N_{field} values of over 50 blow counts (noted as 'Refusal') were deemed to be of very dense consistency or possibly of extremely soft to soft rock strength with a uniaxial compressive strength (UCS) of 0.5–3 MPa. The presence of rock-like indurated material was logged only if core samples were available to confirm its presence, irrespective of the SPT N_{field} values recorded. Where this could be confirmed from core or SPT samples extracted, the thickness, spacing, and depth of indurated beds were recorded.

Based on this approach, zones with similar geotechnical characteristics were identified and provided with an overall description. This process was the first step in the generation of a site-specific geotechnical framework/unit model.

Correlation of In-situ Testing Methods

A numerical correlation between the CPTu-derived SPT N_{60} values, q_c, and the SPT N_{field} values obtained in the paired locations was completed. To obtain a point value to correlate with the SPT N_{field} values recorded, a linear average output value from the CPTu at the start and end of the SPT testing depths was assumed. A graphical presentation of a typical CPT-derived SPT N_{60} trend and SPT N_{field} values plotted with depth for test and drilling location no. 42 is shown in Figure 2.

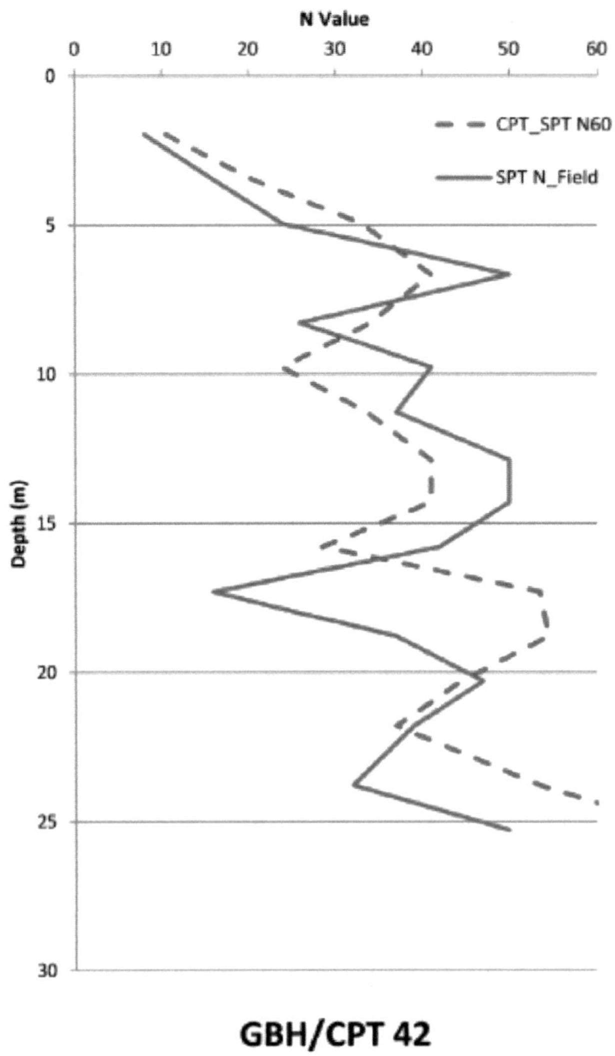

Figure 2. Graphical comparison of CPTu-derived SPT N60 with SPT N_{field} values–Test location no. 42.

Acceptable correlations were found to exist for 58% of the test positions, providing correlation coefficients from 0.5 to 0.94 (Boshoff and Bracken, 2011). Visual assessment of the trends indicated that in several of the poorer correlating cases, trends were offset, but that overall visually the trends are similar.

Correction factors were applied to the SPT N_{field} values, but none of the standard correction factors such as rod energy, overburden, and anvil size, as described by Clayton (1995), could provide a consistent improvement in the correlations obtained compared to using the raw data.

CPTu Interpretations and Borehole Logs vs Laboratory Results

Good correlation was found between the rotary drill sample logging, laboratory results, and the soil behaviour classes as defined by soil behaviour charts developed by Campenella and Robertson (1988) for the CPTu.

The CPTu identified sands and silty sands dominating the profile with some materials, typically those found at the base of the probe holes, classifying as silts or sandy clay-type materials.

The laboratory UCS test results agreed well with the rock strengths identified during borehole logging, with most falling within the strength range (UCS of 0.5 to 3.0 MPa) predicted during logging (Extremely Soft to Very Soft rock). The permeabilities measured in the laboratory for the core samples were in line with permeability predictions based on the CPTu-derived soil behaviour types (sands and silty sand) (Robertson, 1992).

However, some exceptions occurred when correlating CPTu information with samples extracted from the paired rotary drill-holes, indicating that significant variation in induration strength can exist over small distances.

Generation of a Geotechnical Framework/Unit Model

The processes described above led to the definition of a geotechnical unit framework within which further data evaluation could take place. The framework focused mainly on defining zones that would have specific engineering implications for the mining operation.

General internationally accepted guidelines as provided in the literature and summarized in the TRH10 (1994) document compiled by the Committee for State Road Authorities, South Africa (1994) relate CPT q_c to SPT N_{field} values and corresponding density descriptions as shown in Table I below.

Spikes in q_c values could be correlated with very dense to indurated layers located within generally Medium Dense to Very Dense silty sand layers. This finding agreed with observations during logging of the boreholes. It also became clear that the CPTu was also more effective in detecting the presence of very thin indurated layers that would not be identified during drilling and SPT testing. The conceptual geotechnical unit framework was defined considering all the available data and is shown in Table II.

Table I. Internationally accepted consistency descriptors as a function of SPT N and CPT q_c value ranges obtained (adopted from TRH10).

Density description	SPT N	CPT q_c (MPa)
Very Loose	4	2
Loose	4-10	2-5
Medium Dense	10-30	5-15
Dense	30-50	15-20
Very Dense	>50	>20

Table II. Geotechnical unit model derived from Phase 1 fieldwork data and observations.

Zone	Name	Geotechnical description	Distribution
1	Aeolian sands	Free flowing to non-free flowing Loose to Medium Dense sands.	Usually the surface deposit. Occasionally absent.
2	Strandline sands	Non-free flowing clean to organic, occasionally ferruginized, Medium Dense to Very Dense sands.	Usually underlying aeolian sands. Occasionally exposed at surface. Seldom absent, but possible.
3	Strandline sands with indurations	Non-free flowing Medium Dense to Very Dense sands of interbedded with extremely soft (UCS 0.5–1 MPa) to very soft (UCS 1–3 MPa) indurated sands /sandstone.	Lenses generally within Zone 2. Occasionally at top or base. Apparently random distribution. Sometimes absent.
*4	Deposit Base strandline sands, silts and clays	Interbedded sands, silts, and clays of variable consistency.	At base of and occasionally interleaved with Zone2/3. Sometimes absent.
*5	Residual sandy clay	Firm clay, silt or clayey sand.	At base of Zone 2/3 or 4.

*Zone 5 represents the clay basement and is not part of the resource, whereas Zone 4 is a transitional zone between the resource and basement.

Simple principles were developed to ensure a consistent evaluation and optimization of the geotechnical zone boundaries that would constitute the final zone interval database. These include:

- Adopting CPTu data as the more accurate and continuous record of *in-situ* conditions due to generally low sample recoveries in the boreholes
- When delineating and optimizing geotechnical zone boundaries, density would have priority over descriptive properties (such as colour)
- Zone 3 conditions were inferred where significant spikes in the CPTu q_c (30–40 MPa) could be observed, where vibracore drilling was required, or where recovery of core samples confirmed the presence of indurated layers
- If a section was logged in boreholes as Zone 2 and the CPTu data indicated hard layers, this zone would be upgraded or adjusted to Zone 3 in accordance with CPTu data.

3D MODELLING APPROACH

The spatial modelling was completed using implicit modelling software. Implicit modelling uses a rapid 3D interpolation engine to construct 3D models. Implicit modelling techniques have been refined by developers over the past 10 years to suit geological modelling problems, and in this case implicit modelling was also applied to modelling specific geotechnical zones and gathered numerical data in much the same way as modelling geological features and ore grades. The benefit of using implicit modelling is that it allows for the automatic construction of wireframes from scattered downhole data with no need for manual digitization (Cowen *et al.*, 2003).

During Phase 2, the Phase 1 interpretations were verified. Phase 2 also included scrutinizing the geological exploration data gathered to supplement 3D modelling. This in itself provided a challenge as geological logging had been completed from a 'geological and resource modelling perspective', whereas the 'geotechnical zones' were based mostly on the engineering properties of the materials as logged and measured using specific *in-situ* measurement techniques.

Geotechnical zones were modelled as separate 'lithologies'. The geotechnical zone modelling and numerical modelling were approached separately. The measured data was considered hard data, whereas the zone boundary interpretations made were based on a combination of the descriptive and *in-situ* measured data.

Geotechnical Zone Lithology Modelling

Geotechnical zone volumes were generated using spheroidal variogram settings with constant kriging and a range of approximately 300 m. The ellipsoid ratios were optimized to reflect the general dimensions of the volumes being modelled (*i.e.* disk-shaped with small Z relative to ellipse radius).

Zones 1, 2, and 3 were modelled separately; however, Zones 4 and 5 were combined and modelled as one composite region delineating the clayey deposit base consisting of either marine clays and/or residual clayey sands. Areas not defined by the interpolants developed from existing geotechnical data were inferred as comprising of geotechnical Zone 2-type sands.

The following age succession was assumed:

- Zone 3 is considered the youngest, generally cutting indiscriminately through all deposited sands, as cementation took place after deposition of the strandline deposits
- Zones 1 to 5 (excluding Zone 3) were assumed to be of increasing age; however, Zone 2 could be interleaved with Zone 4 type materials (clays and silty sand lenses) due to the nature of deposition of the strandline sands.

Correlation between the geological database generated during exploration drilling and the geotechnical database was considered. After some review, it was confirmed that the merging of the two data-sets would not be ideal and would possibly skew the presentation of the geotechnical information gathered. The two data-sets were therefore kept separate within the modelling software. However, the geological data had to be utilized to some extent during modelling to supplement the limited geotechnical data points available. To correlate indurated conditions captured within the geological database, an approach introduced by the project geologist was used to include an 'induration' interval to identify zones where described/indicated and inferred indurated conditions can be expected (Hees, 2010). This interval could be used to group key lithologies within the modelling space to reflect the distribution of the indurated conditions as per the geological database. This could then be compared with the geotechnical zone boundaries and integrated manually.

In areas where little data was available, the Zone 3 model was extrapolated by utilizing the reviewed and grouped geological data as well as geological cross-sections imported into the modelling software. However, areas did exist where indurated conditions were identified that were not captured in the geological database. Geotechnical data therefore took precedence in most cases when interpreting the Zone 3 material boundaries.

Numerical Data Analyses in 3D

A design chart was developed by the dredge designers matching SPT N number to output in tons per hour for the existing dredge when production issues ensued. For this reason, the first attempts in modelling of geotechnical properties were focused on the SPT N numerical data as known relationships existed. Numerical modelling of the SPT N_{60} and SPT N_{field} values was undertaken and run using similar settings to the lithological model.

A separate 3D interpretation was made using only the numerical data gathered to interpolate areas associated with specific numerical parameter thresholds within specific orebody areas. Numerical data within each modelled zone could also be extracted and analysed statistically. The numerical model can currently provide 3D contour maps of a specific geotechnical parameter modelled at any level or on any section through the orebody volume. The numerical interpretations do not agree in all areas with the zone interpretations, as selected locations were modelled using only descriptive logging information with no numerical geotechnical background data (*e.g.* hydrogeological borehole positions or exploration data) (Boshoff and Bracken, 2012). Gaps in the numerical data coverage still existed; however, the process demonstrated that numerical modelling of a geotechnical parameter can be very useful in this case.

Modelling Results

The model completed for the North West Satellite orebody is provided in Figure 3 as an example.

Of most significance to the mining operation is the spatial distribution of Zone 3, which is the most indurated zone. Modelling of Zone 3 showed it to consist of relatively continuous, but gently undulating, indurated beds and/or lenses across the majority of the Mandena area. It can be absent or underdeveloped in certain locations. The macro appearance of the indurated zones is that of large- to small-scale lensing rather than the existence of distinct indurated horizons, but these macro lenses can be made up of various indurated horizons of variable thickness.

Zone 3 conditions are expected to be concentrated around elevations ranging from +4 m msl to -1 m msl.

Table IV provides the numerical information extracted for each geotechnical zone modelled, providing parameter ranges and typical values of the SPT N_{field} and SPT N_{60} values measured within these areas. Numerical data available for the indurated conditions (Zone 3) is likely to be less representative due to early refusal of the CPTu, and the relatively small number of boreholes drilled also limiting conventional SPT measurement.

Laboratory testing results have confirmed the expected UCS values associated with the most indurated Zone 3 materials. Based on the test results, the maximum strength/hardness of the indurated materials recovered and tested is assumed to be around 2 MPa. Although the properties of Zones 2 and 3 are highly variable, the median SPT N value exceeds 30 for Zone 2 for the majority of the orebody. The Zone 3 median SPT N value is estimated at 53 (possibly higher, if adjustments are made for rock-like materials on which the CPTu and SPT refused – Table IV refers). Current measured SPT N values for Zone 3 ranged from as low as 3 to as high as 100.

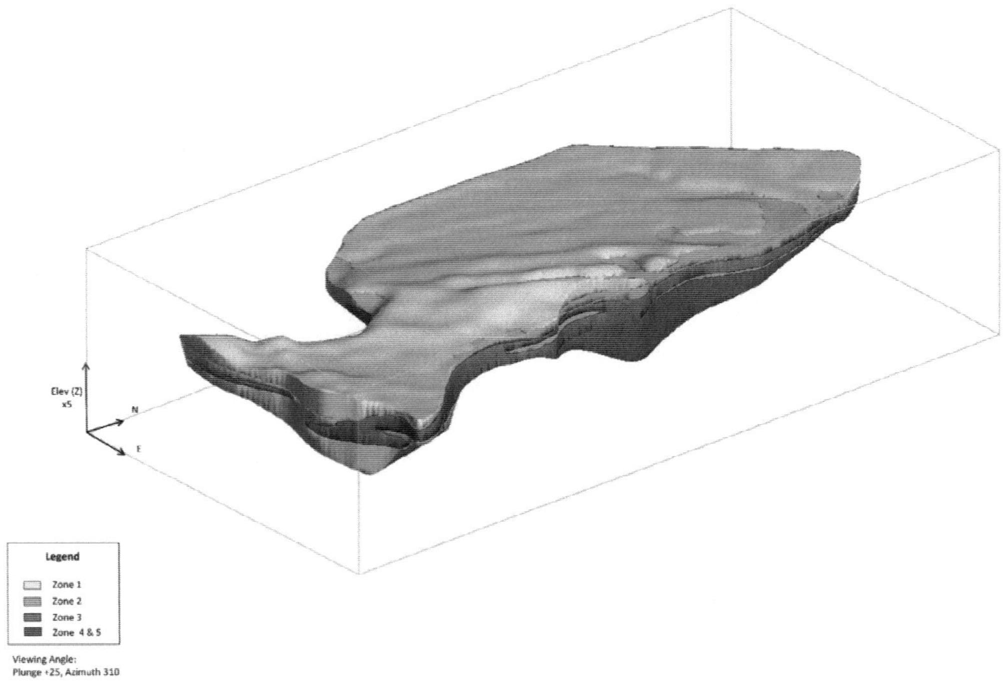

Figure 3. 3D spatial geotechnical model of the North West Satellite orebody ('The Fish'), Mandena deposit, Madagascar.

It is therefore clear from the data evaluation shown in Table IV that the numerical data and the zone model have been successful in delineating distinct geotechnical zones that can be applied to the mine planning and design process.

APPLICATION TO MINING OPERATION

Based on existing design charts, the dredge should be able to achieve an average dredging rate of around 1900 t/h in Zone 2 material and no more than 1300 t/h in Zone 3 material. By applying typical expected weighted SPT N values to certain mining blocks (*e.g.* by determining the weighted average volume that each geotechnical zone represents in a specific mining block), one can more accurately predict advance rates and set realistic production targets and/or plan which areas should rather be earmarked for mechanical preconditioning to obtain the desired throughputs. This information was also used to specify a larger dredge that would be able to deal with most of the geotechnical conditions predicted by the model.

Geotechnical parameters associated with the various zones modelled and the nature of the zones as reviewed in 3D have the following implications for the mining operations:

- The Zone 1 deposits will be easily excavated or dredged
- Zone 4 and 5 materials may result in process inefficiencies due to their more cohesive nature causing blockages of, and build up on, processing equipment and hindering of heavy mineral separation due to increased slimes contents
- All areas will be excavatable using mechanical excavators and bulldozers, although Zones 1 and 2 would be easier to excavate than Zone 3
- Dredging of Zone 1 and Zone 2 is considered possible, but dredging of Zone 3 can result in significant drops in productivity when the stronger indurations are encountered
- In terms of the stability of excavated slopes in the dry mining pits, it is expected that Zone 3 materials can stand vertically over long periods, although not necessarily with a sufficiently high

factor of safety. Zone 2 materials may stand vertically for several metres over short periods, but will collapse to lower angles over time. The aeolian sands are generally free-flowing and excavated slopes will collapse to an angle of 30° to 35°. Due to the variable nature of the deposit (introduced by the combination of the geotechnical zone thicknesses, the presence and scale of the Zone 3 indurations, and localized hydrogeological factors) no single specific slope stability solution or factor of safety (FoS) can be assigned. Slope stability management must be approached as a daily planning and production task requiring ongoing assessment by a competent person.

Table IV. Numerical $SPTN_{field}$ and $SPTN_{60}$ data extracted within each geotechnical zone.

SPTN values inside Zone 1	
Number of Values:	14765
Volume : Zone 1	35669000
Min Value:	1.0
Lower Quartile:	5.0
Median:	8.0
Upper Quartile:	15.0
Max Value:	67.0
Mean:	11.3
Std Deviation:	9.5
Variance:	89.9

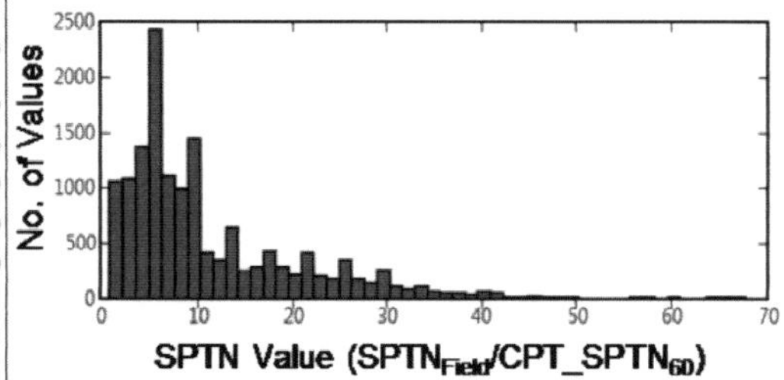

SPTN values inside Zone 2	
Number of Values:	84949
Volume : Zone 2	225290000
Min Value:	1.0
Lower Quartile:	22.0
Median:	30.0
Upper Quartile:	39.0
Max Value:	100.0
Mean:	30.6
Std Deviation:	12.0
Variance:	145.2

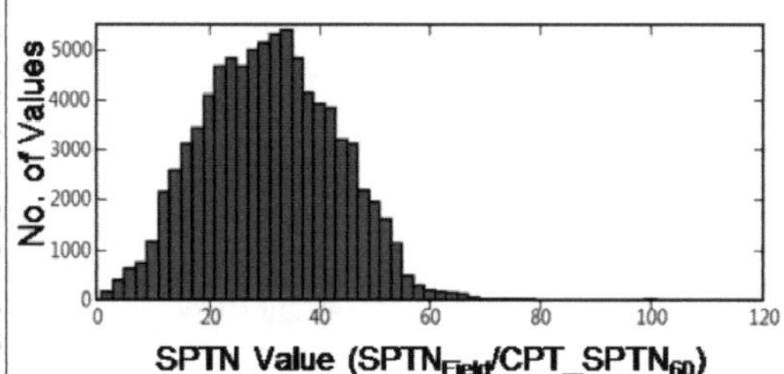

SPTN values inside Zone 3	
Number of Values:	20594
Volume : Zone 3	155500000
Min Value:	3.0
Lower Quartile:	38.0
Median:	51.0
Upper Quartile:	60.0
Max Value:	100.0
Mean:	49.1
Std Deviation:	17.4
Variance:	301.7

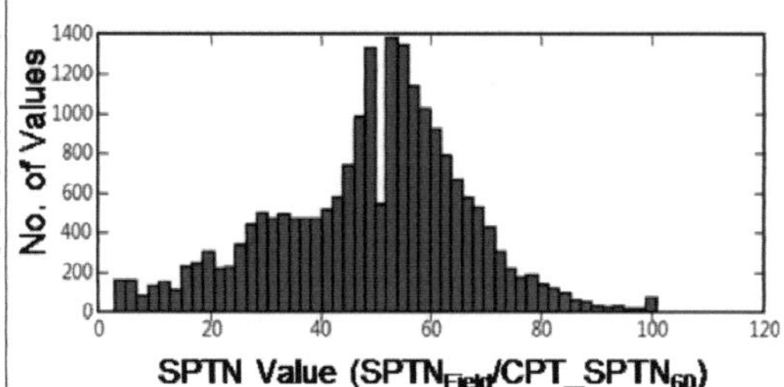

SPTN values inside Zone 4/5	
Number of Values:	4466
Volume : Zone 4/5	43431000
Min Value:	2.0
Lower Quartile:	8.0
Median:	12.0
Upper Quartile:	23.0
Max Value:	100.0
Mean:	17.7
Std Deviation:	15.0
Variance:	225.9

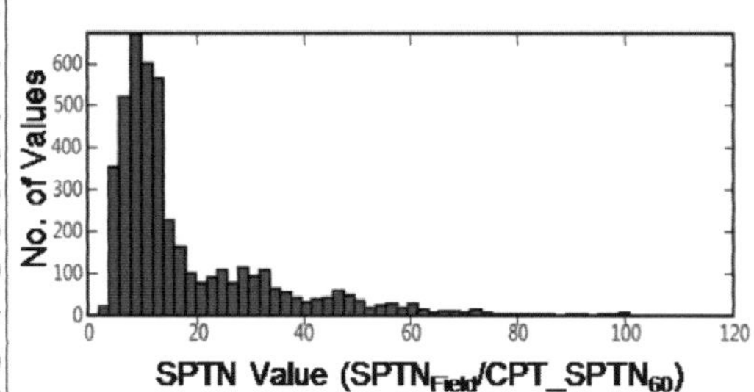

Feedback from the mining operation confirmed that the geotechnical model was utilized as a decision trigger to identify areas destined for dry and dredge mining in 2014. However, dry mining was suspended from November 2014 onwards.

The Zone 3 solid was imported into the Surpac mining software package, and then into the block model attribute to represent how much of a block is in Zone 3 (partial percentage). This helped to estimate the throughput/production rate of the dredge. Surpac Minesched software has a production factor (PRODFACT) attribute, which can be written on a block model. This represents how effective the mining system is compared with its capacity according to orebody condition. The mining rate for each block is then represented by the PRODFACT, which is the mining rate divided by capacity on a given time unit (1 day). This is a function of bench height and cutting rate (which, in turn, is a function of hardness and therefore a function of Zone 3 percentage).

The new dredger utilized as specified following the outcome of the Phase 1 investigations is able to cut Zone 3 material at the reach of the cutter, and preconditioning was planned based on the geotechnical model as part of safety mitigation to prevent unexpected falls of big blocks onto the dredge ladder. Preconditioning was eventually cancelled as the mine was able to cope with the hanging face by managing the risk through a modification of the dredging technique, which basically entails cutting horizontally (*i.e.* laterally) at the base of Zone 3, gradually moving the cutter upwards, so as to minimize the size of blocks that could fall and damage the equipment.

The above methods of planning or anticipating production have therefor offered some benefit to the mine. However, the complexity of the dredging productivity model makes it difficult to confirm the specific contribution of the geotechnical model to efficient production planning, as other constraints in the process also influence how the dredge throughput is planned.

However, by knowing/anticipating mining conditions the mine is able to plan for the way dredging proceeds (*i.e.* planning of cutting faces). An additional benefit to the operation has been the ability to plan how to deal with the downstream impact of Zone 3 mining conditions, as this has an impact on the wear rate of the cutter tooth and hence on inspection frequency, spares management, and preventing overload of the hydraulic system.

The geotechnical characterization of the orebody has therefore provided several ways of optimizing the planning and control of the mining operation.

CONCLUSIONS

The Mandena case study demonstrates the importance of geotechnical investigations in the early phases of a project to ensure understanding of the geotechnical parameters of an orebody and the potential limitations that these could pose to the chosen mining process. Unfortunately, the importance of these investigations in the early phases of mining projects are commonly forgotten, or more focus is placed on civil or resource estimation applications rather than on the understanding of how unusual geotechnical conditions could impact the mining process. It must be added that the conditions encountered at Mandena were unusual compared to the typical well-known heavy mineral sand deposits mined by operators on the South African northeast coastline (*e.g.* Zulti North, Hillendale, Zulti South, and Fairbreeze deposits).

The case study further demonstrates the importance of using an integrated approach that includes various geotechnical and geological techniques to successfully characterize unusual and variable geotechnical conditions.

By using an integrated approach and combining data from geological and geotechnical sources it was possible to generate a geotechnical zone model in 3D into which any of the *in-situ* measured and/or derived parameters obtained from CPTu testing can be imported, modelled, and/or data extracted and analysed statistically for application to the mining requirements. This integrated method of investigating, defining, and presenting large-scale, and unusual geotechnical problems allowed the Mandena operation to enhance its production planning capabilities.

REFERENCES

Boshoff, L.I. and Bracken, A.H. (2012). Project TiO4- Mandena North West and North East Satellite Areas Geotechnical Investigation SRK Report Number 437868/3.1. Draft Report. Rio Tinto Management Services, Johannesburg, South Africa.

Boshoff, L.I. and Bracken, A.H. (2011). QMM Mandena Phase 1 Geotechnical Investigation- SRK Report Number 414630/2. Final Draft Report. Rio Tinto Management Services, Johannesburg, South Africa.

Campanella, R.G. and Robertson, P.K. (1988). Current status of the piezocone test. *Proceedings of Penetration Testing 1988 (),* Orlando, Florida. vol. 1. Balkema, Rotterdam. pp. 93-116.

Clayton, C.R.I. (1995). The Standard Penetration Test (SPT): Methods and use. *Report* 143, Construction Industry Research and Information Association. CIRIA, London. .143 pp.

Committee for State Road Authorities, South Africa. (1994). Technical Recommendations for Highways, TRH10-The Design of Road Embankments. Department of Transport, Pretoria. pp. 62-64.

Cowan, E.J., Beatson, R.K., Ross, H.J., Fright, W.R., McLennan, T J., Evans, T.R., Carr, J.C., Lane, R.G., Bright, D.V., Gillman, A.J., Oshust, P.A., and Tilley, M. (2003). Practical Implicit Geological Modelling. *Australasian Institute of Mining and Metallurgy Publication Series* No. 8/2003. BPA Digital, Australia.

Hees, F. (2010). Orebody knowledge workshop report and presentation. *Biocrete Model Stream 2 Area-Mandena Project.* Rio Tinto Management Services, Johannesburg, South Africa.

Lynn, B. (2008). Summary report on the results of laboratory testing carried out on indurated sand samples. Ref 30141. Rio Tinto Management Services, Johannesburg, South Africa.

Robertson, P.K. (1992). Estimating in-situ soil permeability from CPT and CPTu. Gregg Drilling and Testing Inc., Signal Hill, California, USA.

Lyzandra Isabel Boshoff

Senior Engineering Geologist
SRK Consulting

Lyzandra Boshoff has been involved in the field of engineering geology and geotechnical engineering for the past 14 years. Her key experience includes shallow and deep ground investigations for small to large developments, tailings dams, open cast mine development and planning; 3D geological and geotechnical modelling (Leapfrog 3D Mining) and Stability and Seepage analyses. Lyzandra also has 5 years' experience in the heavy minerals mining and processing industry gained whilst employed as a geotechnical professional and Process Development Metallurgist at heavy mineral mining operations on the north coast of KwaZulu-Natal, South Africa.

Gold and associated industrial heavy minerals in the Icy Cape District: White River to Icy Cape, Alaska, USA

K. Eden[1], A.E. Grosz[2], J.E. Grosz[2], C.J. Giles[2], and J.Y. Foley[3]

[1]Trust Land Office, Anchorage, AK, USA
[2]Global Mineral Sands, Inc. Herndon, VA, USA
[3]Exploration Geologist (Retired), Anchorage, AK, USA

Portions of the Icy Cape placer Au district have been described and exploited since the early 1900s. Industrial studies of Au-bearing garnet-rich intervals promoted a 3.5 million troy ounces resource in the western half of the district. This estimate is not compliant with CIM and SME definition standards for Mineral Resources and Mineral Reserves. We note high variability in the textures of sediments and their heavy mineral (HM) content and composition in the district, probably as a function of depositional environment. Sediments are medium- to fine-grained sands with a valuable HM assemblage that averages 26.0% in a range of 1.1% to 72.6%. Valuable heavy minerals (VHMs) garnet, magnetite, ilmenite, rutile, and zircon, but not including Au, average 7.9% in a range of 1.0% to 37.6% of the HM assemblage. Trace quantities of uraninite, thorite, monazite, xenotime, and others are noted. Analyses show garnets have an average Y content of 726 ppm in a range of 0–2150 ppm. Energy-dispersive X-ray fluorescence confirms compositions commensurate with Ca, Al, and Mg silicates. In this brief report we provide empirical analytical data and note high variability in the textures, HM content, and compositions as a probable function of depositional environment. Icy Cape District land and resources are owned by the Alaska Mental Health Trust Authority and managed by the Trust Land Office (TLO).

INTRODUCTION

The Icy Cape HM District subtends more than 200 km^2 fronting the Pacific Ocean and backing to the foot of the Robinson Mountains at about 1700 m above sea level. We classify the district into geomorphic provinces; the eastern region as deltaic (Figure 1) and the western region as coastal plain (Figure 2) based on geomorphologic expressions. The coastal plain is entirely built up of unconsolidated outwash sediments and ancient beach sand deposits.

The area from the White River to Icy Cape has intermittently produced placer gold since the initial discovery in 1897. Gold production reports don't exist, but it is estimated that at least 16 000 troy ounces of gold have been mined from small-scale placer mining that concentrated on the active beaches and the White River sediments. It was not until the mid-1980s to the early 1990s that mining companies such as Noranda Exploration from Canada and Paraclete Resources from Australia began exploration for large-scale placer gold deposits. Paraclete Resources defined areas for gold placer mining as presented in Figure 2 and delineated a 3.5 million troy ounce resource (Rossetti, 1992). Note that this resource estimate is a historical estimate and is not compliant with CIM and SME definition standards for Mineral Resources and Mineral Reserves.

Placer Au occurs primarily as fine-grained (averaging 0.250 mm in diameter) flat and rounded plates and is strongly associated with garnet-rich layers (inset in Figure 2). Gold recovery tests demonstrated recoveries in excess of 93%, at an estimated 81% purity (Trust Land Office document archives).

In 1996 the land and minerals ownership of the Icy Cape land block was transferred from the State of Alaska to the Alaska Mental Health Trust Authority. The land and mineral resources are managed by the Trust Land Office (TLO). Natural resources between mean high tide and 3 nautical miles (5.56 km) offshore are managed by the State of Alaska. The nature and distribution of HMs in offshore sediments is unknown.

Over the next decade the TLO will develop timber resources in the heavily vegetated region of the coastal plain and delta regions. This will allow synergy with mineral resources development strategies.

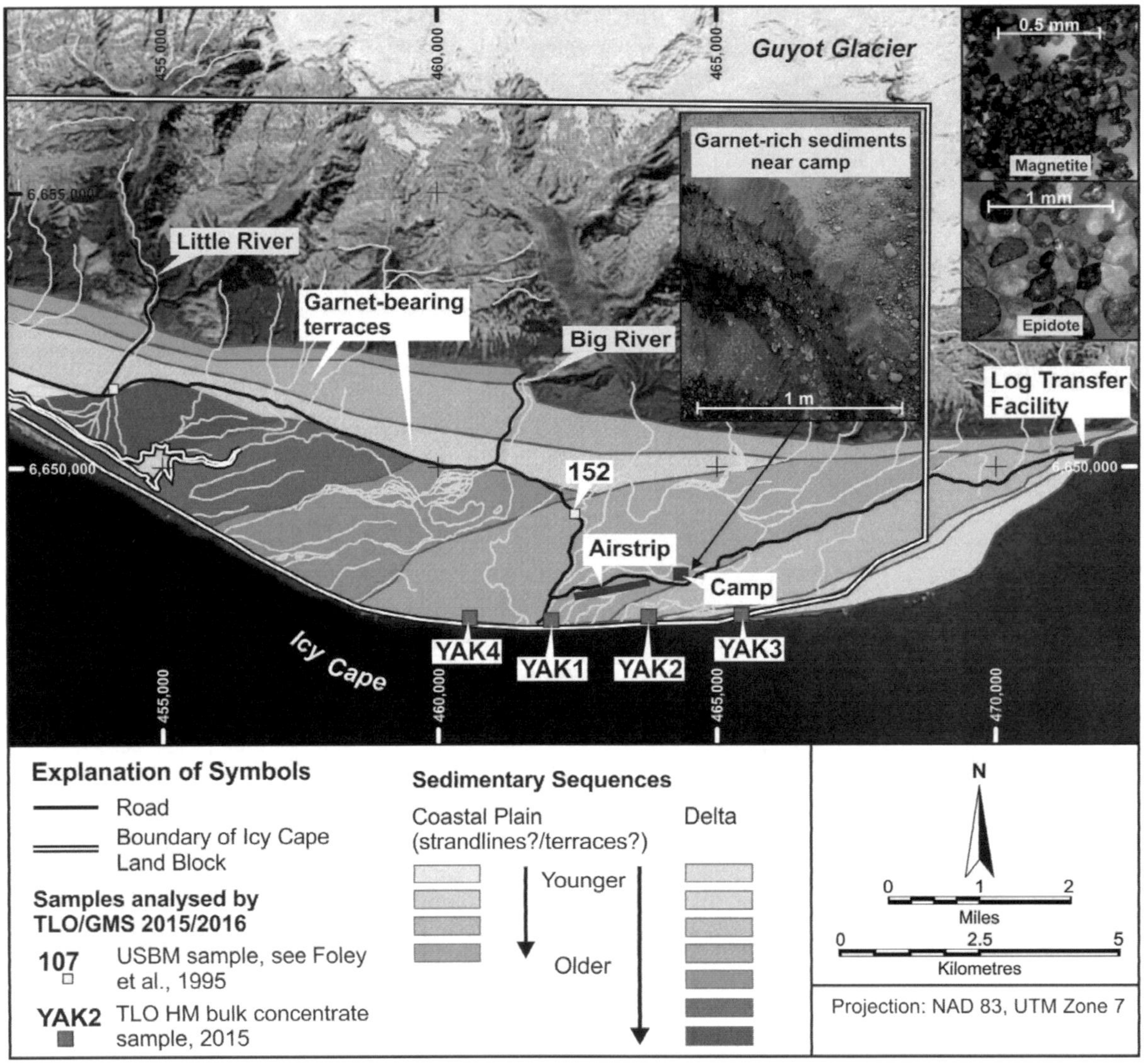

Figure 1. Map showing locations of analysed samples and sedimentary sequences in the eastern Icy Cape District.

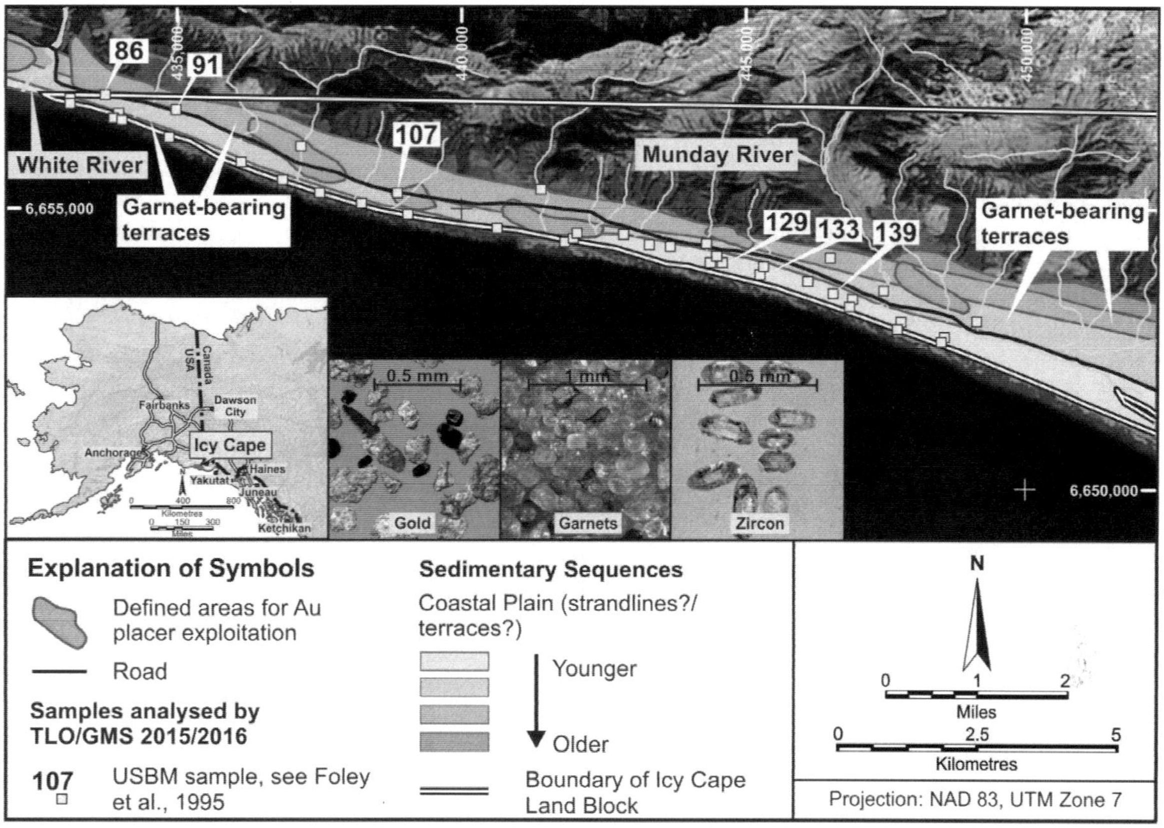

Figure 2. Map showing locations of analysed samples, sedimentary sequences and outlined areas for gold placer mining in the western Icy Cape District.

BACKGROUND

The first published systematic study of HMs in the region that included the district is one of titanium mineral resource potential by Foley *et al.* (1995). Foley *et al.* sampled at about 2 km intervals traversing the modern beach environment using solid-stem power and hand augers in the intertidal zone, beach face, and back beach, and channel sampling wave-cut beach terraces. Samples collected during that study average 1.5 m intervals ranging to 9.6 m in depth. Foley *et al.* (1995) estimate 0.57% valuable heavy minerals (VHMs; 0.49% ilmenite; 0.05% rutile; 0.03% zircon) in a range of <0.1% to 2.9% in modern shoreline sands of a portion of the district west of the Little River. They note that dynamic depositional processes produce large variations in grain size distribution and in HM content. These large variabilities make the determination of statistical central tendencies difficult. Median and mode values are better measures of this tendency, but we present mathematical averages bracketed by ranges.

Foley *et al.* (1995) note that HM fractions contain polymineralic grains, 'rock fragments', interlocked grains of magnetite, pyroxene, plagioclase, sphene, and ilmenite. They note the expansive deltaic sediments at Icy Cape and their reconnaissance Ti, Zr, and Au assays for samples from this area indicate that VHMs are present. Associated with placer gold in the region are trace quantities of platinum group metals (PGMs); (Foley *et al.*, 1995, Table 3).

Formerly confidential industry assessments of Au shown in Figure 2 (TLO document archive) note the sands contain other potentially recoverable minerals. Of these, zircon, garnet, and ilmenite were most likely recoverable, with PGMs regarded as a possibility as their value is recorded and apparently associated with magnetite. These reports also note rutile as present and in need of evaluation as a potential commercial by-product from proposed gold operations.

The stratigraphy of the coastal region (coastal plain and delta) is not well known. Corporate models are based on the major unconsolidated sediment types. They are classified according to lithological character and origin: well-sorted sand and gravel formed at or near the present beach and on former beaches, including strand lines elevated by Holocene tectonic uplift; well-sorted dune sand on or near the present or former beaches; interbedded mud and sand containing much organic debris, which were formed on tidal flats, in bays or lagoons, and in clear lakes and swamps; and interbedded mud and poorly sorted to moderately well-sorted sand and gravel, which were formed on the flood plains or fans of streams on the outwash aprons of glaciers, including terminal and ground moraine and ice-rafted deposits. The thickness of the unconsolidated deposits on the coastal lowland and along the raised beach ridges is not known but is estimated to be of the order of tens of metres.

CURRENT WORK

Our studies relied on methods, procedures, and techniques commensurate with those of the placer Ti-Zr-REE HM industry. Particle-size distribution (PSD) analyses were conducted by use of US Standard stainless steel sieves and are reported for some bulk samples and for their HM content. An average of 120 g was separated in lithium metatungstate (LMT), a heavy liquid with a specific gravity of >2.96. Technical aspects are detailed by Foley *et al.* (1995) and by Grosz *et al.* (1990).

A Frantz barrier magnetic separator was used to separate minerals in the HM concentrate into seven fractions. After a low field-strength (0.05 A) pass through the apparatus set at 15 degrees forward- and 25 degrees side-slopes, the HMs were concentrated according to their magnetic susceptibilities at 0.15, 0.25, 0.35, 0.50, and 1.00 A, yielding a nonmagnetic 1.00 A residue, and their weights were recorded. This approach isolated, for example, over 90% of garnets into the 0.25 A fraction, over 95% of rutile and zircon into the nonmagnetic residue, and epidotes, pyroxenes, and amphibole minerals into intermediate fractions. Because of variations in a mineral's chemical composition and the presence of inclusions and complex intergrowths, a mineral may occur in more than one magnetic fraction. Magnetic fractionation reduces the number of mineral species in each fraction and thus facilitates qualification and quantification of HM species.

Each magnetic fraction was examined by reflected and transmitted light microscopy. Modal mineralogical compositions were determined by point counts of at least 500 grains in combination with visual estimates of field percentage compositions. Percentages of the various minerals were calculated on the basis of their weight in each fraction. The calculated weights of a mineral in the various magnetic fractions were then added together and divided by the total weight of the HMs. The densities of individual mineral species were not compensated for by this method. XRD, XRF, SEM, and microprobe analyses were used to confirm petrographic mineral determinations and to detail compositions.

In this paper, we provide PSD, HM content, modal mineralogical, and adjunct analytical data for samples. We also generated conductor, non-conductor, and trace element determinations that are not included in this brief report.

More than 100 samples from Foley's (1995) study were retrieved from archives at the Geologic Materials Center in Anchorage, AK. From these, 87 samples averaging 275 g were collected by use of Jones splitters, and five samples of garnet-rich beach concentrate collected by the TLO in 2015 were included. The average HM content of the 87 samples is 19.3% in a range of 0.7–72.6%. The PSD of these sands is predominantly less than 0.297 mm and greater than 0.149 mm. With the exception of a few beach samples containing significant very coarse sand- to gravel-sized material, samples were predominantly less than 0.595 mm and greater than 0.105 mm in particle size.

The analytical results for 21 samples representing seven locations as selected for this report are presented in Table I. They represent thicknesses of sediments above mean high tide as modern beach sediments are outside the TLO property boundaries. Each location has defined sampling intervals. For example, SN 86.1 through SN 86.4 denote a core location with successively deeper sampled interval. Sample numbers from Foley *et al.* (1995) are present as map numbers for ease of cross-reference. Table I shows their locations (coordinates in metres, North American Datum 1983, UTM Zone 7), depth intervals in metres, and %HM (SG >2.96) expressed as a percentage of the bulk sample collected. Analyses of these widely distributed power-auger, hand-auger, and channel samples show an average 26.0% HMs in a range of 1.1–72.6%.

Table 1. Analytical results for 21 bulk sediment samples showing sample locations, depth intervals sampled, and HM content.

SN	UTM E	UTM N	From m	To m	Interval m	Wt% HM
86.1	433723	6657013	0.0	2.1	2.1	44.89
86.2	433723	6657013	2.1	4.0	1.9	33.82
86.3	433723	6657013	4.0	5.8	1.8	24.12
86.4	433723	6657013	5.8	7.3	1.5	72.58
91.1	434966	6656749	0.9	2.1	1.2	13.85
91.2	434966	6656749	2.1	4.0	1.9	64.91
91.3	434966	6656749	4.0	5.8	1.8	38.72
107.1	438874	6655268	0.9	2.7	1.8	38.87
107.2	438874	6655268	2.7	5.8	3.1	17.18
107.3	438874	6655268	5.8	6.7	0.9	10.44
129.1	444575	6654017	0.0	2.1	2.1	33.86
129.2	444575	6654017	2.1	4.0	1.9	12.88
129.3	444575	6654017	4.0	5.8	1.8	6.97
129.4	444575	6654017	5.8	7.6	1.8	1.50
129.5	444575	6654017	7.6	9.6	2.0	1.10
133	445248	6653788	0.0	0.8	0.8	21.27
139.1	446545	6653484	0.0	2.1	2.1	48.78
139.2	446545	6653484	2.1	4.0	1.9	10.92
139.3	446545	6653484	4.0	5.8	1.8	8.32
139.4	446545	6653484	5.8	7.2	1.4	2.91
152	462481	6649158	0.0	4.6	4.6	38.99

RESULTS

Medium- to fine-grained sands with locally large coarse- or fine-grained components dominate the coastal plain and delta portions of the district. With the exception of a few samples of relatively very coarse sand to gravel-sized material, mineral grains are 0.6 mm to 0.11 mm sized particles. The PSD of 21 sediment samples and some of their HMs are given in Table II. Particle sizes are given in millimetres. Note the high variability in size distribution of both bulk sediments and their HMs.

Mineralogy

The HMs of coastal plain and deltaic sediments in the district include a large and varied assemblage. Table III shows the distribution of HMs in 21 samples. Magnetite (MAG), ilmenite (ILM), garnet (GAR), rutile (RUT), zircon (ZIR), others (OTH), and VHM (valuable HMs include only MAG, ILM, GAR, RUT, and ZIR) as modal percentages of the HMs are given in Table III. A value of 0.00 denotes <0.01%. Images of minerals are shown as insets in Figure 1 and Figure 2, and they are described in the following section. Small, flat (with irregular and/or curled edges) gold particles as well as uraninite and thorite particles were noted also in many samples (example in inset Figure 1).

XRF analyses confirm chemical compositions commensurate with the presence and relative abundances of Ca, Al, and Mg silicate minerals that have not been examined for potential value as abrasives, water filtration agents, drilling mud components, or other commercial applications.

Table II. Particle size distribution of 21 bulk sediment samples and some of their HMs. Blank cells have no data.

	BULK SIZE DISTRIBUTION (Wt% retained)						HM SIZE DISTRIBUTION (Wt% retained)					
SN	0.595	0.420	0.297	0.250	0.149	<0.149	0.595	0.420	0.297	0.250	0.149	<0.149
86.1	0.7	1.6	10.5	13.9	70.2	3.2	0.1	0.4	4.2	5.3	84.4	5.6
86.2	4.6	3.8	17.2	17.6	53.2	3.6	0.9	0.8	3.6	10.1	79.7	4.8
86.3	10.8	6.1	21.7	17.9	40.2	3.3	2.8	1.8	9.6	7.2	77.1	1.5
86.4	12.5	6.3	21.4	17.5	39.0	3.4	16.9	0.7	2.6	2.1	49.4	28.2
91.1	3.8	2.6	6.8	10.3	65.6	10.9						
91.2	0.6	0.1	1.0	2.8	66.4	29.2						
91.3	2.5	2.6	10.8	10.8	70.5	2.8						
107.1	2.3	0.7	7.5	15.9	66.7	6.9						
107.2	0.5	0.9	11.9	20.2	64.2	2.2						
107.3	6.2	7.3	21.5	18.5	40.8	5.6						
129.1	0.1	0.3	10.2	23.3	65.2	0.8	0.0	0.0	0.9	1.3	95.0	2.8
129.2	2.4	3.2	18.7	25.6	48.3	1.8	0.0	0.0	4.4	8.6	85.0	1.7
129.3	4.0	2.9	17.8	23.0	51.7	0.7	6.2	2.3	4.5	9.0	67.9	10.1
129.4	1.2	1.0	6.7	11.3	52.5	27.2	19.7	3.9	12.1	9.1	53.0	2.2
129.5	0.1	0.4	8.6	17.4	72.2	1.4						
133	0.0	0.1	5.1	19.6	73.1	2.1						
139.1	1.8	3.0	20.4	24.6	47.6	2.7	0.0	0.0	2.0	1.7	3.2	93.0
139.2	1.8	2.3	19.9	24.4	50.6	1.0	2.8	1.4	12.3	10.6	38.6	34.2
139.3	2.5	3.5	21.6	25.1	44.7	2.6	1.5	2.0	2.5	59.9	31.3	2.9
139.4	2.0	0.7	7.2	15.3	68.0	6.8						
152	52.7	16.6	13.1	4.3	8.5	4.8						

Table III. Modal mineralogical composition of 21 samples.

	MINERAL COMPOSITION (Wt% of HM)						
SN	MAG	ILM	GAR	RUT	ZIR	OTH	VHM
86.1	0.73	0.17	10.21	0.47	0.42	88.00	12.00
86.2	0.79	0.17	8.70	0.43	0.31	89.60	10.40
86.3	0.15	0.66	12.70	0.08	0.11	86.30	13.70
86.4	0.14	0.00	1.66	2.89	2.76	92.55	7.45
91.1	1.86	0.07	8.42	0.34	0.23	89.08	10.92
91.2	1.47	0.54	11.75	0.76	0.53	84.96	15.04
91.3	1.70	0.05	6.72	1.94	0.74	88.84	11.16
107.1	1.90	0.18	9.11	0.37	0.45	87.99	12.01
107.2	0.28	0.00	3.94	0.20	0.66	94.92	5.08
107.3	0.48	0.03	3.52	0.32	0.50	95.16	4.84
129.1	0.85	0.00	0.76	0.17	0.02	98.20	1.80
129.2	0.51	0.00	0.42	0.06	0.00	99.01	0.99
129.3	0.44	0.06	0.56	0.02	0.02	98.90	1.10
129.4	0.59	0.08	0.68	0.01	0.04	98.60	1.40
129.5	0.56	0.00	1.24	0.00	0.03	98.17	1.83
133	1.51	0.03	0.34	0.03	0.00	98.09	1.91
139.1	1.93	0.76	34.73	0.07	0.14	62.37	37.63
139.2	1.95	0.03	7.75	0.28	0.08	89.91	10.09
139.3	1.31	0.00	1.82	0.26	0.00	96.61	3.39
139.4	1.43	0.00	0.65	0.06	0.07	97.78	2.22
152	0.40	0.00	0.67	0.04	0.01	98.88	1.12

MAG and ILM are strongly magnetic components of the HMs. They are technically best labelled as oxides as they include Ti-free iron oxides with sphene and amphibole inclusions, haemo-ilmenite, chromite, octahedral and rounded magnetite, and magmatic haematite-ilmenite-magnetite showing exsolution features at 2 µm to >50 µm scale. Other inclusions in the oxide group include sphalerite, pyrite, and REE-rich phases in a range of 2 µm to 10 µm in size. Ilmenite is present in all samples but it is not a liberated 52.7% TiO_2 mineral. The ilmenite appears not to be suitable for pigment manufacture because of trace impurities that include Si, Cr, V, and Al (Foley *et al.* 1995). Figure 4A shows a SEM backscattered image of a typical titaniferous magnetite with titanium-rich exsolution lamellae (thin black bands). Black spots are Al-rich spinel. Figure 4B shows intergrowth of magnetite and ilmenite. Black areas near centre are hornblende inclusions. Oxides average 1.13% of the HM suite in a range of 0.14% to 1.98% (Table III).

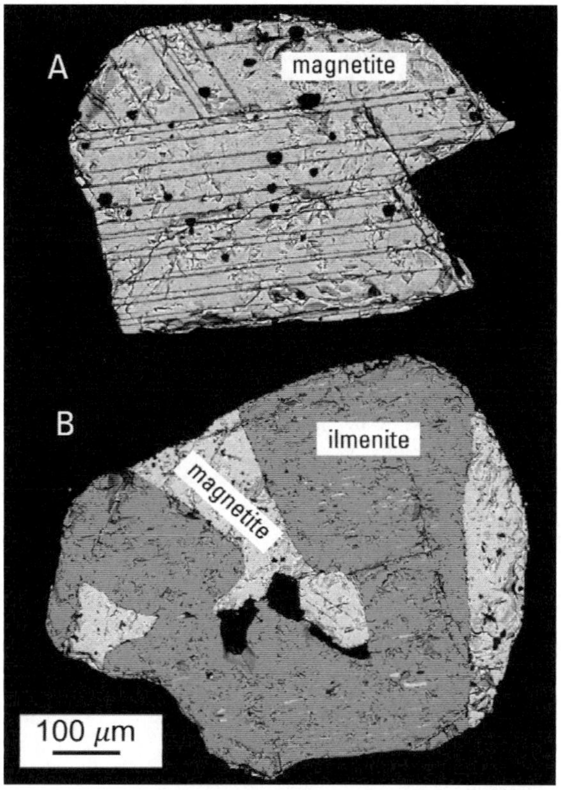

Figure 4. SEM backscattered electron image of oxide mineral particles classified as magnetite and ilmenite (MAG and ILM; Table III).

Garnet

The garnet group was examined in more detail than other HMs as the historically mentioned co- or by-product potential of garnet had not been addressed. Garnet comprises an average of 6.02% of the HM suite in a range of 0.34% to 34.73% (Table III). Our analyses reveal that end-members almandine, pyrope, grossular, and spessartine (not including andradite and uvarovite) comprise 67%, 12%, 6%, and 15% respectively of the garnet group. We distinguish pink, orange, and dark garnets that are mostly euhedral to slightly rounded, with a crushed shard component. Figure 5 shows the principal types: A is pink, B is orange, and C is dark. A is euhedral, B is rounded and contains zircon (white) and quartz (black) inclusions, and C is rounded garnet with quartz (black) and ilmenite (white) inclusions. Many garnets have varying amounts of inclusions that include monazite-(Ce) (up to 25 µm), pyrite (up to 50 µm), ilmenite (up to 50 µm), and xenotime (up to 20 µm). Microprobe analyses show garnets have an average Y content of 726 ppm in a range of 0 to 2150 ppm, with pink and orange types having the highest overall values. The reader interested in Y distribution in garnet-group minerals is referred to Jaffe (1951), Pyle and Spear (1999), references therein, and to subsequent publications.

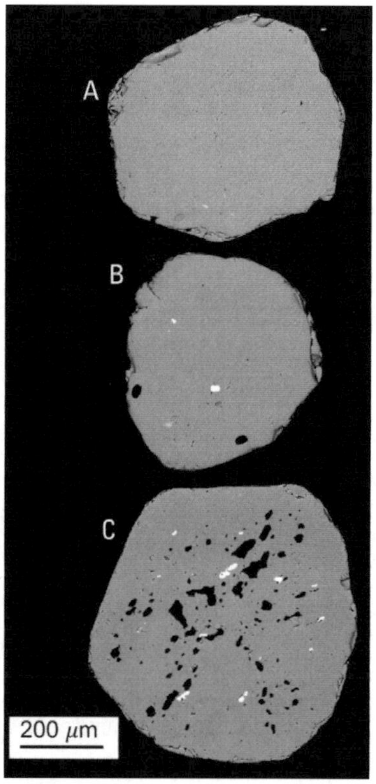

Figure 5 . SEM backscattered electron image of typical garnet particles (GAR, Table III).

Rutile

The rutile group (≥90%TiO_2) includes rutile, anatase, and sphene. This group of minerals averages 0.42% of the HM suite in a range of <0.01% to 2.89% (Table III). Particles appear liberated and are finer-grained than other HM including gold.

Zircon

Zircon is present as fine-grained, clear, doubly terminated euhedral crystals and as crushed shard fragments that are low in U and Th (<300 ppm combined). Figure 6 shows euhedral zircon (A) and a broken fragment of euhedral zircon (B). ZIR comprises an average of 0.34% of the HM suite in a range of <0.01% to 2.76% (Table III). Some zircons contain chalcopyrite inclusions up to 20 μm, monazite inclusions up to 10 μm, uraninite inclusions up to 5 μm, thorianite inclusions up to 2 μm, and thorite inclusions on the order of 1 μm. Arsenopyrite is also occasionally present as inclusions. Zircons are also finer-grained than other HMs, including gold.

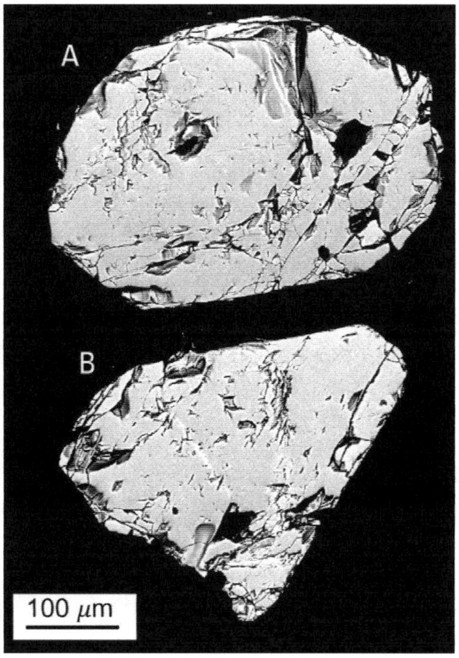

Figure 6. SEM backscattered electron image of typical zircon particles (ZIR, Table III).

Other

For the purposes of this study large and varied groups of HMs are classified as 'others' (OTH, Table III) and retained for follow-up analyses. This group accounts for an average 92.09% of the HM suite in a range of 62.37–99.01%. It includes dominant epidote, zoisite, clinozoisite, pyroxene, amphibole, olivine, serpentine, unidentified opaque and non-opaque sand- and finer-grained particles made up of HM and non-HM fragments. Included in this group are very small but persistent traces of spinels, monazite, cassiterite, uraninite (100 μm particles), chromite (high Al/low Fe), thorite, thorianite, albite, barite, scheelite, and localized carbonate fragments.

VHM

Reported VHM reflect on potential as co- and by-product HMs from gold recovery operations. Excluding gold, VHMs average 7.91% of the HM suite in a range of 0.99–37.63% (Table III). Values are considered underestimates because the OTH group contains a larger, though lower-value suite of industrial minerals than VHMs. Nonetheless, VHMs reach very large values in comparison to traditional titanium-zirconium deposits.

CONCLUSIONS

The Icy Cape District is host to a large and varied body of unconsolidated sediment with documented placer Au resources and associated HM resources that are primarily industrial in nature. We estimate the district to contain in excess of 1.7×10^9 t resource-bearing sediments to 10 m depth. Our analyses confirm potential for mineable HMs with associated Au concentrations. The geological framework, textural, mineralogical, and chemical data indicate a large, but as yet incompletely understood, resource potential for Au, garnet, and other HMs. Very large vertical and lateral variations in PSD and HM content and composition are noted, likely as a function of depositional environment. These, and supportive adjunct formerly proprietary information, refine factors controlling commercial HM viability in the Trust Land area and provide a better understanding of resource potential leading to exploration and development. The TLO will conduct high-resolution aeromagnetic surveys of the district in 2016 and stratigraphic and resources drilling campaigns in 2017.

ACKNOWLEDGMENTS

This work was funded by the Trust Land Office and by ILUKA Resources. The authors would like to thank John Morrison, Wyn Menefee, Leann McGinnis, Paul Slenkamp, and Mike Franger of the Trust Land Office for their enthusiastic support. As well, we note the invaluable assistance and cooperation of ILUKA Resources; specifically of Thomas Hartzog, Adam Karst, Charles Acker, and Doug Holtman. We appreciate the easy availability of data, samples, facilities, and curator support of the Geologic Materials Center of the Alaska Department of Natural Resources in Anchorage, Alaska (http://dggs.alaska.gov/gmc/general-info.php), and in particular the assistance and support provided by Kurt Johnson, Jean Riordan, and Alexandra Busk. James Barker, retired USBM research geologist, provided invaluable information on previous research and exploration campaigns. We also would like to thank Jim R. Tuttle of Sealaska Timber Corporation for his support of this project. Last but not least, we are sincerely indebted to the knowledge, skills, expertise, and time that Harvey E. Belkin and Frank T. Dulong of the USGS provided for their expertise in SEM and microprobe, and XRD, respectively. Their world-class analytical expertise was invaluable to our study.

REFERENCES

TLO archived documents:

Robson, J.M. (1983). Yakataga Beaches Reconnaissance Report, Project 50034-1983. Noranda Exploration Inc., Anchorage, Alaska, USA. 30 pp.

Rossetti, L. (1992). The Yakataga Gold Project, Alaska, USA - Predevelopment Statement. Paraclete Resources Pty Ltd, Beulah Park, SA, Australia. 45 pp., 9 appendices.

Public domain documents:

Foley, J.Y., LaBerge, R.D., Grosz, A.E., Oliver, F.S., and Hirt, W.C. (1995). Onshore titanium and related heavy-mineral investigations in the eastern Gulf of Alaska region, southern Alaska. *Open-File Report* 14-90, US Bureau of Mines. 125 pp.

Grosz, A.E., Berquist, C.R., Jr., and Fischler, C.T. (1990). A procedure for assessing heavy-mineral resources potential of continental shelf sediments. *Heavy Mineral Studies--Virginia Inner Continental Shelf.* Berquist, C.R., Jr., (ed.). *Publication* 103, Virginia Division of Mineral Resources. pp. 13-30.

Jaffe, H.W. (1951). The role of yttrium and other minor elements in the garnet group. *American Mineralogist*, 36 (1), Jan-Feb. 1951, 133-155.

Pyle, J.M. and Spear, F.S. (1999). Yttrium zoning in garnet: coupling of major and accessory phases during metamorphic reactions. *Geological Materials Research* 1. http://www.minsocam.org/gmr/papers/v1/v1n6/v1n6.pdf

Dr Karsten Eden

Minerals & Energy Section Chief
Trust Land Office

Dr Karsten Eden is the Minerals and Energy Section Chief at the Alaska Mental Health Trust Land Office in Anchorage, Alaska and manages the Icy Cape Gold and Industrial Heavy Minerals Project. Karsten is a Certified Professional Geologist with 20 years international experience in the supervision and management of mineral exploration and mine development projects in Alaska, Western Australia, West Africa, Scandinavia and Europe. Throughout his career, Karsten has provided consulting services to many international mining companies and has held several corporate geologist positions. He holds a doctorate degree in exploration geology from the University of Technology in Aachen, Germany.

Mineralogical, geochemical, and processing characteristics of heavy minerals sand deposits of Bangladesh and their commercial aspects

E. Akon

Part-time teacher, Dept. of Geology, University of Dhaka, Bangladesh
Chief Geologist, Premier Minerals Limited, Gulshan, Dhaka, Bangladesh
(Former Chief Geologist & Director, Bangladesh Atomic Energy Commission.)

Systematic exploration in the southeastern coastal belt of Bangladesh for heavy minerals has resulted in the discovery of a number of heavy mineral sand (HMS) deposits in the coastal beaches and offshore islands. The mineralogical, geochemical, and mineral processing characteristics of the heavy mineral sands and their economic potential are presented in this paper. The HMS deposits contain on average 23% total heavy minerals (THM). Heavy mineral sand deposits together contain 1.76 Mt of economically important heavy minerals, comprising ilmenite (1.025 Mt), garnet (223 kt), zircon (158 kt), leucoxene (97 kt), kyanite (91 kt), rutile (70 kt, magnetite (81 kt), and monazite (17 kt). In the heavy fraction the average economic minerals content is zircon 4.20%, rutile 2.04%, ilmenite 26.03%, garnet 6.45%, kyanite 3.92%, leucoxene 2.30%, magnetite 1.87%, and monazite 0.31%. A large deposit of economic heavy minerals has been reported recently in the Bay of Bengal islands in the Bhola and Patuakhali districts and in the river sand bars of the Brahmaputra. Processing of bulk samples has been carried out at the pilot plant of BAEC at Cox's Bazar to test the separation characteristics of individual heavy minerals using gravity, magnetic, and electrostatic separation. Geochemical studies on separated minerals show that the TiO_2 content in rutile is around 90% and ZrO_2 in zircon is around 65%, which are quite satisfactory in respect of commercial grade. The ilmenite contains a low level of titanium dioxide (around 43%) and higher levels of chromium, manganese, and iron. Beneficiation and upgrading will be required to make the ilmenite a marketable commodity.

INTRODUCTION

The valuable minerals in heavy mineral sands (HMS) are predominantly ilmenite, rutile, zircon, and their weathered equivalents. These heavy minerals occur in very low concentrations in a variety of igneous and metamorphic rocks, but being resistant to chemical and physical weathering, and having comparatively high specific gravities, they tend to accumulate in placer deposits along coastal shorelines/beaches or in river channels. The physico-chemical characteristics of the provenance rocks, drainage pattern, and climatic factors played a prominent role in the supply of material for concentration in favourable places. Once liberated from the parent rock, placer minerals become concentrated by hydraulic processes, primarily because of density contrasts. Coastal processes (wave velocity, longshore currents, and wind speed) have their effects on littoral transport, sorting and deposition of placer minerals. A breaking wave takes all the foreshore minerals to the beach but the backwash carries only the lighter minerals back to the sea leaving, behind the heavy minerals.

The Tenth Internation Heavy Minerals Conference
Sun City, 16–17 August 2016
Southern African Institute of Mining and Metallurgy

Repeated wave action results in sorting and the concentration of heavy minerals in a beach placer deposit. After concentration, wind action further enriches the concentration by blowing away the finer and the lighter sand particles, thus leaving the *in-situ* deposit rich in heavy minerals. Emergence and submergence of the coast during the geological past also affected the beach placer formation. Other factors such as coastal geomorphology, neo-tectonics, and continental shelf morphology have contributed for the formation of heavy mineral sand deposits. Beaches are favourable sites for light heavy minerals, *e.g.*, ilmenite, magnetite, rutile, zircon *etc.* whose specific gravity ranges between 4.2 and 6. These heavy minerals have widespread industrial applications. Their exploitation and marketing contribute considerably towards the growth of the national economy. Bangladesh Atomic Energy Commission (BAEC), through its Beach Sand Mineral Exploitation Centre (BSMEC) and Nuclear Minerals Unit (NMU), has been exploring for HMS in the southeastern and southern coastal areas of Bangladesh for more than three decades. A good number of articles have been published on the investigations (Akon and Hossain, 2004; Akon, 2015; Biswas 1983, Howarth *et al.*, 1977; Macdonald 1976; Kabir *et al.*, 2010). Premier Minerals Limited has also engaged in exploration and development of heavy mineral resources since 2007. Through this paper, an overview of the results of exploration activities on mineral sand deposits of Bangladesh and their characteristics is presented.

METHODS

A systematic survey was conducted following a baseline, parallel to the long direction of the deposits, whose general outline was recognized by their topography and occurrence of black sands on the surface. A series of parallel lines was laid out perpendicular to the baseline. A full-scale drilling programme was carried out in the potential areas with a general grid pattern of 500 foot (152 m) intervals between lines and hole spacing of 25 feet (7.6 m) along the lines. The number of lines and boreholes depended on the size of the mineralized area. In total around 7000 boreholes were drilled covering different deposit areas. The sample interval was usually 3 feet (0,9 m). Most of the holes were 5–20 feet (1.5–6.1 m) deep. Samples were taken by hand auger down to the water table and at some points below the water table by casing the hole and using a sludge pump. Sampling was ceased when the heavy mineral grade became too low. A portion of the sample material was panned in the field and the heavy mineral content assessed as trace (< 5%), fair (5–10%), good (10–15%), very good (15–20%), and rich (>20%). Field logs were recorded and the maps and cross-sections produced showing the lateral and vertical distribution of grades.

Before detailed analysis, samples were screened to remove oversize (>1 mm) and undersize fine slime materials (<53 µm). After screening, sample processing included drying, sieving for particle size distribution, estimation of total heavy minerals (THM), and magnetic fractionation of the heavies. The heavy mineral (HM) content was determined from each 50 g sample split by a Carpco laboratory splitter by float and sink method using bromoform ($CHBr_3$; SG 2.89) as a medium. Methylene iodide (di-ido methane, SG 3.30) heavy liquid was used to determine very heavy minerals. A hand magnet was used to remove magnetite from the HM fraction. The nonmagnetic fractions of the samples collected from a line were mixed into a composite sample. The composite sample was sieved into several size fractions and each individual size fraction subjected to petrological and ore microscopy study for mineralogical identification, grain counting, and determining the percentage weight of individual HMs. The density of different heavy minerals was taken into consideration when converting grain counts to weight percentage.

A Frantz isodynamic separator was used for physical separation of individual HMs in different magnetically separated fractions, using progressively higher currents (0.07, 0.2, 0.4, 0.5, 0.7, 1.0, 1.5, and 2.0 A) to separate the minerals incrementally according to magnetic susceptibility. Each magnetic fraction thus obtained was weighed, recorded, and studied for determining the percentage, detailed mineralogy, and chemistry. The dried bulk sample split of several kilograms of >53 µm material was also processed using a laboratory-scale shaking table to produce a HM concentrate for further processing in a Carpco magnet separator, with the nonmagnetic fraction being repassed at increasing currents (0.07–3.2 A) and field strengths (500–17500 G). The percentages of individual separated fractions were determined and characterized mineralogically and chemically. Mineral processing of the bulk sample was carried out in the pilot plant at Cox's Bazar using gravity, magnetic, and electrostatic separators. Selected samples separated in the laboratory and in the pilot plant were chemically analysed using X-ray fluorescence (XRF). The tonnage of ore was estimated by applying the bulk gravity factor to the volume of orebody and the heavy mineral tonnage was calculated by the weighted average grade for the deposit.

RESULTS AND DISCUSSION

Three decades of exploration by BAEC have resulted in the discovery of 17 HMS deposits in the southeastern coastal belt of Bangladesh (Figure 1). Prospective areas for HMs in Bangladesh are shown in Figure 2. Six of the deposits are located along the Cox's Bazar-Teknaf coast (Badarmokam, Sabrang, Teknaf, Silkhali, Inani, and Cox's Bazar), seven deposits in Moheskhali Island, one in Matarbari Island, one in Kutubdia Island, and two in the southern coast (Nijhum Dwip of Hatia and Kuakata of Patuakhali). These deposits range from 500 -10 000 feet (152-3050 m) in length, 50-1000 feet (15-305 m) in width and 3-15 feet (0.9-4.5 m) in thickness. They occur as low linear ridges and dunes (Figure 3). Some deposits extend up to a distance of 3 km inland from the present shoreline. Total heavy mineral percentage varies from deposit to deposit, ranging from 7.30% to 42.20%; and on average these deposits contain 23% HMs.

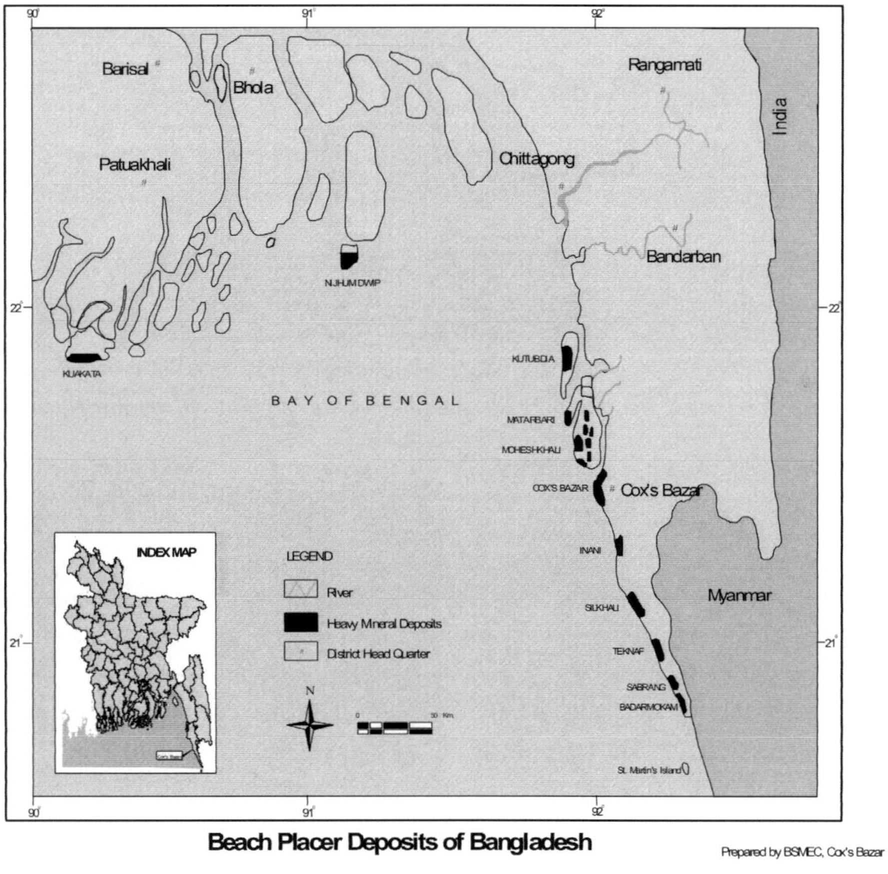

Figure 1. Location of heavy mineral deposits discovered by BAEC.

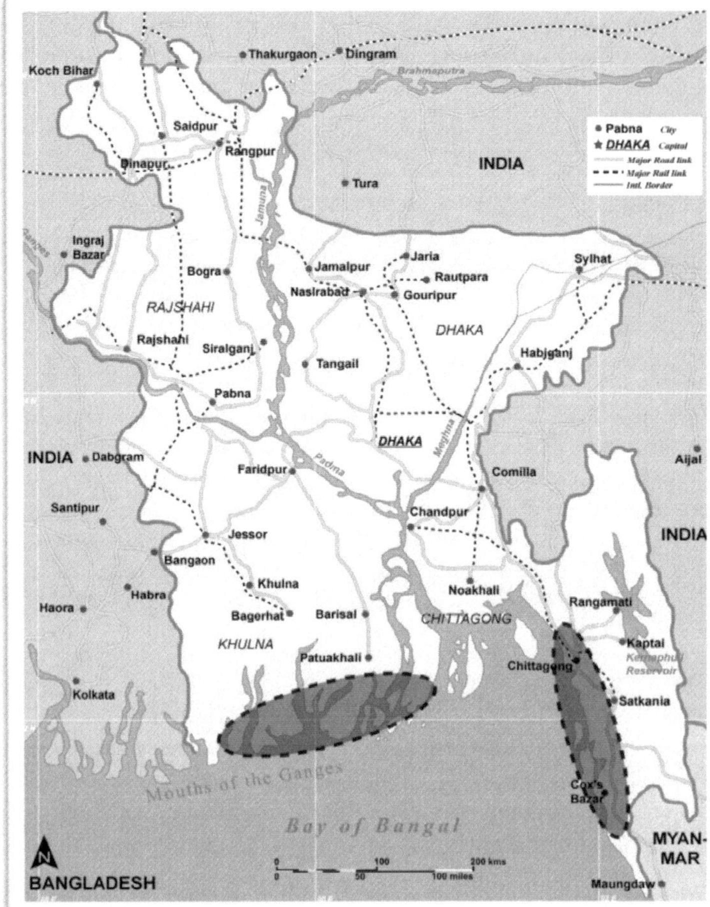

Figure 2. Bangladesh map showing the prospective areas for heavy mineral sand deposits.

Figure 3. Heavy mineral deposition at Teknaf and Silkhali in SE coastal area.

These HM deposits are mostly fine-grained. Resource estimation shows that the HMS deposits altogether contain 4.35 Mt of heavy minerals, among which 1.76 Mt are potentially economic (Table I), which include ilmenite (1.025 Mt), magnetite (81 kt), zircon (158 kt), garnet (223 kt), rutile (70 kt), leucoxene (97 kt), kyanite (91 kt), and monazite (17 kt). Total reserves of economic HMs are depicted in Figure 4. The grain size of the HMs ranges from 0.25 to 0.075 mm, and varies significantly from deposit to deposit. However, in all the deposits the grain size of zircon, magnetite, and monazite is always finer than that of kyanite, garnet, ilmenite, and rutile. Mineralogical composition of the heavy fraction in the deposits is as follows: ilmenite 9.14–46.02%; garnet 4.40–8.05%; zircon 1.20–6.80%; kyanite 0.80%–9.90%; rutile 0.44%–4.00%; leucoxene 0.9%–5.06%; magnetite 0.10%- 4.55%, and monazite 0.01–1.20% (Table II). Mineralogical analyses of the heavy fractions have established that the average content of the important heavy minerals is zircon 4.20%, rutile 2.04%, ilmenite 26.03%, garnet 6.45%, kyanite 3.92%, leucoxene 2.30%, magnetite 1.87%, and monazite 0.31%. The average distribution of the economic heavy minerals is shown in Figure 5.

Table I. Summary of heavy mineral reserves (t).

Deposits	Ilmenite	Garnet	Zircon	Rutile	Magnetite	Leucoxene	Kyanite	Monazite
Main Coast								
Badarmokam	94 530	n.d.	4 932	3 288	10 275	17 002	n.d.	4 932
Sabrang	19 614	3 018	4 184	1 372	1 001	3 740	727	206
Teknaf	163 170	22 424	28 306	13 230	7 208	20 124	14 728	3 045
Silkhali	173 630	39 422	33 300	10 774	3 085	10 970	4 407	3 918
`Inani	53 170	12 810	10 880	4 036	5 545	439	1 404	965
Cox's Bazar	161 000	50 600	26 000	6 440	33 212	10 488	n.d.	2 024
Kuakata	76 015	52 229	9 647	3 911	4 325	9 647	16 800	83.2
Moheskhali Island								
Foreshore Beach	554 961	n.d.	7 145	3 788	824	2 891	Trace	131
Kutubjum	36 440	n.d.	4 568	2 104	1 169	1 158	2 635	202
Fakiraghona	15 746	4 655	1 513	1 458	288	952	5 006	55
Fakirahata	22 157	5 988	5 515	31 870	97	3 677	9 579	194
Baraghoriapara	52 168	11 815	6 017	6 744	1 236	6 416	13 815	927
Panirchara	60 091	13 490	12 264	6 479	3 863	5 519	16 531	531
Hoanak	1 400	n.d.	100	153	127	69	110	4
Other islands								
Matarbari Island	4 962	n.d.	794	295	575	374	Trace	20
Nijhum Dwip	12 978	n.d.	2 052	424	4 384	77	2 592	19.3
Kutubdia	23 796	6 300	3 900	1 908	3 384	2 436	2 592	96

Table II. Economic heavy mineral components (weight%) in heavy fraction of different deposits.

Deposit area	THM%	Zircon	Rutile	Ilmenite	Lecoxene	Kyanite	Monazite	Magnetite	Garnet
Main Coast:									
Badarmokam	23.30	1.20	0.80	23.00	4.38	n.d.	1.20	2.50	n.d.
Sabrang	19.70	6.10	2.00	28.60	5.06	1.06	0.30	1.46	4.40
Teknaf	22.80	6.40	3.00	37.00	4.55	3.33	0.70	1.63	5.07
Silkhali	17.80	6.80	2.20	35.40	2.24	0.90	0.80	0.63	8.05
Inani	24.10	6.20	2.30	30.30	0.25	0.80	0.55	3.16	7.30
Cox's Bazar	18.00	2.50	0.75	17.50	1.14	n.d.	0.22	3.61	5.50
Kuakata	28.95	1.16	0.35	09.14	1.16	2.02	0.01	0.52	6.28
Moheshkhali Island:									
Force-shore beach	42.20	5.98	3.17	46.02	2.42	Trace	0.11	0.69	n.d.
Kutubjum	20.80	4.30	1.98	34.31	1.09	2.48	0.19	1.10	n.d.
Fakiraghona	23.50	2.21	2.13	23.02	1.39	7.40	0.08	0.42	6.80
Fakirahata	21.80	5.70	4.00	22.90	3.80	9.90	0.20	0.10	6.20
Baraghoria-Para	19.20	3.31	3.71	28.73	3.53	7.60	0.51	0.68	6.50
Panirchara	12.80	5.96	3.17	29.35	2.70	8.04	0.26	1.89	6.60
Hoanak (Nalbila)	07.30	1.41	2.15	19.66	0.97	1.55	0.05	1.78	n.d.
Other Island:									
Matarbari Island	22.40	5.22	1.94	32.61	2.46	Trace	0.13	3.78	n.d.
Nijhum Dwip	25.00	2.13	0.44	13.47	0.08	2.69	0.02	4.55	n.d.
Kutubdia Island	29.00	3.25	1.59	19.83	2.03	2.16	0.08	2.82	5.25
Mean	23.00	4.20	2.04	26.03	2.30	3.92	0.31	1.87	6.45

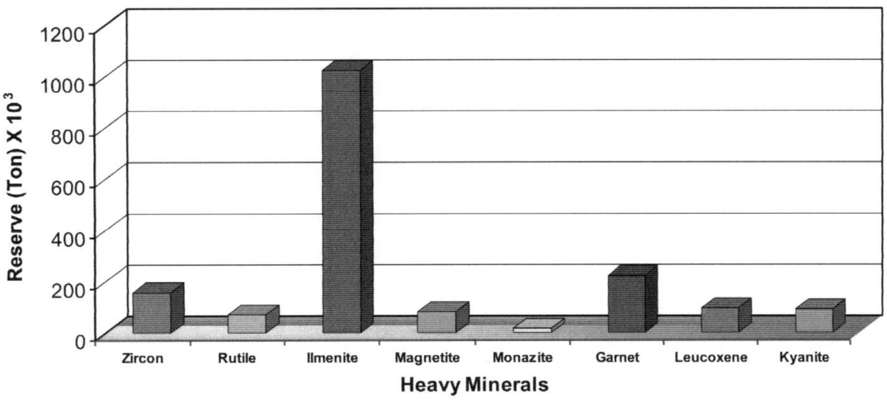

Figure 4. Total heavy mineral reserves.

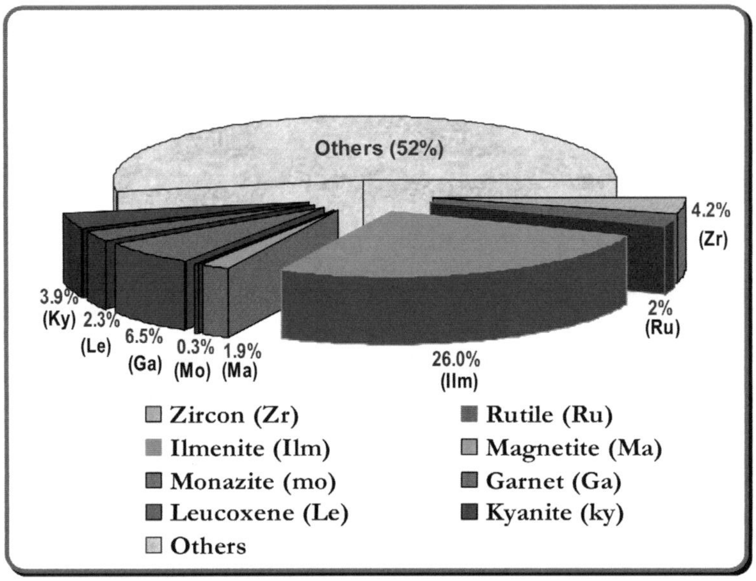

Figure 5. Individual economic heavy mineral percentages in total heavy fractions (based on BAEC estimation).

Bulk samples from the HM deposits were processed in the BAEC pilot plant at Cox's Bazar to determine the separation characteristics of individual minerals using gravity, magnetic, high-tension electrostatic, and induced-roll magnetic separators. Flow sheets of the pilot plant studies are shown in Figures 6 and 7.

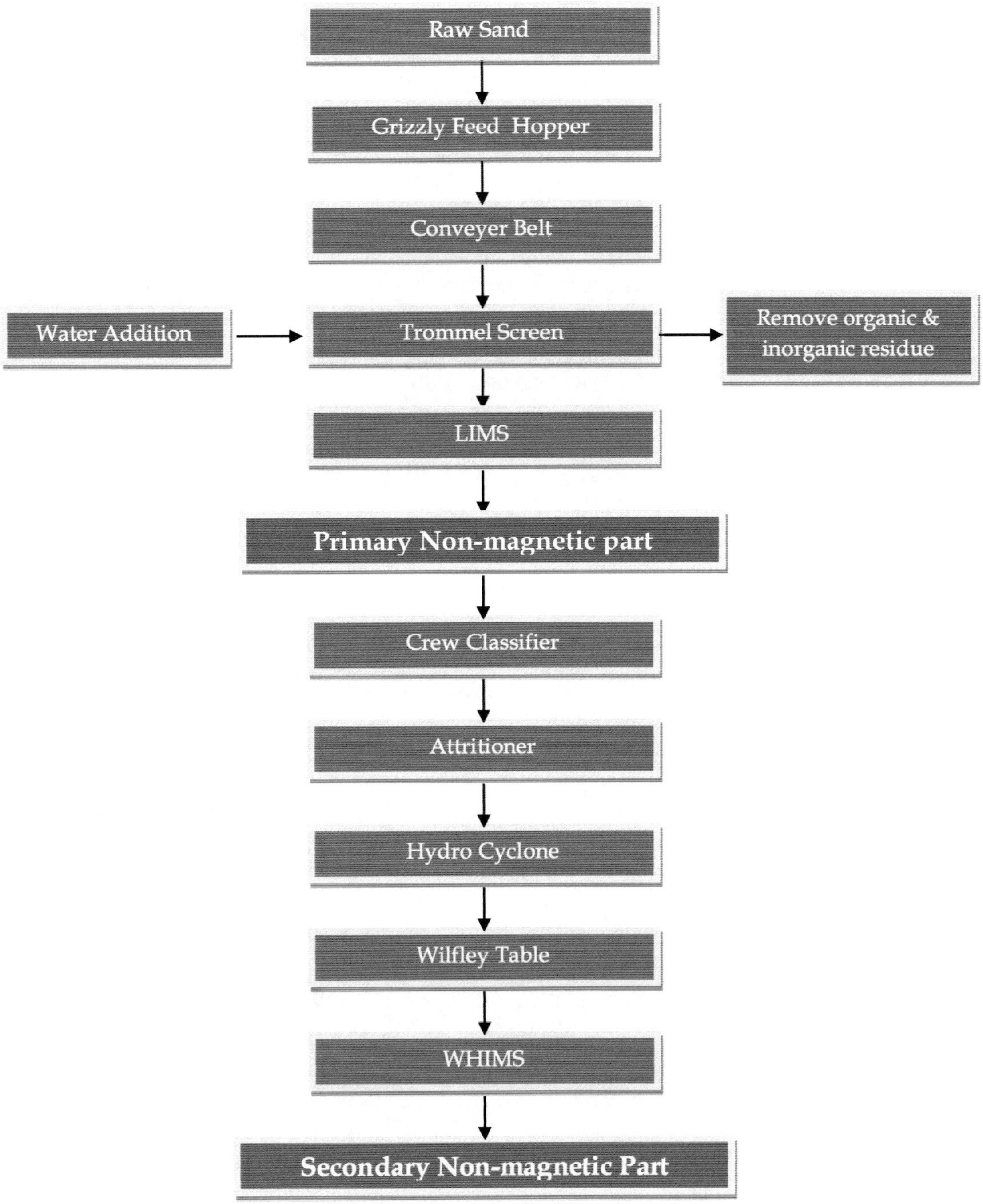

Figure 6. Flow sheet of pilot plant wet section.

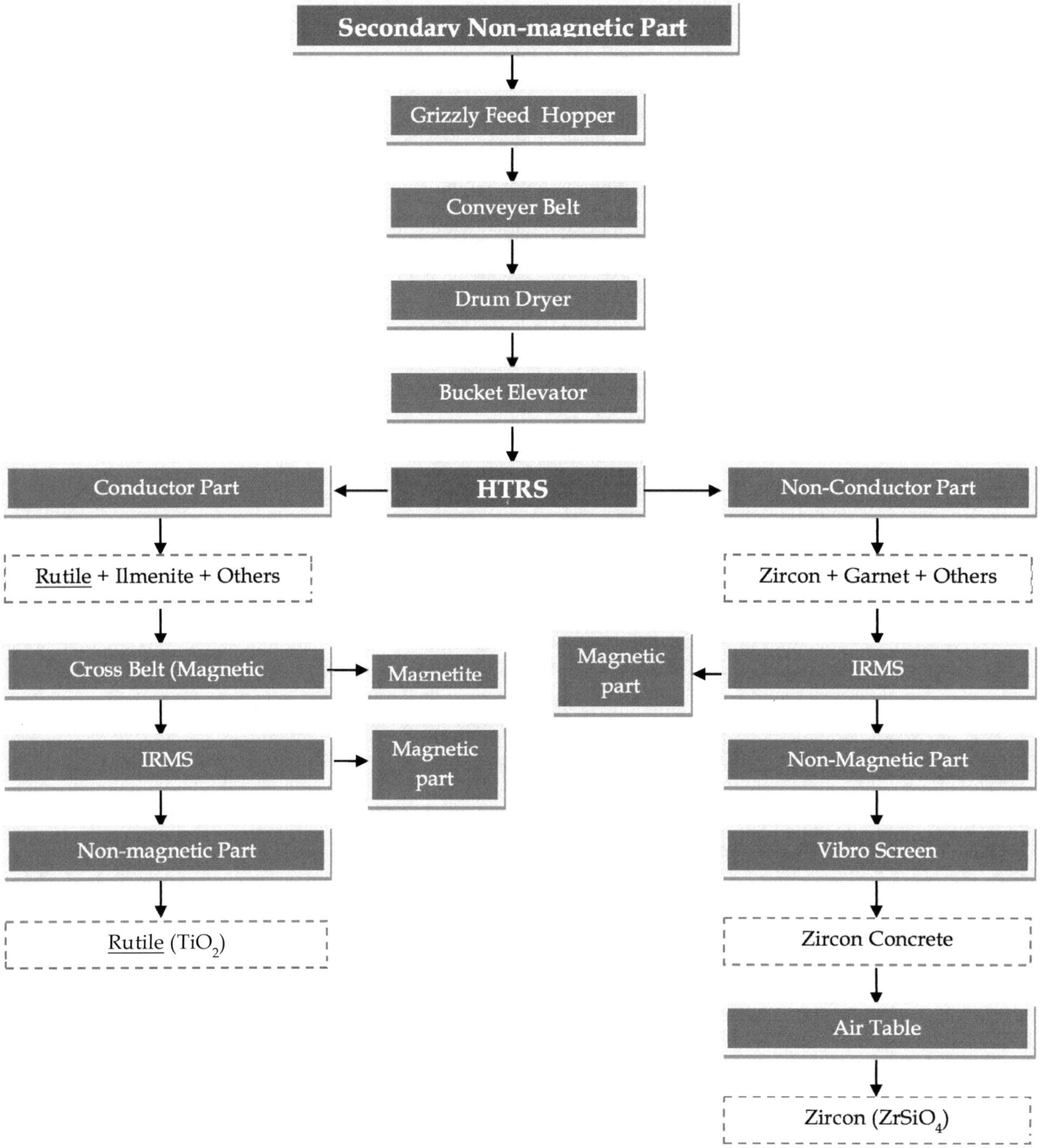

Figure 7. Flow sheet of pilot plant dry section.

Ilmenite and magnetite were separated successfully in the pilot plant with 95% physical purity. Zircon, rutile, and garnet were separated with about 90% purity. Analyses of some of the heavy mineral products separated at the pilot plant are shown in Table III. ZrO_2 in zircon is around 65%, which is quite satisfactory for a commercial product. An important value addition activity for the zircon concentrates is dry grinding of the concentrates to zirflour for application as an opacifier in the ceramic industries. In Bangladesh there is a high demand for zirflour.

The uranium content of minerals such as zircon and monazite is a key parameter in the economics of mineral sand production, as high uranium contents may significantly downgrade the value of the product (Duddy and Kelly, 1999). The presence of uranium orthorium is not considered a problem until the combined content is greater than 500 ppm. More than 500 ppm (U+Th) is observed in some separated fractions of Badarmokam mineral sand deposits (PML, 2012). During production, the uranium content of various products has to be monitored to adjust processes to separate desirable low-uranium mineral fractions and to maximize the quality of the refined product of commercial grade.

Although the desired physical grade was achieved for ilmenite, its chemical characteristics are not favourable. The ilmenite contains a low level of titanium dioxide (around 43%) and higher levels of chromium, manganese, and iron, and thus fails to meet industrial requirements. As ilmenite is the most dominant constituent in the beach placer deposits, extensive research and development work is planned to upgrade the ilmenite through different beneficiation processes such as oxidation-reduction, leaching, and smelting, which add substantial value by upgrading the low-value titanium mineral into a high-value product, synthetic rutile - a much sought-after value-added feedstock for the production of pigment. TiO_2 in rutile is around 90%, and in commercial grade it needs to be more than 90%. Undesirable silicon content is higher. The upgrading of ilmenite and rutile will be required to make these minerals commercially viable. So far, more than 2000 t of economic HMs produced in the pilot plant have already been utilized by different industries in Bangladesh.

Table III. Chemical analyses of a few separated heavy minerals.

Constituent	Zircon (90%)	Constituent	Rutile (90%)	Constituent	Ilmenite (97%)	Constituent	Garnet (90%)	
ZrO_2	65-66%	TiO_2	90%	TiO_2	42.37%	Fe_2O_3	27.80%	
SiO_2	30.6-32.8%	ZrO_2	4-5%	Fe_2O_3	25.62%	Al_2O_3	24.80%	
Fe_2O_3	0.18-1.1%	Fe_2O_3	5.9%	FeO	28.75%	TiO_2	2.48%	
TiO_2	0.13-0.9%	SiO_2	1.45%	SiO_2	1.20%	Cr_2O_3	1.45%	
S	0.03%	Al_2O_3	0.40%	P_2O_5	0.03%	CaO	0.25%	
P	0.03%	Nb_2O_3	0.49%	MnO	1.30%	MnO	2.45%	
		S	0.02%	Cr_2O_3	0.57%	SiO_2	35.80%	

OTHER EXPLORATION PROGRAMMES

Since 2007, Premier Minerals Limited (PML), a Singapore-based international mineral development company, has been engaged in mineral sand exploration activities in Bangladesh. Extensive exploration activities have been carried out on its 7986-hectare license in the Badormokam-Shilkhali area of Teknaf in the Cox's Bazar district. More than 800 percussion drill-holes were drilled to a depth of 3–24 m, with an average borehole depth 10 m. A total of 739 lithologic logs were prepared reflecting the subsurface conditions of the areas and a total of 9980 samples were collected and subjected to various tests and analyses. A requisite number of representative samples were analysed and tested in the certified laboratories in Bangladesh and in the advanced laboratories of CRL, Auspac, AML, and Titanatek in Australia. The mineral sands of the explored area are very fine to fine-grained. The average particle size distribution of mineral sands of Badormokam-Shilkhali is shown in Table IV (PML, 2012). The HM content varies from 15- 31%. The fractional distribution of the composite samples from Badormokam and Shilkhali deposits, along with the XRF analyses of individual fraction, is shown in Tables V and VI (PML 2012). Fractions of valuable minerals can be identified through interpretation of the chemical data.

Based on exploration activities PML has ascertained significant resource of valuable heavy minerals - 185 kt of zircon, -52 kt of rutile, and 150 kt of garnet. The heavy fractions contain 4.95% zircon, 4% rutile, 16% ilmenite, 7.15% leucoxene, and 6.45% magnetite. Having carried out a feasibility study, PML has submitted an Exploitation Scheme to Bureau of Mineral Development (BMD), Bangladesh to develop mineral sands resources in its license areas. PML has not yet been awarded the Mining Lease. Based on current reserves, PML plans to establish a small commercial plant of 400 t/h with an initial project life of 10 years. However, the life can be increased up to 20 years when adjacent areas become available for exploration and mining. PML has also carried out preliminary surveys in a number of Bay of Bengal islands in the southern part of Bangladesh under the Bhola and Patuakhali Districts. A surface sample collected from Dhal Char contains 43% HM and sample from Vashar Char contains 50% HM on average (PML, 2013). Recently, a titanium-rich HM deposit is reported in the Brahmaputra River basin containing 10–15% THM (Rahman *et al.*, 2014.). Ilmenite from the deposit contains up to 52% TiO_2, which is higher that the ilmenite of the coastal beach.

Table IV. Average particle size distribution of mineral sands of Silkhali-Badarmokam.

Size (μm)	250	212	180	150	106	75	53	-53
Average%	2.85	5.04	9.39	15.85	42.95	15.06	5.31	3.55

Table V. Fractionation of Badarmokam composite sample and XRF analysis of individual fractions.

Fraction	Mass%	TiO_2	ZrO_2	Fe_2O_3	SiO_2	Al_2O_3	Cr_2O_3	MgO	P_2O_5	U+Th (ppm)
HS Mag	8.49	21.80	0.27	74.70	2.09	0.89	0.71	0.39	0.04	19
Mag 1	44.54	17.30	0.21	37.60	22.40	12.50	1.23	2.57	0.20	396
Mag 2	32.66	6.93	1.76	15.00	34.80	20.70	0.57	3.54	0.57	1273
Nonmag +150 Mic.	4.51	25.40	4.24	1.61	30.10	24.30	0.04	0.66	2.56	217
Nonmag - 150 Mic.	9.80	23.30	36.30	0.76	26.30	5.22	0.04	0.16	1.56	1309
Total (calc.)	100.00	15.25	4.44	28.14	25.45	14.01	0.80	2.38	0.55	732

Table VI. Fractionation of Silkhali composite sample and XRF analysis of individual fractions.

Fraction	Mass%	TiO_2	ZrO_2	Fe_2O_3	SiO_2	Al_2O_3	Cr_2O_3	MgO	P_2O_5	U+Th
HS Mag	4.40	22.10	0.37	64.10	7.66	2.59	0.91	0.87	0.09	55
Mag 1	45.21	7.53	0.15	24.80	33.50	17.20	0.57	4.26	0.14	293
Mag 2	36.16	7.65	0.82	12.80	28.30	19.00	0.13	4.00	0.33	342
Nonmag +150 Mic.	4.28	14.70	1.16	1.41	40.50	26.40	0.03	0.83	3.11	93
Nonmag - 150 Mic.	9.95	26.50	10.20	1.09	34.50	10.50	0.03	0.38	3.80	442
Total (calc.)	100.00	10.41	1.45	18.83	30.88	16.94	0.34	3.48	0.70	307

CONCLUSIONS

Exploration by Bangladesh Atomic Energy Commission (BAEC) at the beaches and offshore islands in the southeastern coastal belt of Bangladesh has established 17 high-grade heavy mineral sand (HMS) deposits with 23% THM on average and containing about 2 Mt of economically important heavy minerals (ilmenite, zircon, rutile, leucoxene, kyanite, garnet, magnetite, and monazite). Ilmenite is the dominant heavy mineral of economic importance in these deposits. The quality of the ilmenite is not favourable, but there is scope for upgrading to satisfy the requirements of end-users. The quality of the zircon is satisfactory which industrial requirements. Premier Minerals Limited has also identified a significant heavy mineral resources in Teknaf Peninsula. Heavy mineral sand deposits constitute one of the important mineral resources of Bangladesh and they are extremely promising for commercial exploitation.The heavy minerals found in the beach sands, river sand bars, and islands in the Bay of

Bengal need to be mined sustainably and utilized for the socio-economic benefit of the country. Due attention has yet to be given by the government agencies to the development of these heavy mineral resources, in spite of their substantial economic potential. These resources, if exploited, will certainly bring significant economic benefit to the country and contribute towards employment generation and the establishment of mineral-based industries. A detailed feasibility study should be undertaken as early as possible to decide on the development processes and to determine the real economic potential of these.

ACKNOWLEDGEMENTS

The author has worked at Beach Sand Mineral Exploitation Centre and Nuclear Minerals Unit of Bangladesh Atomic Energy Commission (BAEC), Bangladesh for 32 years. Most of the investigations were done during his BAEC service and he is greatly obliged to the BAEC authority for providing the facilities to carry out the investigations. The author is thankful to his colleagues, particularly the geologists of BAEC, for their assistance during field, laboratory, and pilot plant studies. The heavy mineral investigation was also supported by Premier Minerals Limited. Thanks are also due to PML, and particularly to Craig Figtree for his help with regard to the work.

REFERENCES

Akon, E. and Hossain Shakhawat, A.T.M. (2004). Prospect of mineral sand deposits of Bangladesh. Abstract, *32nd International Geological Congress*, Florence, 20-28 August 2004.

Akon, E. (2015). Status of mineral sand resources of Bangladesh: prospect and challenges for its development. *Recent Advances in Mineral Development and Environmental Issues.* Das, M., Singh, P., Sahoo, H.K., Mohapatra, B.K., and Faruque, B.M. (eds.). Capital Publishing, New Delhi, India. pp. 77-92.

Biswas, M.A.B. (1983). Heavy mineral variability in fore-shore sediments of Teknaf- Inani beach of Chittagong. *Nuclear Science and Applications, Series B*, 14-15, 88-95.

Duddy, J.R. and Kelly, P.R. (1999). Uranium in mineral sands. *Bulletin* no. 26, Australian Institute of Geoscientists.

Howarth, W.J. Allen, D.W. Spry, A.H., and Moeskops, P.J. (1977). Mineral Sands Study, Cox's Bazar, Bangladesh. *Amdel Report* no. 1168, Australian Mineral Development Laboratories, Frewville, South Australia. 81 pp.

Kabir, M.Z., Deeba, F., Rajib, M., Karim, M.M., and Zaman, M.M. (2010). Radioactivity and heavy mineral concentration at Coxs Bazar beach area, Bangladesh. Beach Sand Minerals Exploitation Center, Cox's Bazar, Bangladesh. BSMEC/TR-1/Dec-2010, 16 pp.

Macdonald, E.H. (1976). Final project report, Cox's Bazar pilot plant project, Bay of Bengal Bangladesh, for BAEC and the Australian Development Assistance Agency.

PML (2012).Technical report. Premier Minerals Limited (Singapore). 76 pp.

PML (2013). Bangladesh heavy minerals project. Premier Minerals Limited. 46 pp.

Rahman, M.A., Pownceby, M. I., Haque, N., Bruckard, W.J., and Zaman, M.N. (2014). Characterisation of titanium-rich heavy mineral concentrates from the Brahmaputra River basin, Bangladesh. *Transactions of the Institution of Mining and Metallurgy Section B: Applied Earth Science* 123 (4), 222-233.

Eunuse Akon

Part time teacher, Geology Department
Chief Geologist, Premier Minerals Limited, Gulshan, Dhaka, Bangladesh
(Former Chief Geologist & Director, BAEC, Bangladesh)
University of Dhaka, Bangladesh

Dr. Eunuse Akon is the former Chief Geologist and Director, Bangladesh Atomic Energy Commission (BAEC) who worked in Beach Sand Mineral Exploitation Centre and Nuclear Mineral Unit of BAEC for 32 years in various capacities. He did his B. Sc. Honors and M. Sc. degrees in geology from the Geology Department of Dhaka University, Bangladesh in the years 1973 and 1974 respectively; and obtained Ph. D on mineralogy and geochemistry from Bangalore University, India in the year 1984. He is an experienced geologist with more than 35 years extensive professional experience in the field of heavy minerals, atomic minerals, coal, Precambrian rocks etc. He has published over 30 technical papers in these areas. He was awarded fellowships, scientific visits and advanced trainings in a number of countries namely Japan, Korea, Thailand, Austria, India, Pakistan, Vietnam, China and Australia on mineral exploration, mining and optimization of mineral resources and thus gained knowledge and expertise in the field of mineral development. Presently, he is working as the Chief Geologist of Premier Minerals Limited, a Singapore based mineral development company engaged in the development of heavy mineral sands resources of Bangladesh. He is also a part time teacher of the Department of Geology, University of Dhaka, Bangladesh.

Expanding the geotechnical horizon at the Zulti South heavy mineral deposit, South Africa

P. de Kock[1], F. Hees[1] and A. Louw[2]

[1]Rio Tinto Iron and Titanium, Johannesburg, South Africa
[2]Richards Bay Minerals, Richards Bay, South Africa

Zulti South is an orebody located in Richards Bay and is being developed by Richards Bay Minerals. At the start of the Pre-feasibility Study in 2010, a number of different mining methods were contemplated, of which dredging was the prime contender. Dredging can be a cost-effective method of mining mineral sands deposits. However, in comparison to other mining methods, dredging is capital-intensive and therefore a high production rate is required if it is to attain the lowest life-cycle cost. For the Pre-feasibility Study to correctly specify a dredge that could achieve high production rates, the geotechnical character of the orebody had to be known, and thus a geotechnical exploration programme was conducted. The primary geotechnical determinant of a dredge's performance is the hardness of the orebody as determined by the Standard Penetration Test. The presence, location, extent, and nature of indurations such as calcretes can adversely affect a dredge's performance. By combining information from several different exploration techniques it is possible to cross-validate the data, identify distinct domains of like geotechnical attributes, and construct a spatial model of those domains. A reliable geotechnical model leads to technically and economically sound mining project decisions.

INTRODUCTION

The Zulti South mineral sands orebody located in Richards Bay, South Africa (Figure 1) is being developed by Richards Bay Minerals (RBM). Although other mining methods were considered, dredge mining was, at the onset of the Pre-feasibility Study in 2010, considered to be a favoured option as RBM has, very successfully, been dredge mining the dunes on the Tisand and Zulti North leases (Figure 1) for the past 40 years. Although dredging has proven to be a cost-effective method of mining mineral sands deposits, it is capital-intensive in comparison to other mining methods, and therefore a high production rate is required to attain the lowest life-cycle cost. For the Pre-feasibility Study to correctly specify a dredge that could achieve high production rates, the geotechnical character of the orebody had to be known, and thus a geotechnical exploration programme was conducted.

This paper first elicits the geotechnical factors required for productive dredging and then describes the exploration campaign that was designed and executed to measure those geotechnical factors. How the data was interpreted and modelled is discussed, and finally the use of the geotechnical model and data in specifying a dredge is described.

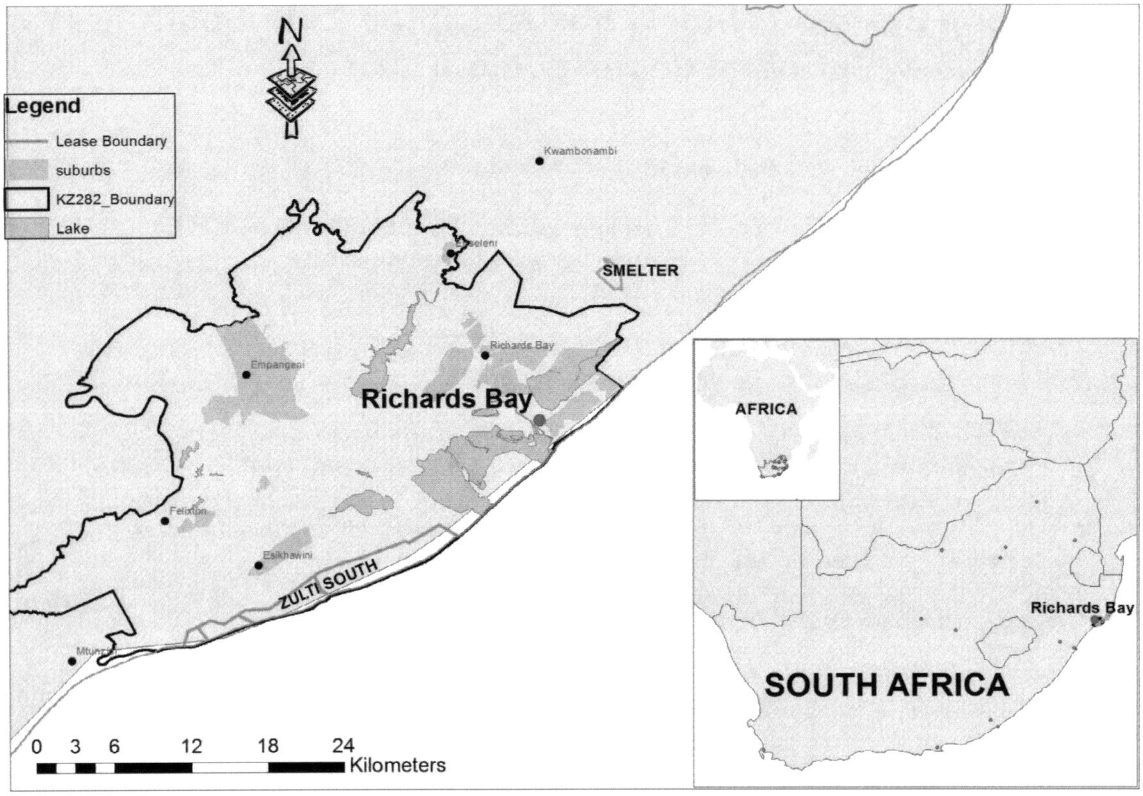

Figure 1. Location of Zulti South orebody.

PRODUCTIVE DREDGING

The dredging process can be broken down into four sub-processes (Vinke, 2012), each of which could become the constraint on production. The four sub-processes are: cutting, suction, discharge, and hauling.

The suction and discharge sub-processes are not the concern of the geotechnical programme as their specifying parameters, such as density, particle size distribution, and dredging depth, are better determined by the resource definition and metallurgical testing programmes.

Cutting is the process of breaking the structure of the ore and mixing it with water. The amount of power supplied to the cutting process is the primary determinant of its production rate. There are other lesser influences on the productivity of the cutting process, such as shape of the cutter head or wheel and the type of cutting teeth, but they are optimized in relation to the installed cutting power (Vinke, 2012).

The cutting power required to achieve a given production rate is dependent on the hardness of the orebody. The primary measure of hardness for soils is the Standard Penetration Test (SPT). Miedema (1999) has computed the maximum production rate per 100 kW of cutting power in relation to SPT. Miedema's production rate curves represent a theoretical upper limit - in reality the production rate per unit of cutting power will be less because of effects like clogging and bulldozing. The major dredge manufacturers and contractors each have their own proprietary, empirically derived curves relating production rate, cutting power, and SPT (de Vries, 2013).

Hauling is the process of seeking ore to cut. Imagine a bucket wheel rotating while stationary in one position. There is a void left where the teeth of the bucket wheel cut the orebody, and to maintain production either the ore must flow to the bucket wheel or the bucket wheel must be moved to fresh ore. That is the hauling process. Hauling is achieved by swinging the dredge from side to side, and by stepping forward. If the propensity of the ore to flow is known, then a minimum swing speed of the dredge can be specified.

Indurations such as biocretes, ferricretes, and calcretes can impede the flow of ore to the dredge cutter, adversely affecting production. Production strategy in response to indurations can vary depending on the nature, extent, and location of the indurations. Figure 2 contemplates four different scenarios of induration location.

In Figure 2a there is no induration, and ore is free to flow onto the dredge cutter. In this case the minimum requirements for dredge cutting power and swing speed will suffice.

In Figure 2b induration is found near the base of the orebody where the cutter must operate. This induration will affect the cutting process and not the hauling process. There are three possible responses: (1) accept a lower production rate; (2) increase the cutter power to deal with the induration; or (3) incur an orebody loss and cut above the induration.

In Figure 2c the induration is located just above the elevation of the dredge cutter. The normal flow of ore above the induration layer is impeded. To compensate, the dredge swing speed or cutter power can be increased. In this situation greater than normal ore losses can be anticipated.

In Figure 2d the induration is located high in the face. The majority of the ore is not impeded, so the direct impact on production should be minimal. However, this may pose a safety hazard to activities such as anchor moves.

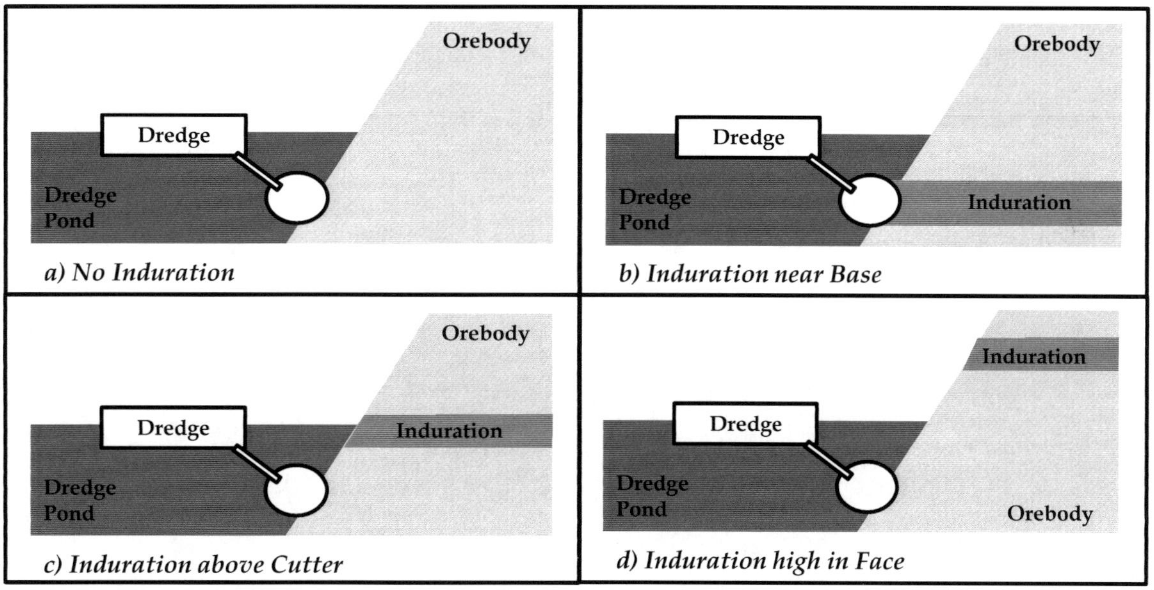

Figure 2. Different scenarios of induration.

In summary, to achieve a given production rate the dredge must have sufficient cutting power and swing speed. The cutting power is determined by the hardness of the orebody as measured by the SPT. Indurations can modify the cutting power and swing speed required depending on their nature, extent, and position in the mining face.

GEOTECHNICAL CAMPAIGN

Design of a dredge that will meet its desired production target requires orebody knowledge in two main aspects, namely: hardness as measured by the SPT; and a spatial model of indurations or hard layers.

In this section of the paper the pre-existing geotechnical information is assessed and then the exploration campaign to address the geotechnical orebody knowledge needs is discussed.

Pre-existing Information

Exploration drilling at Zulti South commenced in 1989 with various phases of reverse circulation (RC) drilling. In 2005 RBM trialled sonic drilling, a new technology at the time, that proved to yield very reliable, representative samples. All of RBM's resource classifications today are based on sonic drill data.

In 2006 a pre-feasibility sonic drilling campaign was carried out at the Zulti South project to provide the data required for a geological, geotechnical, and resource model and to define the parameters likely to influence the selection of a mining method.

The 2006 work was particularly aimed at understanding the coastal stability of the sea-facing dunes and comprised:

- Geotechnical investigation of the materials that make up the general stratigraphy of the resource, *i.e.* the Sibayi, KwaMbonambi, and Kosi Bay Formations
- Geohydrological study of the entire lease area to develop a numerical groundwater model.

Eighteen sonic boreholes during the 2006 resource drilling campaign were dedicated to the geotechnical investigation. Clusters of boreholes were drilled at approximately 1600 m spacing along the length and on the coastal foredunes of the Zulti South resource.

While the sonic resource borehole data included descriptive data of a geotechnical nature, the specific measurement of SPT was limited to 18 boreholes clustered on the sea-facing dune outside of the bulk of the mineable orebody.

2010 Pre-feasibility Geotechnical Campaign

A geotechnical exploration programme was designed to gather sufficient SPT data and to enable mapping of geotechnical domains. Multiple exploration techniques were selected to ensure depth of information and to enable cross-validation. The 2010 pre-feasibility campaign comprised the following:

- 72 rotary mud boreholes
  - Collection of SPT data at depth intervals ranging between 1 m and 2 m
  - Installation of piezometers for long-term groundwater monitoring
- 118 cone penetration test (CPT) holes
- 50 sonic boreholes
  - Collection of SPT data
  - Detailed geological logging
- Three exploration trenches
  - Unconsolidated drained shear
  - *In-situ* sand replacement densities
  - Grading and material analyses
  - Consolidation tests
  - Sonic/SPT/CPT verification
  - Sand flow tests
- Supplemented by 554 sonic resource boreholes primarily intended for resource and geological modelling, but logging included descriptive geotechnical data.

The combined 2006 and 2010 drilling and CPT programmes resulted in an overall grid spacing of approximately 300 m, as shown in Figure 3, which is believed to sufficiently cover all the geomorphological domains across the lease area.

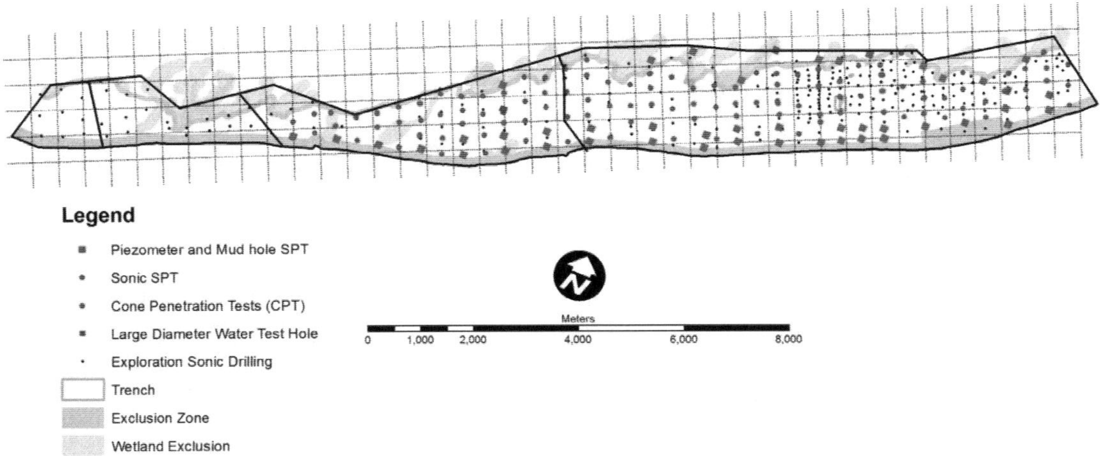

Figure 3. Geotechnical drilling campaign (Louw and Cawthorn-Blazeby, 2013).

The SPT, CPT, and sonic resource borehole geotechnical logging protocol will be further explained.

Standard Penetration Test (SPT)

The SPT employs a sample tube, which is driven into the ground at the bottom of a borehole by blows from a slide hammer, which drives the sample tube 150 mm into the ground. The number of blows needed for the tube to penetrate each 150 mm up to a depth of 450 mm is recorded. The sum of the number of blows required for the second and third 150 mm of penetration is termed the 'standard penetration resistance' or the '*N*-value'. In cases where 50 blows are insufficient to advance the tube through a 150 mm interval, the penetration after 50 blows is recorded. Correction factors are then applied to the field results (N_{field} or uncorrected data), such as rod energy, hammer efficiency, depth (Clayton, 1990, 1995; Rogers, 2006), to produce the final *N* results. The blow count provides an indication of the density of the ground, which has an effect on the productivity of a dredge (ASTM, 2011).

A significant correlation between sonic borehole SPT maximum values was found to occurr at depth intervals of 3 m. The sonic outer casings were installed in 3 m lengths and it is thought that the sonic drilling vibrations during the addition of a length of casing densified the sand, leading to the higher SPT values. Although the sonic SPT values were discarded for this study, the samples provided invaluable data in terms of lithology, hardness, and slimes content, as well as the presence of calcrete or ferricrete indurations. For the determination of the slimes content, the ASTM sieve size 325 was used (<45 μm fraction), including particles sizes in the silt and clay fractions.

Cone Penetration Test (CPT)

In the CPT test method, a conical-shaped penetrometer is used to determine the point resistance during penetration (Robertson and Campanella, 1986) as it advances into subsurface soils at a steady rate. As the instrumented cone is pushed into the ground at a controlled rate of between 1.5–2.5 cm/s by hydraulic force, it measures and records the resistance to the hydraulic force at 20 mm intervals, and converted the data into SPT *N*60 values for correlation with borehole SPT *N* values. The procedures for performing CPTs are standardized in ASTM D-5778 and D-6067.

One advantage of CPT over the SPT is a more continuous profile of soil parameters (Robertson, 1990). The CPT data can be used to interpret subsurface stratigraphy, and through use of site-specific correlations, can provide data on engineering properties of soils intended for use in mining design.

The CPT data was generally considered to be consistent and repeatable; however, the method was unable to penetrate beyond a limited depth in the higher dune area in the north of Zulti South.

Sonic Resource Borehole Geotechnical Logging

The four parameters hardness, clay, calcrete, and ferricrete content were described at the field, through 'sample cores' collected by sonic drilling and/or split spoon. These are descriptive parameters, and can be considered as containing some degree of subjectivity. Nevertheless, the sample slimes content was later determined by laboratory analyses, which could be used to double-check the consistency of the other parameters described during sample logging.

Hardness was measured using a plasterer's knife and was represented by an index between 0 and 3, where 0 is very soft, increasing to a maximum of 3, which is very hard. Clay content was represented by an index between 0 and 4, 0 being recorded as non-sticky (normally the material cannot form a shape, no plasticity), versus index 4, recorded as very sticky (normally forms shapes like sausages or circles, without cracking). This provides an indicator of slimes (silt + clay) in the sand. A distinction exists between 'clay' which is from field observation of the core, and 'slimes' which is a laboratory measurement (particle size <45 μm).

Figure 4 presents the results found for the hardness codes and for the slimes codes respectively. Most of the intervals sampled (63%) were classified in the field as very soft, with less than 5% classified as very hard. The field classification for clay demonstrated that approximately 50% of the intervals sampled were classified as non-sticky sediment, also considered as clean to very clean sand. In general, intervals associated to the Sibayi or KwaMbonambi formations presented hardness and slimes codes changing from 0 to 1, increasing to 1 and 2 in the Kosi Bay Formation.

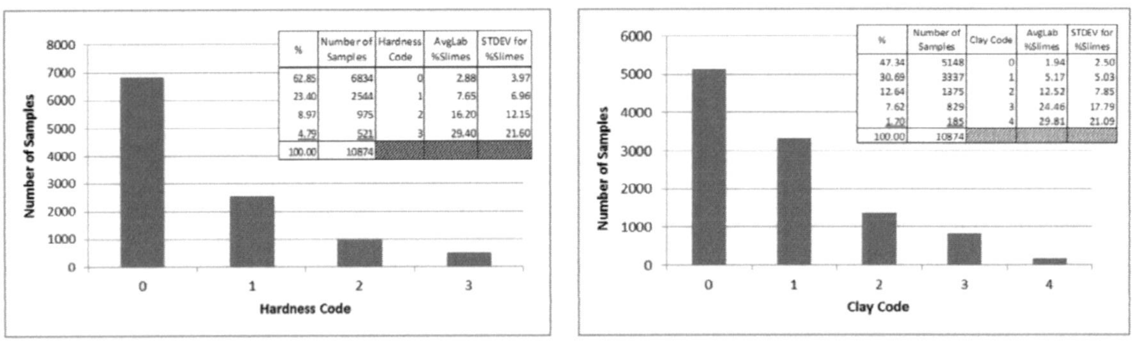

Figure 4. Results for (hardness (left) and clay (right) codes used in the field.

Calcrete was represented by an index between 0 and 3, where 0 is no calcrete and 3, considered as intact cemented core or containing large calcrete cobbles (>2 cm). Ferricrete was represented by an index between 0 and 4, where 0 is considered ferricrete absent and 4 as intact hard core or containing large ferricrete pebbles (>2 cm).

Figure 5 presents the results found for the calcrete and for the ferricrete codes respectively. Approximately 90% of all intervals described were classified in the field as either lacking calcrete or having only some of calcrete (Figure 5a), and the vast majority of samples described had no ferricrete (Figure 5b). It seems that calcrete is present generally in the intervals associated Sibayi Formation.

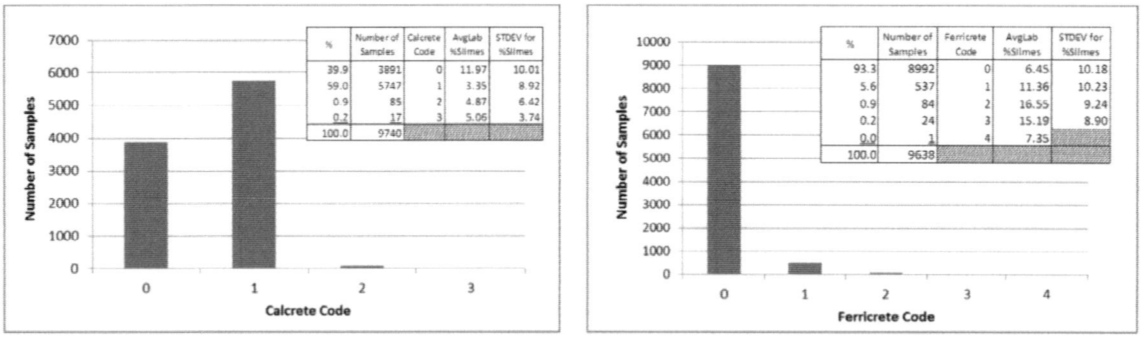

Figure 5. Results for calcrete (left) and ferricrete (right) codes used in the field.

Figure 6 presents the results of laboratory analyses conducted for slimes determination, compared to the ferricrete, calcrete, hardness, and clay codes used in the field to describe the samples. The results indicate that there is a strong correlation between ferricrete or hardness codes and the slimes content of measured intervals. The only exception was observed by the intervals in which calcrete was described (Figure 6c). In this case, although slimes content increases slightly with the calcrete codes used, high slimes intervals (average 12%) are not associated with calcrete formation.

The laboratory results for the slimes determination demonstrated a strong correlation with the hardness codes used in the field. This correlation is directly proportional, *i.e.* the higher is the hardness code the greater the slimes content. Similarly, the distribution of ferricrete codes, although the vast majority of intervals present code 0, is also proportional to the presence of high slimes (Figure 6a). On the contrary, intervals with presence of calcrete (code 1, 2 and 3) are generally associated with clean sand, containing low slimes (<5%).

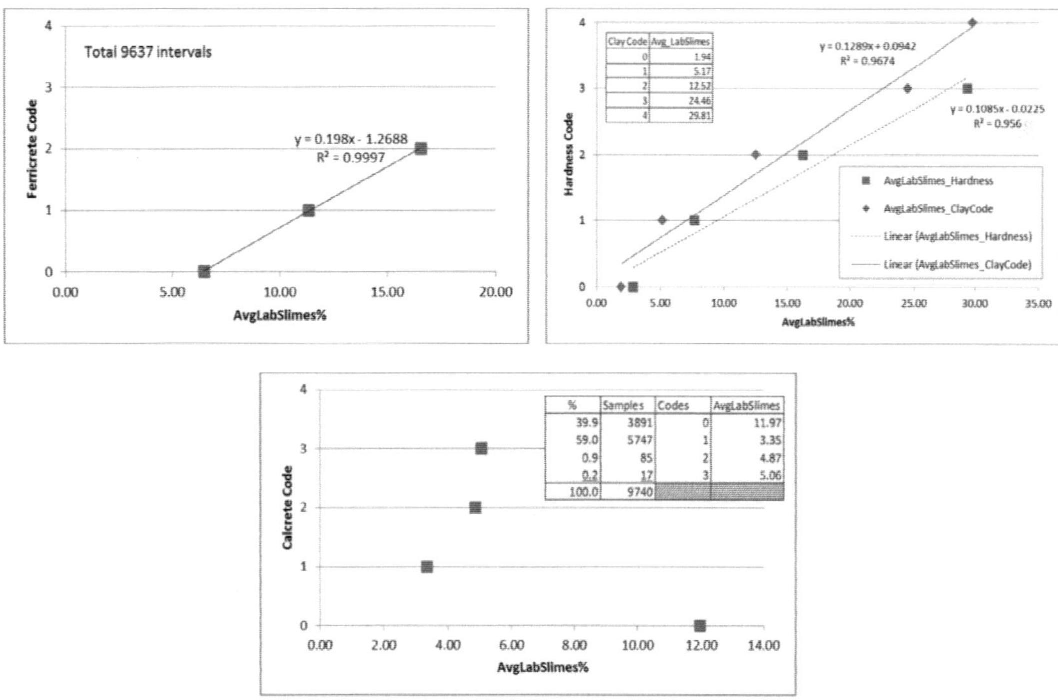

Figure 6. (Top left) laboratory results for slimes content compared to (top right) ferricrete and hardness and clay codes. (Bottom) laboratory results for slimes content compared to calcrete codes.

DATA INTERPRETATION AND MODELLING

The data interpretation was aimed at identifying locations, and later the respective volumes, containing similar geological-geotechnical properties, which would indicate possible hard or soft mining conditions.

In order to provide such information, domains were defined where anomalies in hardness, clay and slimes, calcrete, and ferricrete content (from geological sonic drilling and the respective logs), as well as with high CPT and SPT counts, were identified (geotechnical drilling and respective data). Such anomalies were plotted in several sections, in which the continuity of these domains was interpreted along and across the sections. When such areas were found at roughly the same elevation in nearby holes, they were interpreted as continuous and part of the same domain.

The indication of possible hard layers, initially identified by the geological logging (clay, hardness, calcrete, and ferricrete), later related to the slimes content obtained by laboratory analyses, had another important component to be analysed; namely, the results acquired by CPT and SPT.

Figure 7 provides an example of two geotechnical drill-holes (CPT and SPT) located less than 3 m from each other, at a very similar collar elevation. When plotted, these results were used to identify possible hard layers and to indicate the intervals for which the geological description would be carefully investigated.

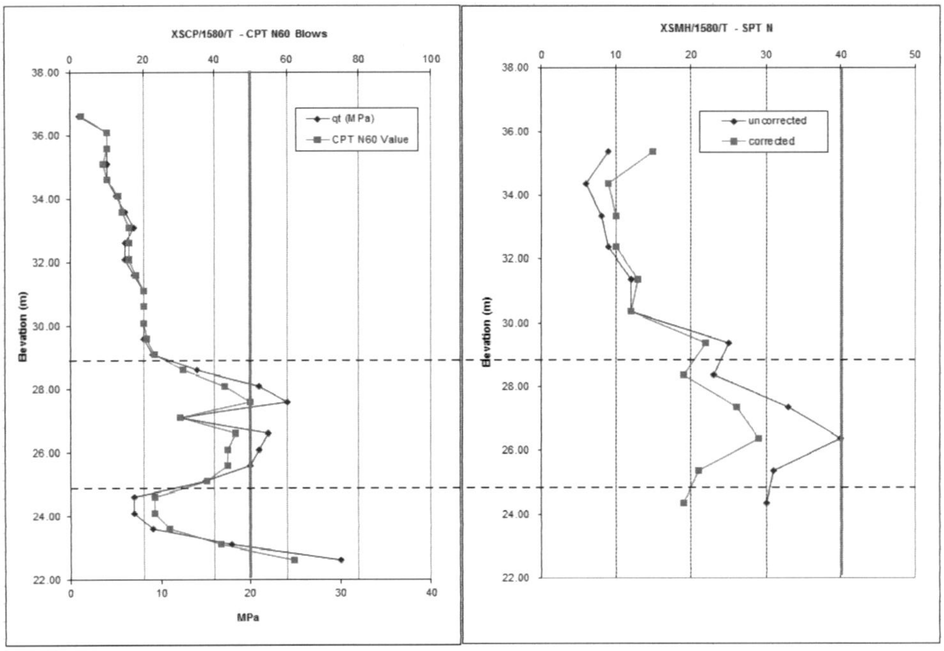

Figure 7. Results from the same location for CPT (left) and SPT (right).

Thus, integrating the geological information gathered from sample logging with the geotechnical results obtained by CPT and SPT drilling allowed the identification of parameters that were used to separate domains with similar results. CPT *N*60 values were flagged at 50 as being significantly hard, and SPT values were flagged at 40 for significant hardness. Interpretation was based on a combination of the above parameters, with specific Surpac string numbers/colours assigned to the interpretation:

> String 8 - Red: used where the geological hardness rating is from 2 to 3, clay codes are 3 to 4, and where CPT *N*60 values greater than 50 correlate with the geological hardness, or where SPT values are above 40 and correlate with the hardness readings.

String 7 - Orange: used where the CPT *N*60 values are greater than 50, or SPT values are greater than 40, but where there is no apparent correlation with any of the other parameters used, such as geological hardness, slimes, or calcrete/ferricrete.

String 6 - Yellow: where there is a geological hardness of 2 and clay codes rating of 2, thus there is likely to be some difficulty with mining, but not to the same extent as for the red strings.

String 5 - Green: friable calcrete. Where there is a hardness rating of less than 2, and a calcrete rating of 2, this indicates the presence of calcrete, which is unlikely to create problems unless below or at the water table.

After modelling the 2D strings in sectional views, 3D solids were then extrapolated between the strings to provide a geotechnical model of Zulti South domains. These domains were coloured according to the same conventions as the 2D strings. Thus red represents areas where high CPT *N*60, and SPT values were acquired, correlating with high geological hardness ratings. Finally, these domains were provided to the mine planning team, indicating the areas and respective volumes where some mining difficulty may be experienced.

DREDGE SPECIFICATION

The majority of the orebody, more than 90% by volume, is unaffected by indurations, as seen in Figure 8. This allowed for the specification of a base case dredge that would achieve the desired production rate in the majority of the orebody.

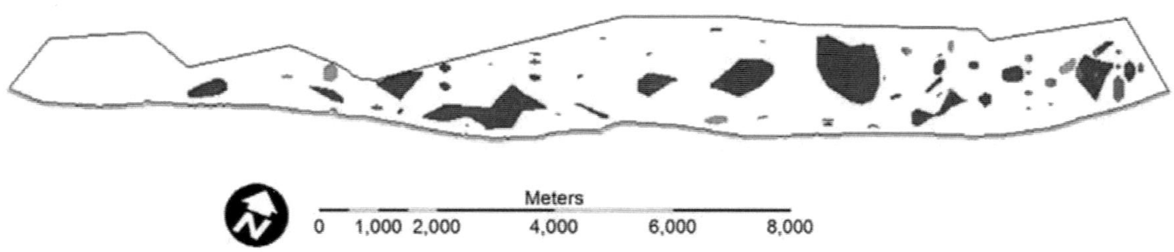

Figure 8. Geotechnical domains in plan view.

For the minority of the orebody affected by induration, a trade-off between three options is apparent, namely: (1) utilize a larger dredge with more cutting power; (2) avoid mining the indurated ore and incur an orebody loss; (3) accept a lower production rate when mining with the base case dredge.

The option to mine the induration-affected ore with the base case dredge and accept a lower production rate is the preferred option. The higher capital and operating cost of a more powerful dredge did not warrant its purchase. The location of the induration in the vertical plane meant that it was not practical for a dredge pond to rise over, and mine the ore above, the induration.

CONCLUSION

In order to design a dredge that can achieve a desired production rate the hardness of the orebody must be known so that sufficient cutting power can be installed. Most dredge manufacturers determine cutting power requirements using the SPT measure. Indurations such as calcretes and ferricretes can have an adverse impact on dredge production. Knowledge of the occurrence of indurations, their nature, and horizontal and vertical extents allows a project to improve a dredge's design.

Multiple exploration techniques and multiple measures are recommended because:

- No single exploration technique can provide the full depth of information required
- Any particular technique can involve technical challenges. For example: CPT has a limited depth of penetration, SPT measured from a sonic rig yields invalid results at regular intervals, and observational logging data can be subjective
- Comparing data from different techniques allows calibration of observation data and validation of the data.

Geotechnical and geological characterisation was used to identify distinct domains and develop a geotechnical model of the distribution of domains for the Zulti South heavy mineral deposit. This characterization was developed through the use of CPT, geotechnical drilling, and SPT testing, and by combining information from the geological logging and characterization acquired from several different exploration techniques.

A clear and reliable geotechnical model allows for the identification of mining options, and provides confidence in the results of trade-off studies.

REFERENCES

ASTM (2010). D6067-10, Standard Practice for Using the Electronic Piezocone Penetrometer Tests for Environmental Site Characterization, ASTM International, West Conshohocken, PA.

ASTM. (2011). D1586-11. Standard Test Method for Standard Penetration Test (SPT) and Split-Barrel Sampling of Soils. ASTM International, West Conshohocken, PA.

ASTM. (2012). D5778-12. Standard Test Method for Electronic Friction Cone and Piezocone Penetration Testing of Soils , ASTM International, West Conshohocken, PA.

Botha, G. and Porat, N. (2007). Soil chronosequence development in dunes on the southeast African coastal plain, Maputaland, South Africa. *Quaternary International*, 162-163, 111-132.

Campenella R.G. and Robertson P.K. (1988). Current status of the piezocone test. *Penetration Testing 1988*, vol. 1. Balkema, Rotterdam. pp. 93–116.

Carter M. and Bentley, S.P. (1991). *Correlations of Soil Properties*. Pentech Press, London. 130 pp.

Clayton, C.R.I.. (1990)., SPT energy transmission: theory, measurement, and significance. *Ground Engineering*, 23 (10), 35–43

Clayton, C.R.I., (1995). The Standard Penetration Test (SPT): methods and use. *Report* 143, Construction Industry Research and Information Association (CIRIA), London. 143 pp.

De Vries, G. (2013). Vuyk Engineering Rotterdam. Personal communication.

Louw, A. and Cawthorn-Blazeby, A. (2013). 3-D domain modelling of geotechnical data. Zulti South pre-feasibility study. Richards Bay Minerals.

McGregor, J.A. and Duncan, J.M. (1998). Performance and use of the standard penetration test in geotechnical engineering practice. Center for Geotechnical Practice and Research, Virginia Tech.

Miedema, S.A. (1999). Considerations on limits of dredging processes. *Proceedings of the 19th Annual Meeting and Technical Conference of the Western Dredging Association*, Louisville Kentucky, 16–18 May, 1999.

Peck, R.B. (1973). *Foundation Engineering*. 2nd edn. Wiley.

Porat, N. and Botha, G. (2008). The luminescence chronology of dune development on the Maputaland coastal plain, southeast Africa. *Quaternary Science Reviews*, 27, 1024–1046.

Robertson, P.K. and Campanella, R.G. (1986). Guidelines for Use and Interpretation of the Electric Cone Penetration Test. 3rd edn. Gaithersburg, MD: Hogentogler & Co. 196 pp.

Robertson, P.K., 1990, Soil classification using CPT. *Canadian Geotechnical Journal*, 27 (1), 151–158.

Rogers, J.D. (2006). Subsurface exploration using the standard penetration test and the cone penetrometer test. *Environmental and Engineering Geoscience*, 12 (2), pp. 161–179.

Vinke, T. 2012. Introduction to Suction Dredger Production Optimisation. *Training Institute for Dredging*, Kinderdijk The Netherlands.

Paul de Kock

Principle Mining Engineer
Rio Tinto Iron and Titanium

The past twelve years have been in the mineral sands industry involved in a wide spectrum of activities from tactical and strategic mine planning to projects. Project experience covers the entire pipeline from Concept to execution. Prior experience was predominantly in planning at large opencast coal mines.

Flavio Justiniani Hees

Project Manager
Rio Tinto Iron and Titanium

Flavio Hees has 25 years' experience in minerals exploration with focus on heavy mineral sands and rock ilmenite exploration, geological and structural mapping, geological and geotechnical modelling, field and laboratory data management and resources estimation. In the last 15 years, his work has included project management mainly to manage teams to deliver multidisciplinary studies. In his current position, Mr Hees provides support to different projects and mine operations, going from stakeholder engagement and government liaison to technical studies and management.

Adriaan Pieter van Niekerk Louw

Superintendent Geological Drilling
Richards Bay Minerals - Rio Tinto

Adriaan is an exploration geologist with 15 years' experience in the mining industry on a number of commodities including diamonds, platinum, nickel, bauxite and 6 years in heavy mineral sands. He was the geological resource project lead for Zulti South prefeasibility study.

Mine rehabilitation using *Eucalyptus* at the Tronox KZN Sands Hillendale mine

C. Smith[1], B. Cocks[2], J. Beukes[2], and M. Vlok[2]

[1]Paperbark Forestry Consulting, Pietermaritzburg, South Africa
[2]Tronox KZN Sands, Empangeni, South Africa

Mining operations to extract valuable heavy minerals from ancient red sand dunes at the Hillendale mine of Tronox KZN Sands (Pty) Ltd near Richards Bay commenced in 2001 and were completed in 2013. Since Tronox KZN Sands is currently expanding its mining operations to areas under *Eucalyptus* forest plantations, it is important to restore the dune system back to productive land and to take measures to prevent erosion. *Eucalyptus* trees can also play an important role in the restoration of the dune system; stabilizing the soil and restoring soil structure through the dense root system of the trees. Research was therefore conducted to determine whether *Eucalyptus* could be successfully established on the restored dunes and to evaluate the most suitable rehabilitation methods for establishing *Eucalyptus* under conditions of reduced fertility and poor soil physical conditions caused by the hydraulic mining process. A trial was established in December 2012 comparing various land preparation methods and fertility treatments. The trial was designed so as to test a gradient of different physical and fertility treatments in combination with each other. The results show that *Eucalyptus* trees can be successfully established on sites disturbed by mining provided that land preparation is conducted properly and that soil fertility is improved by retaining and re-spreading topsoil and applying fertilizer regularly.

INTRODUCTION

The Hillendale mine is located close to Richards Bay, adjacent to the town of Esikhaweni. The orebody, consisting of ancient red sand dunes, was removed to depths of up to 65 m by hydraulic mining. The ore was beneficiated in a gravity separation circuit to produce a heavy mineral concentrate (HMC), with sand tails being returned to reconstruct the mined dunes and the majority of the fines fraction disposed in a residue storage facility (RSF). Hattingh *et al.* (2007) described the development of the soil reconstitution process for the Hillendale mine. This process involved the mixing of some fines (<45 μm material high in clay) with sand to produce a material approaching the texture of the pre-mining soils, and depositing this material onto the sandy backfilled dunes. Hillendale's targeted end land use is the establishment of commercial sugar cane on the majority of the mined area. The focus of development of rehabilitation methodologies was thus aimed at dryland sugar cane cultivation. The Hillendale mine reached its end of life at the end of 2013 and is currently in the closure phase.

Tronox KZN Sands has recently established the Fairbreeze mine south of Mtunzini as the replacement mine for Hillendale. The majority of the Fairbreeze mine area, as well as the Port Durnford area for which Tronox holds a prospecting right, is currently used for commercial forestry. It was thus recognized that, as was done for sugar cane establishment, trials would be required for optimizing the methods for establishing *Eucalyptus* in rehabilitated soils. As the geology and mining process used at Hillendale and Fairbreeze are similar, it was decided to conduct the trials on a section of Hillendale that has been backfilled and covered with the capping layer of sand and fines. The knowledge gained from the trials would direct the application of rehabilitation methods employed at the Fairbreeze mine on areas earmarked for *Eucalyptus* forestry. The trial sites were established and planted towards the end of 2012 by Tronox in collaboration with Paperbark Forestry Consulting.

TRIAL AREA AND PROPERTIES

The trial was established on a relatively level section of the Hillendale mine known as Paddock 18. This paddock had a capping of reconstituted soil material deposited over reshaped base dune sand. The particle size distribution of the reconstituted material consisted of an average of 22% clay, 8% silt, and 70% sand, corresponding to a sandy loam/sandy clay loam texture with a relatively high bulk density of between 1.60 $Mg.m^{-3}$ ^{and} 1.66 $Mg.m^{-3}$. The water available for plant growth in the soil as defined by the available water capacity (AWC) was moderate; between 118 $mm.m^{-1}$ and 125 $mm.m^{-1}$. Measurements taken in May and June of 2012 indicated that soil water contents throughout the reconstituted soil profile were so high that soil aeration was below critical levels for plant growth. Soil pH, soil salinity, and exchangeable sodium percentage of the reconstituted soil were similar to natural soils developed on the dunes. However, the cation exchange capacity (CEC) was higher and exchangeable Mg levels were much higher in the reconstituted material. The combination of high Mg levels and low Na levels in the presence of the low electrolyte concentration of the soil solution probably resulted in the poor soil physical conditions observed, as shown in Figure 1.

Figure 1. Poor physical structure of the reconstituted soils manifested by 'slaking' and crusting. This is partly due to the chemical properties of the soils and the physical damage to the soils during the mining and recombining process.

TRIAL DESIGN AND PREPARATION

A field trial was designed to test appropriate land preparation and fertility treatments that would allow successful establishment of fast-growing *Eucalyptus* species. The trial site was previously planted to Sun hemp, which was disked-in prior to other treatments being implemented. Land preparation treatments included rotovating, disking, ripping, and ridging in various combinations and soil fertility treatments included topsoil addition, inter-row planting of Sun hemp and *Casuarina* trees, and a multiple fertilization treatment. Figures 2 and 3 show the trial area being prepared and seedlings being planted on ridged sections. The trial was designed so as to test a gradient of these physical and fertility treatments in combination with each other. The trial was designed as a randomized block design with four tillage treatments (whole plots) and four fertility treatments (sub- or split-plots) as depicted in Figure 4. Each treatment was replicated four times, resulting in four blocks oriented along the length of Paddock 18 in a north-south direction. Each plot consisted of 8 × 8 trees with 4 × 4 measured and two guard rows. The trees were planted at a spacing of 2.0 × 2.5 m, resulting in a stand density of 2000 stems per hectare. With the planted trial plots as depicted in Figure 5 (a photograph of the trial area shortly after being planted), the physical differences caused by differing land preparation methods can be seen. The trial was planted to a *Eucalyptus* clonal hybrid (*Eucalyptus grandis x urophylla)* on 28–29 November, 2012. Control plots were established the same time at the nearby Maholoholo plantation with a similar *Eucalyptus grandis x urophylla hybrid.* The Eucalyptus grandis x urophylla clone was selected as it is one of the most widely planted in the Zululand region and is considered a 'broad performer', doing well in a wide range of climatic and soil conditions.

Figure 2. Disking of the trial area in preparation of planting Eucalyptus seedlings.

Figure 3. Planting of Eucalyptus seedlings on a ridged trial plot.

Figure 5. View of the planted trial plots at Hillendale Mine.

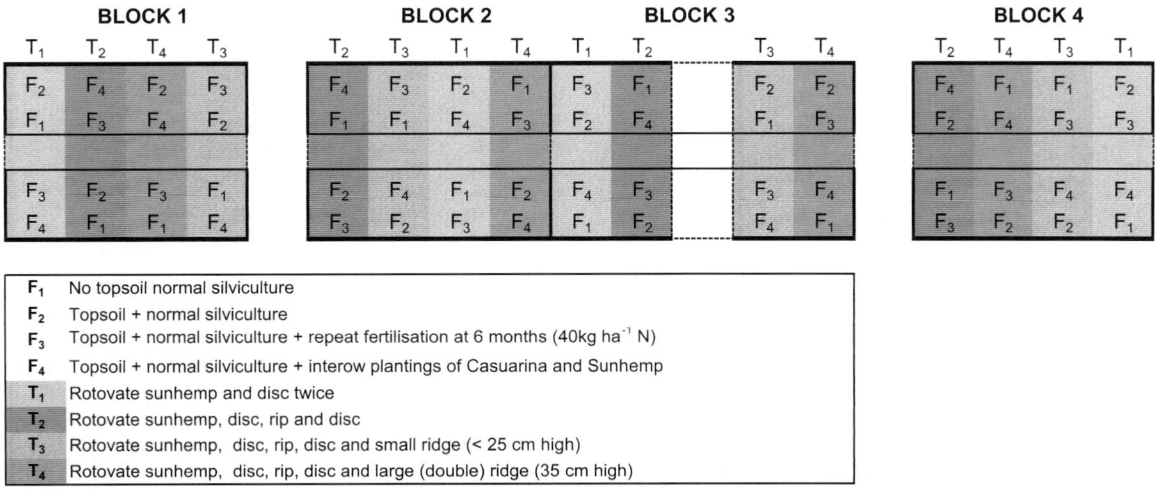

Figure 4. Trial design layout showing the matrix of soil treatments as well as fertility treatments investigated.

VARIATES MEASURED AND DATA ANALYSIS

Tree survival, height, and diameter breast high (dbh - when the trees were large enough) were measured every six months after the trial was planted. The data was analysed as a randomized block design with split plots by an analysis of variance (ANOVA) using standard statistical software. In order to normalize the data, survival data was square-root transformed. At two years the volume of the trees was calculated and used as a variate in the analysis.

RESULTS AND DISCUSSION

Experience gained from establishing the trial highlighted that mulching of forest residues after harvesting is desirable to promote breakdown of the nutrient rich-material. Removal of topsoil prior to mining is best carried out by excavators stripping soil from between existing stumps and piling in mounds prior to transport to stockpiles or areas in the final stages of rehabilitation. Preference is given to placing fresh topsoil and minimizing stockpiling for extended periods. Spreading of topsoil on reshaped dunes needs to be undertaken by low-ground-pressure bulldozers in the dry season if possible. A significant feature affecting the whole rehabilitation process is the high water content in the reconstituted material (from the underlying sand to the sand/fines mix), which affects access, effective land preparation and subsequent crop growth. Figure 6 shows some of the best-performing trees at one year of age.

Figure 6. Tree growth one year after planting.

After the second year of growth it was observed that the effect of tillage diminished while the effect of fertility was highly significant. Growth of the best treatment was further compared to a nearby site with similar clonal type and soils, and after two years stand volumes were one-third of the control plot. Figure 7 shows the considerable differences in tree growth at two years of age for two different treatments, F3T3 and F1T1.

Figure 7. Trees in the foreground with no topsoil treatment and minimum land preparation (F1 T1), compared with trees in the background where soils were ridged and fertilized (F3 T3), at two years of age.

Figure 8 presents the stand volumes for different treatments after three years of growth. There were no significant differences in stand volume between tillage treatments, but it is of interest that ridging (T3 and T4) had the highest stand volume of the tillage treatments. This was due to the early advantage that ridging imparted on the soil due to the extremely wet initial conditions in the paddock.

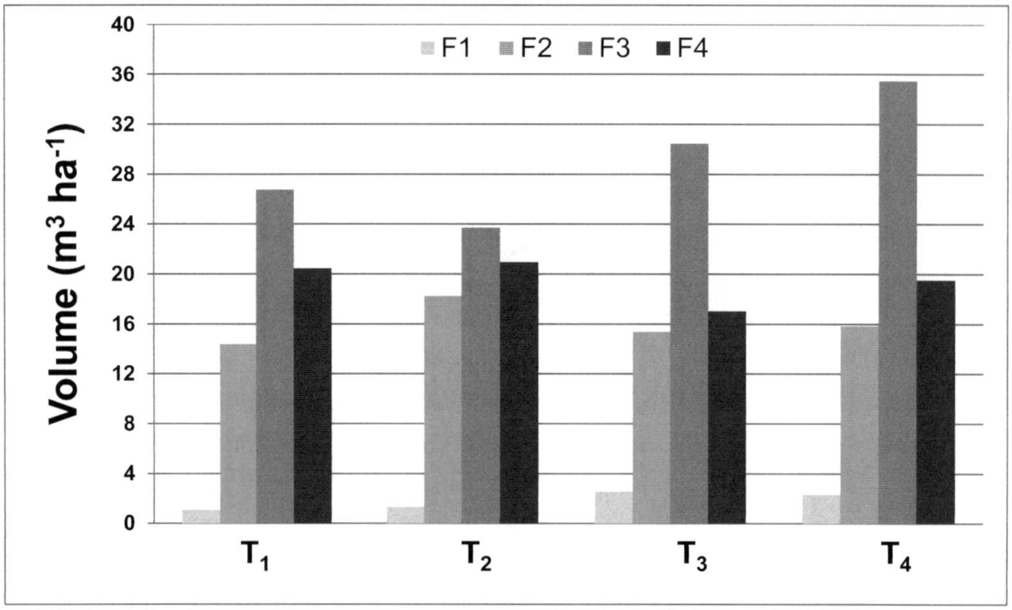

Figure 8. Effect of tillage and fertility treatments on the stand volume of a three-year-old E. grandis x urophylla clonal hybrid (GU 7) at Hillendale Mine.

The effects of fertility treatment on stand volume were highly significant, as shown in Figure 9.

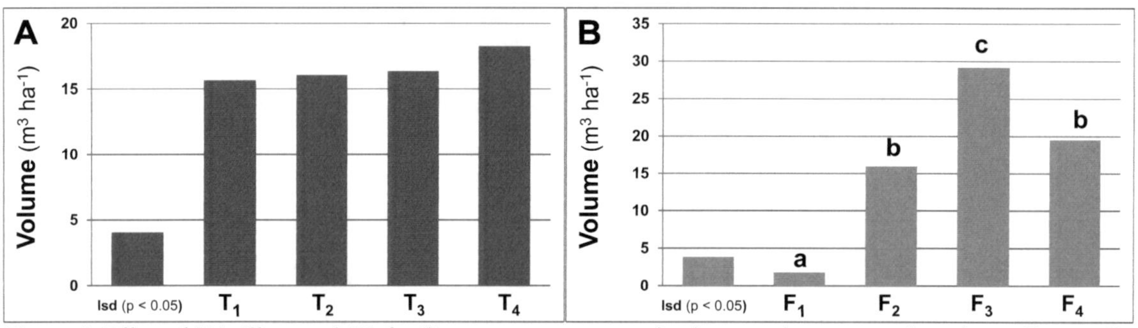

Figure 9. Effect of (A) tillage and (B) fertility treatment on stand volume at three years of age for an E. grandis x urophylla clonal hybrid (GU 7) at Hillendale. Different letters above each bar indicate treatments that are significantly different from each other at the p < 0.05 level based on the least significant difference (lsd). No significant differences were detected between tillage treatments.

Multiple fertilization combined with topsoil retention resulted in by far the highest stand volume (F3), being nearly 50% greater than topsoil alone without fertilization, and topsoil with cover crops and planting with *Casuarina* in the inter-row (F2 and F4; Figure 9). Not retaining topsoil coupled with no fertilization resulted in stand failure (F1; Figure 9). The results clearly demonstrate the importance of improving soil fertility through applications of fertilizer and retaining topsoil.

When stand volumes are considered, the best growth was achieved in treatment plots that were ridged combined with topsoil addition and multiple applications of fertilizer (F3T4, Figure 8). The average stand volume of the best treatment at the Hillendale trial of 35.5 m^3 ha^{-1} in treatment F3T4 was exactly half that obtained with a similar clone grown in a nearby plantation at Maholoholo (70.5 m^3 ha^{-1}) at three years of age. The annual increment over the third year for the trial stand was, however, 18.1 m^3 ha^{-1} while for the control stand it was 19.4 m^3 ha^{-1}. An increased annual increment, compared to the first two years of growth, was thus observed for the trial plot (F3T4). Even though the growth rates are lower than the control plot they represent rates that can be considered commercially viable.

After three years, soil fertility has emerged as the main driver for performance, which is explained by the impact of the hydraulic mining process in causing massive nutrient depletion. However, as the plantations establish and the process of nutrient recycling starts, the need for additional nutrient inputs will decline. Figure 10 shows an aerial view of the trial plots after three years of growth. The blocks representing failures as well as the well-performing blocks can be clearly differentiated.

Figure 10. Aerial view of the tree trial plots after about 3 years, indicating the different growth rates of the treatment plots in the trial.

CONCLUSION

After three years of monitoring of the tree trial at Hillendale Mine it can be concluded that *Eucalyptus* can be successfully established following the appropriate land preparation methods and fertilization regimes. The trial shows conclusively that appropriate tillage is important for good survival and growth early in establishment, but treatments improving soil fertility promoted by far the largest contribution to sustained tree growth. In this respect, retention and re-application of the topsoil coupled with fertilization, particularly nitrogen, are the keys to successful rehabilitation with *Eucalyptus*. Growth rates were behind those on undisturbed soil in an adjacent forest plantation with a similar clone but, in the best treatments, corresponded to lower productivity rates obtained commercially in KwaZulu-Natal plantations.

The trial will continue and allow for further evaluation of the performance of the stands over a typical full rotation. At this stage of the trial, conclusions regarding the main drivers of success are already apparent and information is available to guide establishment on a larger scale as part of the rehabilitation process. A further benefit of the trial is that it constitutes a benchmark that can be used for future trials on different areas of the mining operation. Recently, establishment of *Eucalyptus* on the Hillendale residue storage facility was also investigated, and preliminary measurements after the first year indicated growth rates that are comparable to the best treatments in the trial.

ACKNOWLEDGEMENTS

The authors wish to thank Alexander Bekker and Denis Oscroft for their role in establishing the trial and for ongoing maintenance and measurements of these special trees.

REFERENCE

Hattingh, R.P., Beukes, I., Krause, B., and Kotze, H.C. (2007). Soil reconstitution as a key driver for successful rehabilitation at Hillendale Mine. *Proceedings of the 6th International Heavy Minerals Conference 'Back to Basics'*. Southern African Institute of Mining and Metallurgy, Johannesburg. pp. 31–34.

Jan Adam Beukes

Principle Process Improvement Engineer
Tronox KZN Sands

Jan is a Principle Process Improvement Engineer at Tronox KZN Sands and in this role is responsible for the rehabilitation planning at the Hillendale mine. Jan has also previously been involved with the development of Tronox' s new Fairbreeze mine in the role of Client Manager. As a metallurgical engineer Jan has held several technical positions at Tronox KZN Sands. Apart from these Jan had roles at the then Iscor Research and Development facilities in Pretoria, and also at the BHP Billiton Olympic Dam operation in South Australia.

The effect of net positive suction head (NPSH) on pump and stacker cyclone performance

J.D. Singleton[1] and L. Sarisar[2]

[1]Weir Minerals Africa (Pty) Ltd, Johannesburg, South Africa
[2]Weir Minerals East Africa Ltd, Mwanza, Tanzania

Concentration of mineral sands from the primary ore at Base Resources' Base Titanium Kwale operation in Kenya is carried out in two sections; the wet concentration plant (WCP) and the dry plant (mineral separation plant, or MSP. The sand and clay from the wet concentration process are mixed and disposed of hydraulically onto the mine tailings dam using two streams of three-stage series pumps, Warman® High Pressure 8/6 FF AHP, discharge taking place via Cavex® 650 Stacker hydrocyclones.

Weir Minerals supplied the Kwale mineral sand operation with 55 Warman centrifugal slurry pumps and 46 Cavex classification cyclones in different duties. The pump supply included two pump trains, each consisting of three Warman 8/6 FF AHP high-pressure metal pumps in series to dispose of the tailings. Each tailings pump train feeds a 650CVX Cavex tailings stacker cyclone for tailings dam construction.

The two streams of Warman 8/6 FF AHP pumps receive feed from a common hopper and dispose of it through a 2 km, 300 NB pipeline, with the Cavex 650CVX stacker cyclones at the end of each tailings line. Operational concerns arose during commissioning, due to the high-pressure tailings pump bearings overheating and the stacker cyclones performing inconsistently. This paper documents an on-site study conducted by Weir Minerals to determine the root cause of these problems. The study scope included effects of net positive suction head available (NPSHa) against net positive suction head required (NPSHr) of the pumps, effect of pipeline design and elevation differences on series pumping, effect of pump feed hopper design on pump performance, effect of pump performance on cyclone performance, as well as the effect of stacker cyclone installation on cyclone performance while evaluating the effect of cyclone feed pressure on performance.

INTRODUCTION

The term 'heavy mineral sands' is generally used to refer to placer-type deposits of zircon, ilmenite, rutile, and leucoxene, but also other minerals that may occur in the same or similar deposits such as staurolite, garnet, and chromite (Moore, 2011).

Base Resources' Kwale processing plant is designed to recover three separate products - ilmenite, rutile, and zircon. Ore is received at the wet concentrator plant (WCP) from a conventional dozer trap mining unit (DMU) via a slurry pipeline. The WCP removes slimes at a particle size <45 µm, concentrates the valuable heavy minerals (ilmenite, rutile, and zircon), and rejects most of the non-valuable, lighter gangue minerals. The WCP incorporates a number of gravity separation steps using spiral concentrators. The heavy mineral concentrate (HMC), containing 90% heavy minerals, is then processed in the mineral separation plant (MSP) (Base Resources, 2016).

The final tailings are discharged into a single hopper, from where two streams of Warman® 8/6 FF AHP high-pressure pump trains feed two Cavex® 650CVX stacker cyclones. The stacker cyclones are used for tailings dam construction and are fed via a 2 km 300 NB pipeline.

PROBLEM STATEMENT

Figure 1 depicts a sketch of the tailings hopper, tailings pumps, and tailings stacker cyclones set-up.

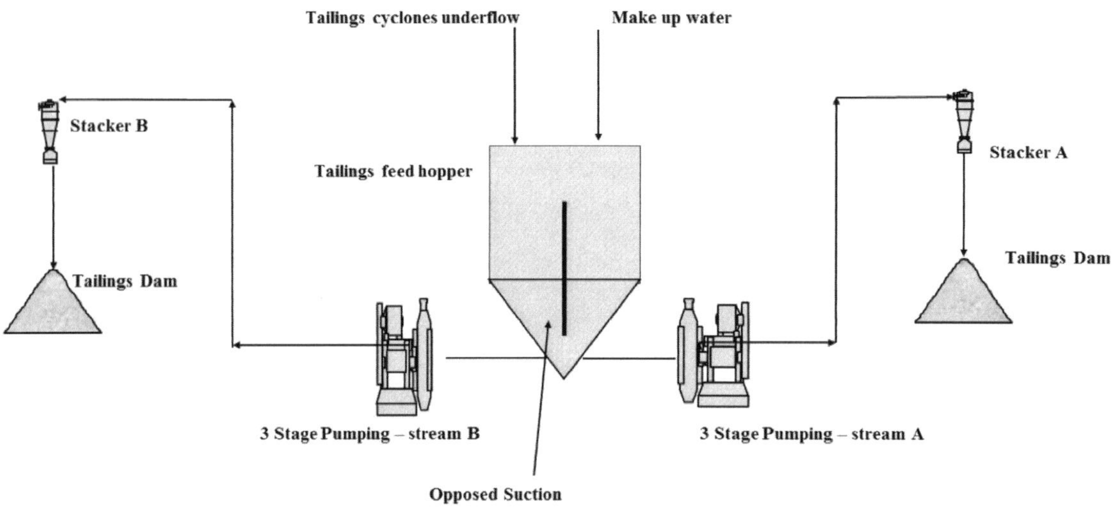

Figure 1. Kwale tails hopper and tailings pump trains.

During commissioning, operational concerns arose due to the high-pressure tailings pump bearings overheating and the stacker cyclones performing inconsistently. An on-site study was conducted by Weir Minerals to determine the root cause of the bearings overheating and the inconsistent performance of the tailings dam stacker cyclones on stream A (eastern stream) as compared with stream B (western stream).

The design specifications for each stream required a total dynamic head (TDH) of 185 m at 620 m^3/h, and a cyclone feed pressure of 101 kPa and pulp feed density of 1.45 t/m^3. This data was used to specify the pumps and to determine the pump speeds. During design, the two pipelines and suction conditions were assumed identical. The actual installation resulted in some system TDH deviation between the two streams. Stream A had an actual higher system head compared to stream B. This resulted in the need for different variable speed drive (VSD) settings for the two streams. The total system head changed from time to time as the stackers moved along the dam wall, either away from (increasing total head) or back towards (decreasing total head) the plant as the tailings dam wall was built.

With the resultant fluctuations in flow rate, the performance of the stacker cyclones was erratic, resulting in inconsistent cyclone underflow densities which in turn resulted in poor tailings dam wall construction. The cyclone performances varied between roping, good operation, and excess water in the underflow.

KEY CONCEPTS

Total Dynamic Head (TDH)

Weir Minerals' slurry pump handbook (2009) defines total dynamic head, H, as the head that is required by a given system to maintain a given flow rate, Q, through the system. H varies with Q; the relationship of H with Q is known as the system resistance and may be expressed algebraically or graphically. The main components of TDH are:

- Total discharge head
- Total suction head.

The relationship is:

TDH = Total discharge head - Total suction head:

$$H = (Hd) - (Hs), \text{ or } H = (Hgd + Hvd) - (Hgs + Hvs) \quad [1]$$

(A list of nomenclature is provided at the end of this paper.) The values Hvd and Hvs are always positive. Hd is usually positive above the pump centreline. Hs may be positive (above pump centreline) or negative (below pump centreline). When Hs is positive:

$$H = (Hd) - (Hs), \text{ i.e. } H = Hd - Hs \quad [2]$$

When Hs is negative:

$$H = (Hd) - (Hs) \text{ i.e.: } H = Hd + Hs \quad [3]$$

Total Discharge Head (*Hd*)

The basic simple formula for total discharge head is given by:

$$Hd = Zd + Hfd + Hve \quad [4]$$

Zd may be positive or negative. If applicable, additional terms must be included in the formula to account for an increased value of Hd due to any contractions (for example, nozzle friction loss) and enlargements; friction loss in a flow-measuring device; and exit into pressure-fed equipment, for example, a hydrocyclone (Weir Minerals, 2009).

Total Suction Head (*Hs*)

The basic simple formula for total suction head is:

$$(Hs) = (Zs) - Hi - Hfs \quad [5]$$

Hs and Zs may each be positive or negative. If applicable, additional or substitute terms must be included in the formula to account for increased or decreased values of Hs due to any contractions, enlargements, and flow measuring device. These are as follows:

- Liquid supply surface being under pressure, Hpr, or under vacuum, Hva
- Differential column head loss, Zc.

One factor limiting the suction performance of a centrifugal pump is the net positive suction head (NPSH) required at the pump intake to avoid cavitation (Weir Minerals, 2009).

Net Positive Suction Head Required (NPSHr)

The NPSHr by a centrifugal pump, at any given point on the head/flow rate (*H/Q*) curve, is the minimum net amount of energy (expressed in metres of head of the actual mixture being pumped above absolute zero pressure) required in the fluid at the entrance to the impeller to avoid cavitation. During cavitation, vapour bubbles form at points where the net positive head falls below the vapour pressure of the liquid. The subsequent collapse of these bubbles as they flow with the liquid into a zone of higher head may cause severe erosion of the impeller. The lowest head in a centrifugal pump occurs behind the leading edge of the vanes in the 'eye' of the impeller. Formation of vapour pockets at these points has the following adverse effects on the pump performance:

- The head developed decreases
- The efficiency drops
- Rumbling or crackling noises and vibration are produced, often resulting in mechanical failure
- The impeller can be subjected to excessive erosion.

The net positive head (NPH) at a point in a pipeline is the absolute pressure head at that point, plus the velocity head, less the vapour pressure. Thus, if a pressure head gauge reading is obtained at a point in a pipeline, the NPH at that point is equal to the gauge head reading, plus atmospheric pressure head, minus the liquid vapour pressure head, plus the velocity head. Gauge readings above atmospheric pressure are taken as positive and those below atmospheric pressure as negative. The NPH at the suction inlet of a pump is called the NPSH and the minimum NPSH required to avoid cavitation is usually shown on pump performance curves as 'NPSH required' (NPSHr) (Weir Minerals, 2009).

NPSH Available (NPSHa)

For a particular pump installation, the NPSH available must be determined from the system conditions and liquid characteristics. The NPSH available (NPSHa) must exceed the NPSH required by the pump at the duty point to prevent cavitation from occurring (Weir Minerals, 2009).

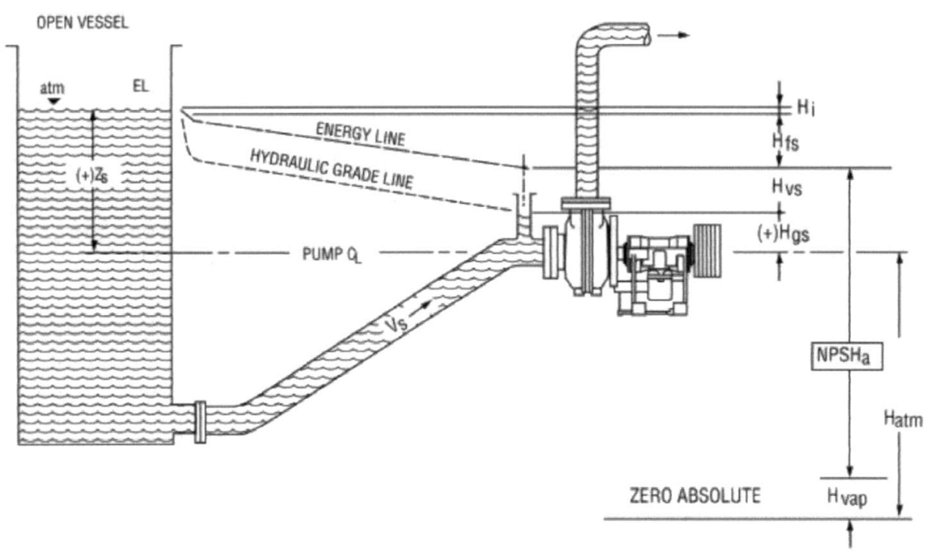

$NPSHa = Hatm - Hvap \pm Hgs + Hvs$	(1)
$NPSHa = Hatm - Hvap \pm Zs - Hi - Hfs$	(2)

Figure 2. NPSHa for positive suction conditions (Weir Minerals, 2009).

Series Pumping

Many pumping duties require slurries to be transported over long distances and/or against very high discharge heads, for example, against total heads well in excess of heads that can be developed by a single centrifugal slurry pump. Typical examples include many requirements for the pumping of concentrates, tailings, power station ash, and underground fill Weir Minerals, 2009

Single Pump

Figure 2 represents a single centrifugal pump operating at duty point 'A' and at pump speed n_1. At the required flow rate Q_1, the pump will develop a head H_1 at an efficiency em_1, and at a power absorbed of P_1.

Two-stage Pump Unit

Figure 3 represents two identical pumps, configured in series so that the entire flow discharged from the first-stage pump is piped under pressure through a short length of piping, directly to the suction flange of the second-stage pump into the discharge pipeline. If both pumps are operated at the same speed, n_1, and as both are handling the same required flow rate, Q_1, each will develop the same head, H_1, at the same efficiency, em_1, and absorb the same power, P_1. The total head developed by the two-stage pump unit combination is:

$$(H_1 + H_2) = 2H \times H_1 \qquad [6]$$

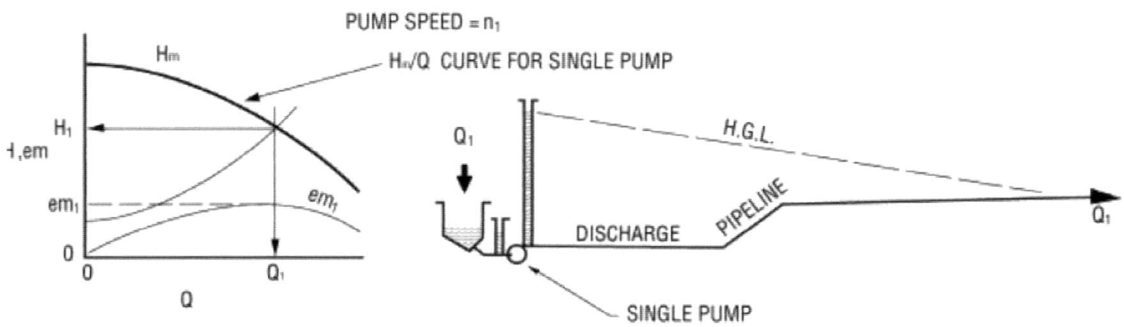

Figure 3. Single pump (Weir Minerals, 2009).

Accordingly, both pumps will be operating under the same conditions, except that the suction head and the discharge head of the second-stage pump will both be higher by the value of the discharge head of the first-stage pump (less small losses in the interstage piping) as presented in Figure 4 (Weir Minerals, 2009).

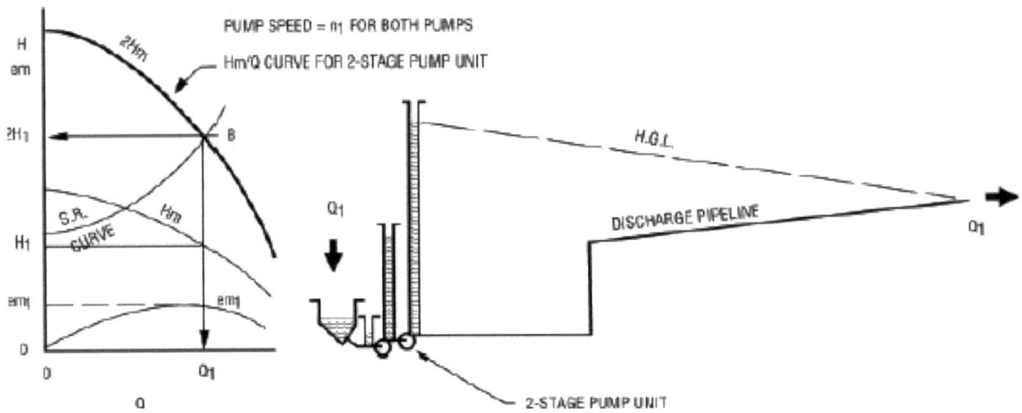

Figure 4. Two-stage pump unit (Weir Minerals, 2009).

If the total head developed by each pump varies from one pump to another due to different speeds or different effects of wear, the total head developed by the multistage unit will be the sum of the individual total heads developed by each of the pumps. Similarly, the total power absorbed will be the sum of the individual powers absorbed by each of the pumps as illustrated in Table I (Weir Minerals, 2009).

Table I. Calculation of power and head for multistage sets (Weir Minerals, 2009)

Stage	Suction head for stage (m)	Total head developed by pump	Power consumed by each pump	Discharge head for next stage
1st stage	X	H_1	P_1	$X + H_1$
2nd stage	$X + H_1$	H_1	P_1	$X + 2H_1$
3rd stage	$X + 2H_1$	H_1	P_1	$X + 3H_1$
4th stage	$X + 3H_1$	H_1	P_1	$X + 4H_1$
4- stage unit: total head developed = $4H_1$ total power absorbed = $4P_1$				

Slurry pumps

Two types of pumps are used in the hydraulic transportation of solids: centrifugal pumps and positive displacement pumps. The centrifugal pump produces a head and a flow rate by increasing the velocity of the liquid through the pump due to the rotating vane impeller. A positive displacement pump operates by alternatively filling a cavity and then displacing a given volume of liquid. A positive displacement pump delivers a constant volume of liquid for each cycle, independent of discharge pressure or head (Engineering Toolbox, 2016).

Slurry pumps can handle solid particle sizes from microscopic to large rocks up to 300 mm (Weir Minerals, 2009). An important aspect of the slurry pump is the material of construction of its casing, which is subjected to the final discharge pressure. There are two main types of casings: those with internal replaceable liners and those with solid metal volute casings. The Warman AH pumps are centrifugal horizontal split case types, designed for a wide range of erosive and/or corrosive applications, used for slurries containing high concentrations of erosive solids or where an extremely heavy-duty pump is required. The Warman AHP model has a thicker casing and heavy-duty bearing assembly, with a short impeller overhang. This unit is designed for high-pressure applications. The liners originally selected for the Kwale tailings pumps were changed from metal to polyurethane, which resulted in longer wear liner life.

Hydrocyclones

The principle and basic design of conventional hydrocyclones was first patented in 1891, but only found significant application in industry after the Second World War (Svarovsky, 1984). Hydrocyclones, commonly referred to as cyclones (Arterburn, 1982), have become the accepted standard method of classifying slurries in the mineral industry since the mid-1950s (Plitt, 1976). Classification hydrocyclones separate solid particles on the basis of size. A classification cyclone uses water as carrier and separating medium, and is different from dense medium (DM) cyclones, which use magnetite or ferrosilicon mixed with water as carrier and separating medium. A stacker cyclone is a classification cyclone fitted with a Linatex® discharge regulating valve to regulate the underflow density for concentrate and tailing stackers. The overflow pipe must break to atmosphere more than 1.8 m below the spigot to form the perfect vacuum for the water to report to the overflow via the vortex finder, and the solids to report to the underflow via the spigot. A Linatex sock is added to the end of the overflow pipe to ensure the pipe is seen by the cyclone as a water column, which creates the required vacuum to ensure optimum underflow densities.

RESULTS AND DISCUSSION

Observations

In addition to the problem experienced when comparing the design flow rates with the actual installation, some system TDH deviation was observed between the two streams. The total system head changed from time to time as the stacker cyclones moved along the dam wall, either away from or towards the plant as the tailings dam wall was constructed. Other physical observations included high levels of pump vibration, considerable feed aeration, and oscillating flowmeter readings. The eastern side flowmeter exhibited greater extremes, varying between 630 to 770 m^3/h. It was also observed that stream A (eastern stream) was exhibiting the above problematic fluctuations more than stream B (western stream). The analysis compared the design data, as-built data, the pipeline data, and all the other system details with the current running conditions. Preliminary findings included:

- Actual flow rates from plant SCADA = 630-660 m^3/h
- Eastern pumps (pumps 62, 94, and 96) were running at between 95 and 97% VSD speed setting, at a discharge pressure of 1911 kPa, and pulp density of 1.30 t/m^3
- Western pumps (pumps 63, 95, and 97) were running at between 64 and 66% VSD speed setting at a discharge pressure of 1099 kPa, and pulp density of 1.32 t/m^3
- Metallurgical specifications were taken as D_{50} = 285 µm
- Maximum particle size of 1250 µm, solids specific gravity (SG) of 2.65
- The tailings pipeline details were provided as: first 1000 m from the pumps was 355 HDPE, PN25; the next 600 m was 355, PN20; the next 900 m was 315, PN16; and the last 700 m was 315, PN10. Corresponding internal diameters (IDs) were 254 mm, 273 mm, 256 mm, and 276 mm respectively
- The pumps are at 107 m above sea level (mAMSL) and the stackers are at 83 mAMSL
- The cyclone underflow represents the tailings used to construct the tailings dam wall. Cyclone feed pressures were estimated to be about 100 kPa
- Considerable aeration of the pumps' feed material occurred in the suction hopper. The conical hopper tank level was relatively low.

The differences in the duty points of stream A and stream B can be attributed to the following factors, which result in preferential suction conditions:

- Sump design (conical section with too small a volume cylindrical section)
- Sump capacity

- Point of water addition in sump
- Pipe lengths to the various stackers
- Pipe diameter differences in various pipelines
- Difference in elevation of stacker cyclones, which would result in static differential heads.

Results

Based on the operational and installation details, the implied actual duty point on the A stream (eastern) was plotted on the pump performance characteristic curve, as shown in Figure 5. All pump curves depict the configuration of three Warman8/6 FF AHP high-pressure pumps in series.

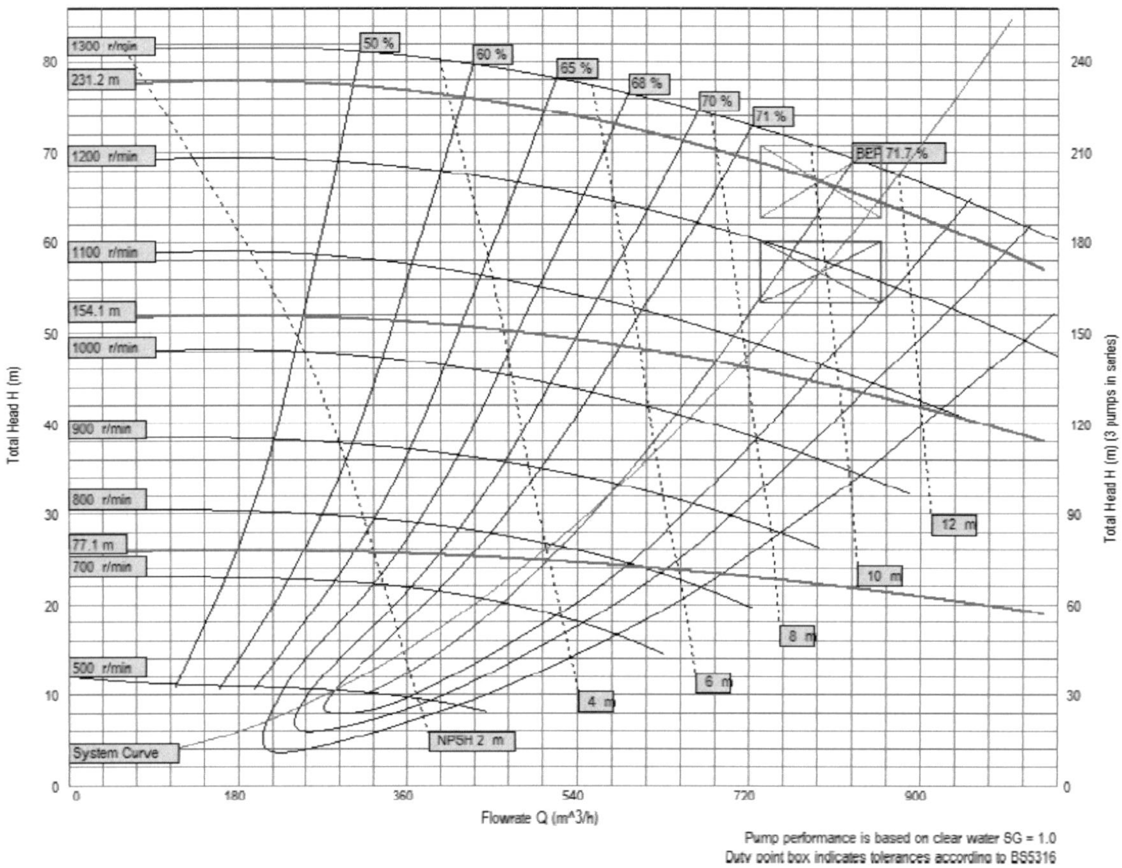

Figure 5. Stream A (eastern) pump train – actual duty point.

The blue duty point on the pump curve is the under-rated (water) duty point. The red duty point represents the derated duty point, which accounts for the D_{50} average particle size, the solids SG, the slurry concentration by volume, the pump impeller diameter, as well as efficiency ratio (efficiency of slurry/efficiency of water) and head ratio (head of slurry/head of water). The blue curve is the pump best efficiency line. The red curve is the system curve of the individual pipeline. For every pipeline design and flow rate, there is a corresponding total system head. The plot of this flow rate against the corresponding total system head gives the system curve (the red curve). At each duty point, the system curve intersects the pump speed line. The intersection of the system curve and the pump speed line is the pump duty point. For each duty point there is a corresponding NPSH required by the pump, thus, the NPSHr curve is identical to the system curve (the red line). The NPSHa is essentially the net summation of the suction side heads, static suction head, plus the vapour pressure of the liquid for the applicable temperature, plus the friction and other minor losses. For a particular installation, if the sump level is not dropping or rising, then the NPSHa curve essentially is a straight horizontal line. When the NPSHr curve (system curve) intersects the NPSHa line then there is insufficient NPSH, which will result in pump cavitation.

The duty point, resulting in a 97% VSD, 1290 r/min actual pump speed, on stream A (eastern), shows that the NPSHr at the operating duty point was 10.13 m. The NPSHa, however, was only 7.97 m. This indicated an inadequate NPSHa for the pump, which leads to instability on the suction side. Figure 6 illustrates the NPSHa against the NPSHr against flow rates for the different streams.

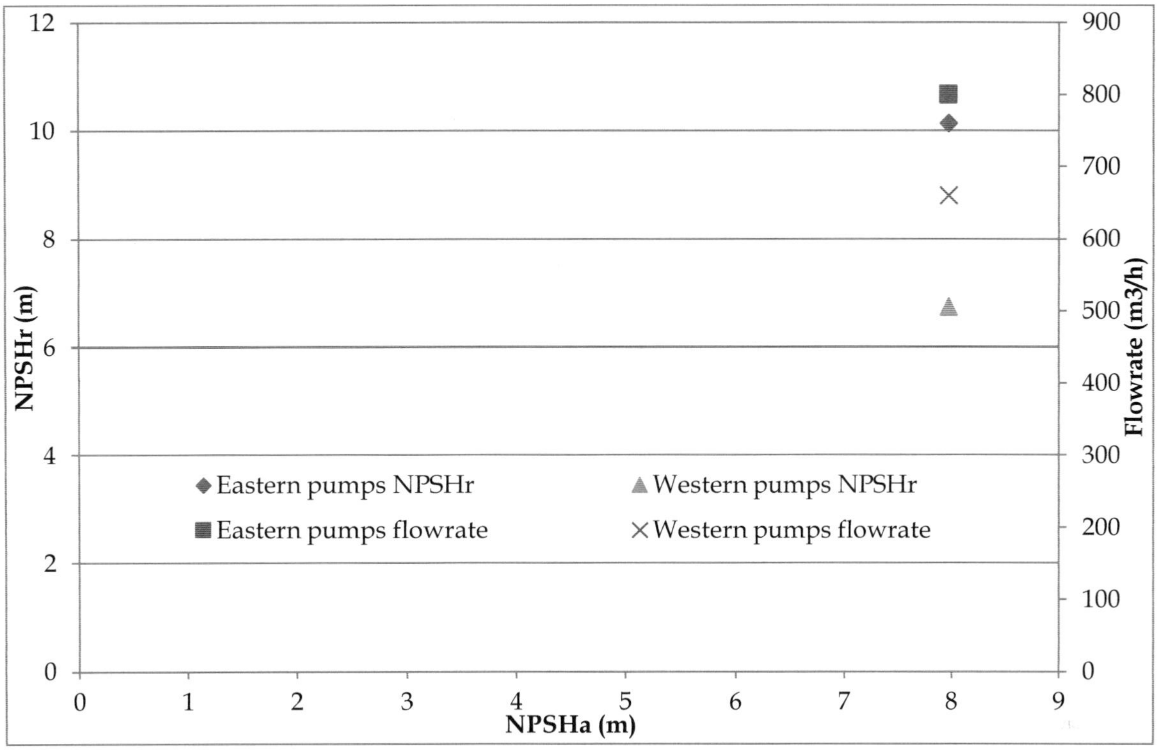

Figure 6. NPSHa and NPSHr of eastern and western pumps.

The actual duty point on stream B (western) was also plotted on the pump performance characteristics curve, as shown in Figure 7.

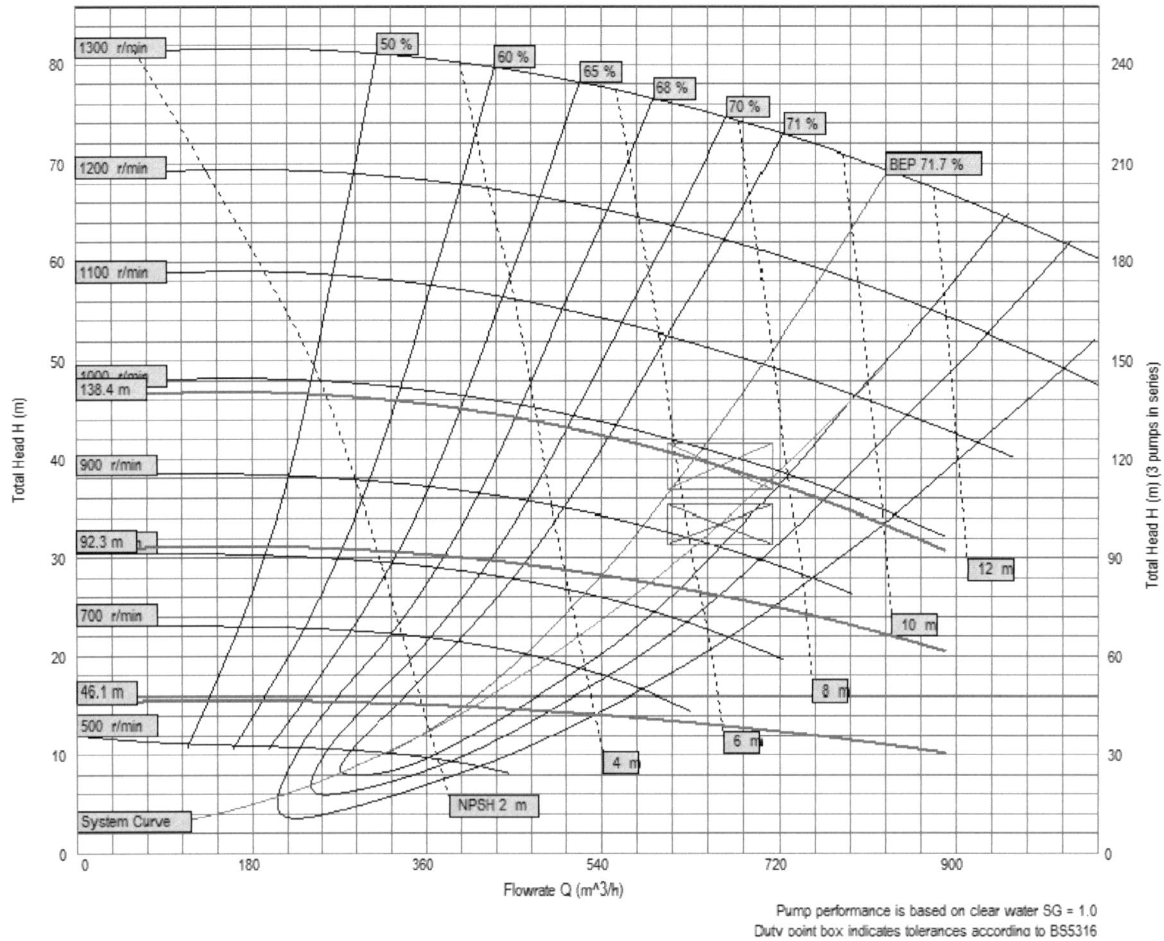

Figure 7. Stream B (western) pump train – actual duty point

From the second duty point, the NPSHr by the pumps was 6.74 m. The implication is that there is adequate NPSH available for the pumps. As a consequence of the above observations, a further study was conducted to establish the relationship of the pump speed to actual pump performance (duty point, and hence NPSHr) with the pump bearing performance and temperature.

Temperature Measurement

The temperatures were simultaneously monitored on both the eastern and western pumps. Temperature readings were taken at the drive end (DE) and the non-drive end (NDE) of the bearings. The western pumps were running at 67% VSD (operating at about 900 r/min) and the eastern pumps at 97% VSD (approximately 1290 r/min). The readings were taken and trended as shown in Figures 8 to 11.

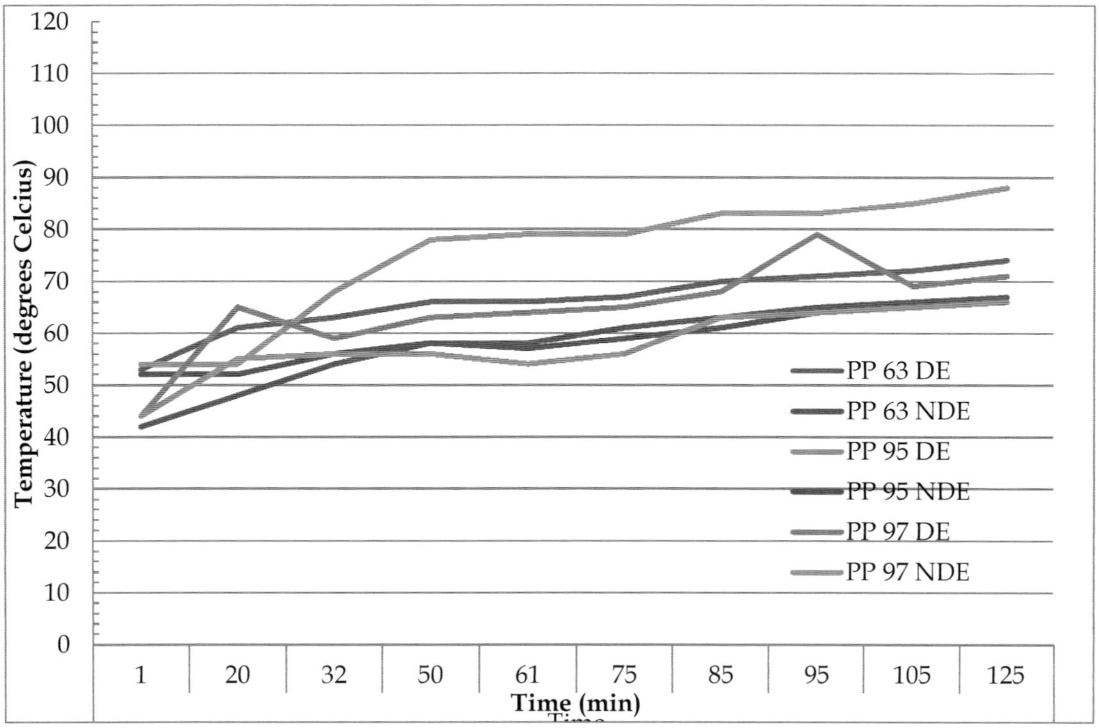

Figure 8. Temperature (°C) trends for western pumps (B).

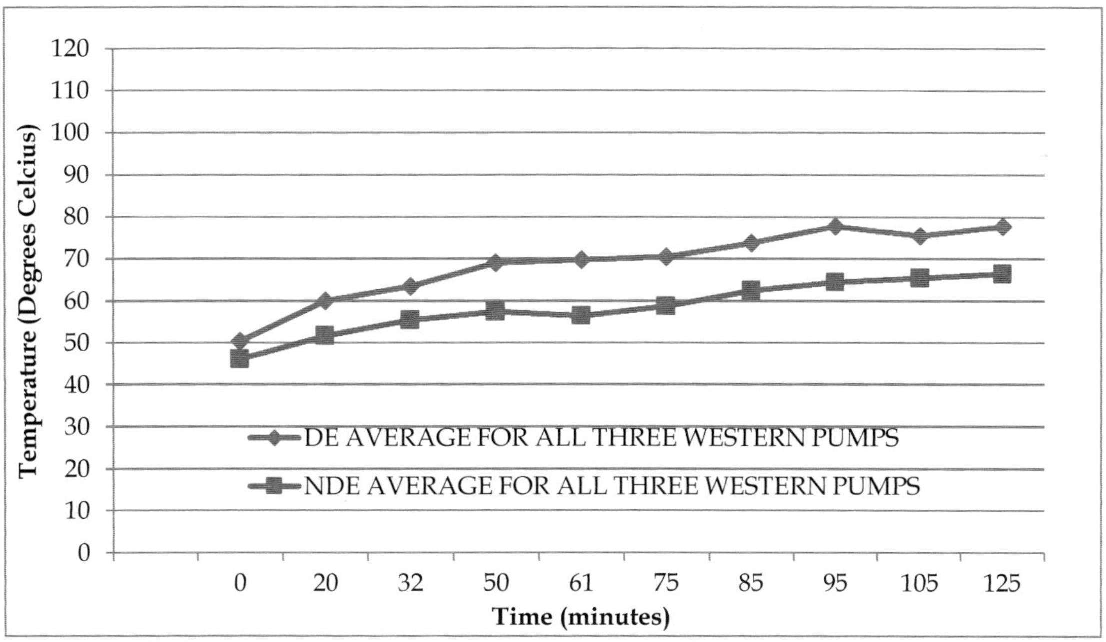

Figure 9. Average temperature (°C) trends for western pumps (B).

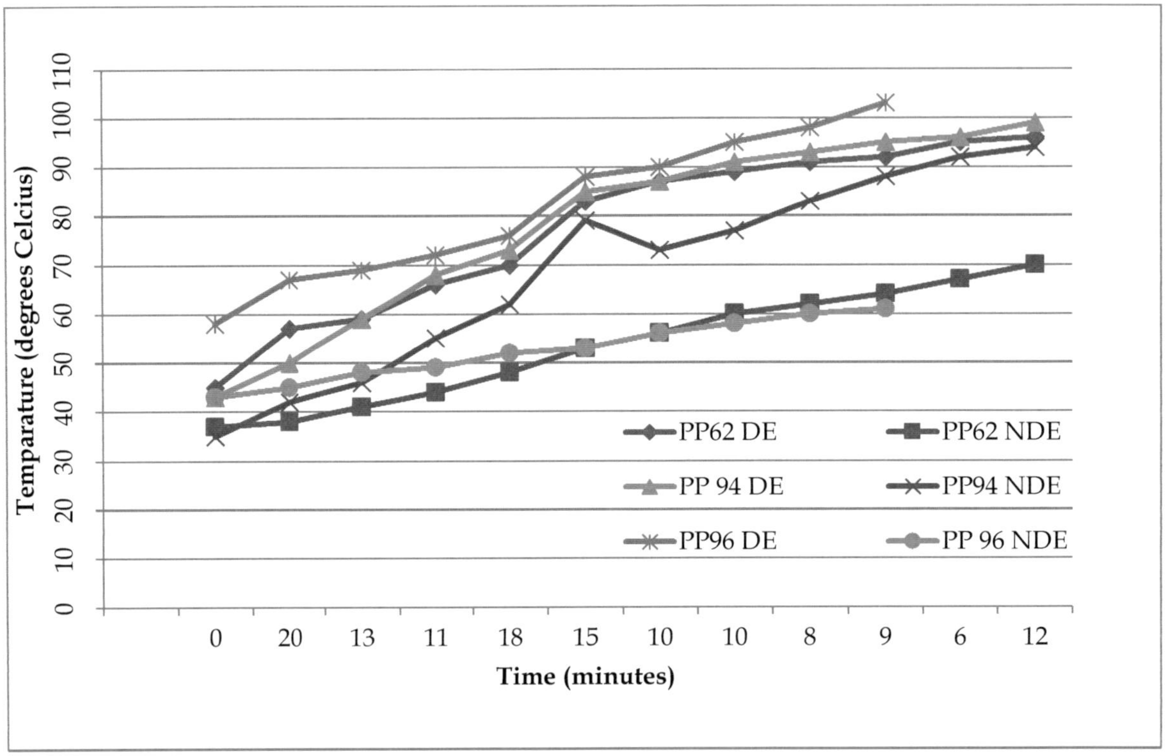

Figure 10. Temperature (°C)) trends for eastern pumps (A).

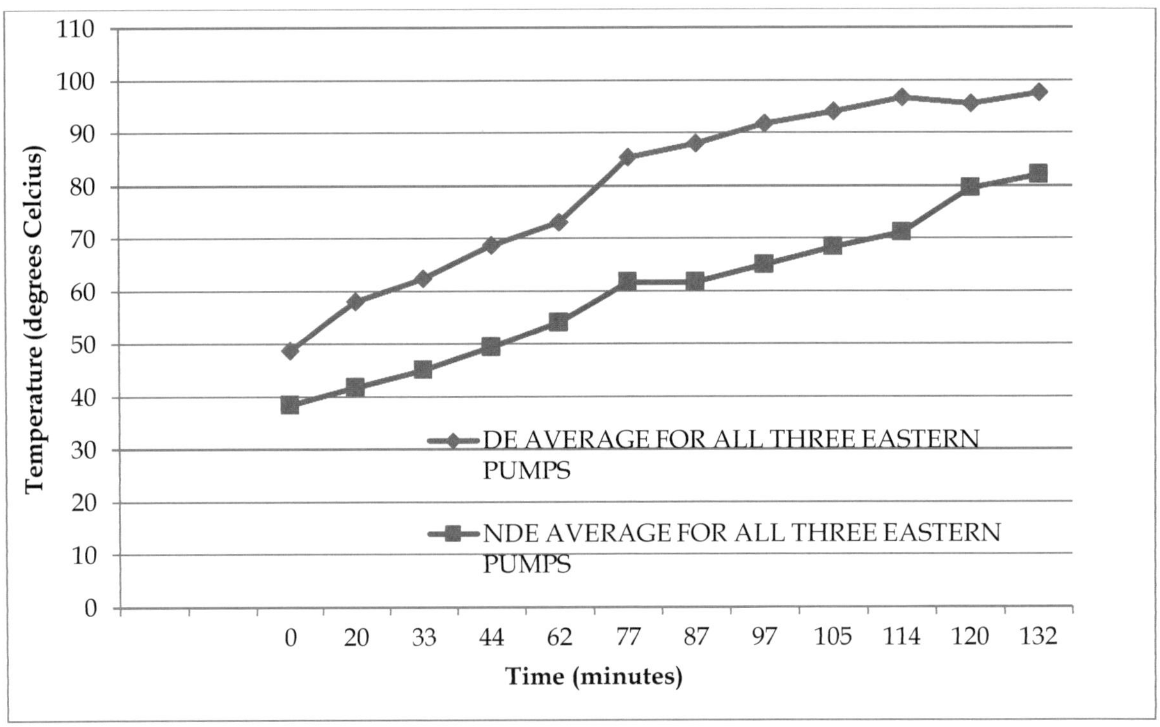

Figure 11. Average temperature (°C) for eastern pumps (A).

The higher pumps speeds of 1290 r/min on the eastern pumps resulted in the bearing temperatures continuing to increase in excess of 100°C within the experimental timeframe of 2 hours, whereas for the western pumps running at 900 r/min, the bearing temperatures levelled out at about 75 to 80°C. The pumping situation of the eastern pumps results in the pump running into cavitation mode, aggravated by feed aeration and slumping surging. This situation results in high axial and radial thrusts on the bearings, as well as high vibration levels, causing the heating of these bearings.

The cyclone operational problems could be addressed only after feeding the cyclones at consistent flow rates. It was observed that the stacker cyclones were not operating under vacuum due to the incorrect overflow pipe arrangements. Inconsistent feed to cyclone stackers resulted in variable operating pressures of cyclones, which directly impacted solid split and underflow densities. When the feed pressures are too low (less than 50 kPa), there is no split to the overflow. When the feed pressure to the cyclone stacker cyclone is too high (greater than 100kPa) the vacuum is broken, which results in a diluted underflow.

STUDY FINDINGS

The pump speeds were largely a consequence of the TDH, which varied with the operational configuration (position of the stackers) from time to time. It was therefore recommended that a fourth stage of pumping be considered for the extreme TDH situations in order to ensure that each pump ultimately runs at a lower speed of 900 r/min at the operational configuration. This will result in lower NPSHr, lower wear rates and consistent feed to the stacker cyclones. When the system TDH reduces due to new operational requirements, the series stages can be slowed down or idled as necessary.

The suction conditions of the installation had a distinct detrimental effect on the system. The cylindrical/conical sump design, opposed suction, and the dilution water line from the bottom of the tank contribute significantly to the aeration. Corrective actions need to be implemented to increase the sump height; maintain the baffle plate between the two suctions; closely control the sump levels; and minimize pump speed variation. These variations have undesired effects on the downstream cyclone stackers' performance. Dilution water should be added from the top of the tank and not at the bottom of the tank, as was taking place at the time.

The control philosophy on the sump should cater for tank level and retention time of the vessel. It is recommended to control the sump from a minimum 600 m^3/h flow rate to a maximum 630 m^3/h flow rate, with the balance of the volume being contributed by dilution water. The Cavex stacker cyclones will operate efficiently within this operating window, despite the feed density to the cyclone. This operating window will allow the pumps to run slower, thus decreasing the bearing temperature. An increased tank level is required to increase the NPSHa, aiding the implementation of this revised control philosophy.

The following cyclone configuration changes should be carried out:

- Amend the installation of the stacker cyclones so that the overflow pipe discharges at more than 1.8 m below the spigot.
- Install a 'Linatex sock' to the end of each overflow pipe to restrict flow rate so that the overflow pipe remains full of water during operation, which will ensure that a continuous vacuum is present
- Enlarge the vortex finder on the stacker cyclones from 290 mm to 305 mm to lower the feed pressure range to between 80 kPa and 95 kPa in order to ensure optimum stacking ability. The change in vortex finder would allow for the stacker to accommodate changes in the TDH without having to change the cyclone's internals.

CONCLUSION

An understanding of the entire pump system curve is essential to ensure effective plant operation. This includes analysis of the system curve from suction conditions to discharge conditions, pipeline set-up, dynamic head understanding, pump and cyclone operational limitations, pump performance effect on cyclone performance *etc*. At the Kwale operation it was observed that due to a low NPSHa and the other detrimental suction conditions, the bearings on the Warman 8/6 FF AHP pumps were constantly overheating in excess of the allowable limits (stream A eastern pumps). The inconsistency in the feed rate resulted in instability in the tailings line, which adversely impacted the performance of the two 650CVX Cavex tailings stacker cyclones. When the flow rates were stabilized, the cyclone's performance was predictable within a feed pressure range of 85–100 kPa.

Moving the dilution water to the top of the tank assisted in a more even split between the two pump trains. The closer the cyclone stackers operated to the plant, the lower the TDH, which reduced the bearing temperatures. The cyclone underflow density has been consistent subsequent to the installation of the cyclone overflow pipe arrangement and the enlargement of the vortex finders from 290 mm to 305 mm. The solids in the stacker underflow were consistently above 78% w/w and an overflow density of less than 1.01 t/m^3 as the stacker cyclones operated at less than 100 kPa feed pressures. The high, consistent stacker cyclone underflow densities resulted in faster construction of the tailings dam wall.

RECOMMENDATIONS

Projects are ongoing to implement the following recommendations to further improve the situation:

- Conversion of pumps to Warman WRT™ liners, which will reduce the NPSHr
- Redesign and modification of feed hopper to increase volumetric capacity and residence time, and as a result increase NPSHa
- Implementation of a fourth pumping stage
- Regular review of the operating philosophy in relation to stacker position and tailings densities

ACKNOWLEDGEMENTS

Weir Minerals is grateful to Base Titanium Kwale Operations for their collaboration and permission to publish this study.

NOMENCLATURE

Absorbed power - *P*
Bearing drive end - DE
Bearing non drive end - NDE
Differential column head loss - *Zc*
Dozer trap mining unit - DMU
Flow rate through the system - *Q*
Head - *H*
Liquid supply surface being under pressure - *Hpr*
Liquid supply under vacuum - *Hva*
Metres above sea level - mAMSL
Mineral separation plant - MSP
Net positive head - NPH
Net positive suction head available - NPSHa

Net positive suction head required - NPSHr
Pump speed - *n*
Suction head for stage pumping - *X*
Total discharge head - *Hd*
Net positive suction head - NPSH
Total dynamic head - TDH (H)
Total suction head - *Hs*
Variable speed drive - VSD
Wear resistant technology - WRT
Wet concentrator plant - WSP

REFERENCES

Arterburn, R A. (1982). The sizing and selection of hydrocyclones. Krebs Engineers, Menlo Park, CA. pp. 1-19.

Base Resources. (2016). http://www.baseresources.com.au/projects/kwale-mineral-sands-operations/, [accessed 17 April 2016].

Engineering Toolbox. (2016). http://www.engineeringtoolbox.com/classification-pumps-d_55.html [accessed 17 April 2016].

Moore, P. (2011). Heavy minerals. *International Mining*, February 2011, 60-68.

Plitt, L.R. (1976). A mathematical model of the hydrocyclone classifier. *CIM Bulletin*, December 1976.

Svarovsky, L. (1984). Hydrocyclones. Holt, Rinehart and Winston, Pennsylvania.

Weir Minerals. (2009). Slurry Pump Handbook, 5th Edition, Weir Group plc

J.D. Singleton

General Manager: Comminution
Weir Minerals Africa

Since May 2015 JD holds the position of General Manager: Comminution at Weir Minerals Africa where he is overseeing the integration of Trio Products into Weir Minerals in Africa and the Middle East after acquisition. JD is responsible for all aspects of profit and loss, HSE, product management, applications, sales, services, engineering, inventory control, project management and aftermarket supply and support of the comminution product lines. After joining Weir in 2009, JD fulfilled the role of Cyclone product manager role where he was responsible for setting up the Cavex cyclone business into a sustainable business unit and between 2013 and 2015 he also fulfilled the role of Process manager at Weir Minerals Africa where he was responsible for applications/process engineering of all products.

Lerionka Sarisar

Country Manager
Weir Minerals East Africa

Lerionka started his career as Trainee plant engineer at Tata Chemicals Magadi, Kenya, where he was involved in Lean Six sigma Green belt dust extraction projects to improve on safety and ash recovery at the grinding and screening section of the SAM plant at Tata chemicals. Lerionka later fulfilled the positions of plant and power plant maintenance planner as well as mechanical and electrical maintenance shift team leader and acting maintenance planning manager at Tata chemicals Magadi.

Lerionka joined Weir Minerals East Africa in August 2011 as mechanical engineer and was promoted to country manager for Tanzania in March 2014. Lerionka is responsible for sales and slurry pump applications in East Africa for Weir Minerals.

Rare earths: applications, mineral resources, thorium, and the balance problem

D. Kennedy[1] and T. Blench[2]

[1]Honorary Research Fellow, University of Birmingham, UK.
[2]Chairman, Steenkampskraal Thorium Ltd, Johannesburg, South Africa.

Rare earth (RE) metals are crucial for the performance of a wide variety of technologies and have been identified as 'critical materials' in the EU, the USA, and Japan. They have only been available in commercial quantities as pure, individual elements for the last 60–70 years due to their mutual similarities in chemical properties and the requirement for complex processing to separate the pure metals. Mineral compositions vary, but the crude generalization is that if one RE metal is in demand then applications must be found for the remaining scandium, yttrium, and lanthanides that will also be present in the mineral. Uses for the REs have fluctuated through the years, with the cost burden for production settling on different materials (those in demand) such as early uses for cerium in lighter flints through yttrium, europium, and terbium for colour television screens, samarium for the first RE permanent magnets, and for neodymium and dysprosium in the most recent RE permanent magnets. The quantities required are minor by commodity standards, but their critical impact is significant in modern technology. The methods of processing, the control and use of actinides (thorium), and in particular the applications of REs in permanent magnets will be examined as the driving force for the RE industry.

INTRODUCTION

The rare earths (REs) are categorized by the International Union of Pure and Applied Chemistry as the lanthanides (element numbers 57 La to 71 Lu inclusive) plus yttrium and scandium.

RE metals are not intrinsically rare - crustal abundance ranks them with the more familiar metals tin and lead. They are unfamiliar to most metallurgists for a number of reasons: they have not been in high demand, their oxidation potential is extreme, they are heavy, they have similar chemical properties and are hard to separate, they occur together, and have only been separated in economic quantities since the 1960s. The RE sector can be characterized as small in scale, involving expensive process techniques, and of limited opportunity, though the unique properties and technological impact of RE metals are drawing increased attention.

The Tenth International Heavy Minerals Conference
Sun City, 16–17 August 2016
Southern African Institute of Mining and Metallurgy

To place the size of the industry in perspective, by the early 21st century approximately 120 kt/a of rare earth oxides (REO) were being processed. This compares with approximately 1.8 Mt/a for nickel and about 70 kt/a for cobalt. This 120 kt/a of REO is spread over 15 different metals. While the bulk chemical properties of the REs may be similar, they are fundamentally different metals that have specific applications; only in the crudest of applications does the natural mixture of the metals find some uses.

With a history of economic exploitation spanning only 60 years, applications that exploit the properties of RE metals have been emerging and are now becoming more interesting to a wider audience. The outer valence shells determine that they are generally of valency three, but the progressive filling of the 4f shell through the series determines that the REs are physically dissimilar in behaviour and can exhibit useful physical and chemical properties. Due to these unique physical and chemical behaviours, certain of the metals are difficult to substitute in important technological applications such as in phosphors and permanent magnets. The perpetual problem in RE processing is that the natural occurrence of the metals in raw material feed does not match demand: if there is a requirement for one metal, then a market for the remainder has to be found.

The metals occur together in varying ratios depending on the host mineral, and while some may be in demand, others are not; this constitutes the 'balance problem' (Falconnet, 1985). The most common minerals contain up to 15 metals; of the 17 in the IUPAC scientific definition, promethium has no stable isotope and does not occur in nature and scandium is obtained as a by-product of uranium extraction. Generally speaking, it is the demand for REs in permanent magnets (neodymium, dysprosium, praseodymium, samarium, and terbium) that is driving the whole RE industry at the current stage of its development.

There is growing interest in the potential for recycling RE materials; this is driven by the possibility of helping to balance supply and demand as it tends to address the materials in common usage, it avoids the necessity of processing large quantities of surplus elements, and the materials have already been separated from their radioactive relatives. In the recycling of REs, this may be a case of competitive recycling that decreases manufacturing cost rather than typical metals recycling that preserves intrinsic material value, such as in platinum and gold recovery.

Radioactivity and the REs are intrinsically connected. Monazite minerals, which have has been processed for many years, initially as by-products from heavy mineral sand processing and from underground mining, were originally exploited for their thorium (Th) content. Monazite may contain from 5% to 15% Th of the total RE/Th content of the monazite. Thorium was an extremely important material for lighting; it was used for gas mantles since its incandescent properties are unsurpassed. The RE by-product, which was commonly called 'technical cerium', contained the entire set of REs from the source mineral; applications for technical cerium were found in glass polishing (cerium rouge) and as mischmetal (well known as an alloy with iron for lighter flints) when extracted from mixed chloride by fused-salt electrolysis.

Although radioactive elements are now generally considered a waste from RE processing, since the 1960s there has been an active interest in the use of thorium for power generation. Most famously, a molten salt reactor using thorium ran at Oak Ridge National Laboratory (USA) from 1965 to 1969. There are current generations of thorium fuel on trial, where one of the key factors may be the lack of fission by-products that can be used in nuclear weapons, whereas the production of weapons-grade actinides was a key factor in the dominant use of uranium for power generation. Thorium is significantly more energy-efficient in application (1 kg of Th is equivalent to 200 kg of uranium) and its residual radioactive isotopes have much shorter half-lives than those of uranium, and therefore require storage for shorter duration.

OVERVIEW OF APPLICATIONS FOR RARE EARTHS

As described, the core problem with REs is the imbalance between demand and the natural distribution of elements from the various deposits. Each element has specific physical properties and few applications use a mixed compound. Table I represents a guide to the principal applications for the various REs; this does not cover the residual quantities of materials that are left unused, such as the significant quantities of cerium that are surplus to requirements.

Table I. The principal classes of application for rare earths from global production (Kennedy, 2014).

Class of application	Comment
Permanent magnets	Consumes approx. 22% of separated REs
Metals and alloys	Consumes approx. 20% of separated REs
Catalysts	Two distinct groups, together consuming approx. 19% of RE production
Polishing compounds	Approx. 13% of RE production as a mixed compound
Glass additives	Approx. 8% of RE production.
Phosphors	Estimated at 7% of RE production; mostly Y, Eu. and Tb.
Ceramics	Approx. 6% of RE production.
Other	Remaining 5% of RE production.

Mineral Resources

To gain perspective on the balance problem it is necessary to consider the supply chain. Table II lists the principal minerals that have been exploited for their RE content, with one notable omission, the 'ion adsorption clay' of southern China, which is described separately.

Table II. Commonly processed types of rare earth minerals.

Mineral	Chemical formula	Comment
Monazite	$(RE, Th)PO_4$	Phosphate. Significant levels of thorium and varying ratios of REs depending on source, tending to LREE
Bastnaesite	$(Ce, La, Nd, Y)FCO_3$	Fluorocarbonate. Low levels of Th; high proportions of LREE
Xenotime	$(Y, Yb)PO_4$	Phosphate. Principally yttrium phosphate with significant substitution of Dy and Tb
Loparite	$(Ce, Na, Ca)(Ti, Nb)O_3$	Oxide. Rare earth containing 'perovskite' type of mineral

The commonest minerals processed are monazite and bastnaesite. The RE ratios in these minerals are dominated by the light rare earth elements (LREE). The ratios of REE are not consistent between different deposits and may also vary within a deposit.

Within the RE industry there is a common delineation between 'light rare earth elements' (LREE) and 'heavy rare earth elements' (HREE). This is a somewhat arbitrary division between the lighter elements La to Eu and the heavier elements Gd to Lu plus Y. This is a reflection of how the materials are processed in industry, with yttrium concentrating with the HREE.

Possibly the most economically important RE deposits are the ion-adsorption clays of southern China. These are variable in composition, but carry disproportionate quantities of the scarce HREE. Though the grade of these deposits is relatively low, in the region of 0.3% to 1.0% total rare earth oxides (TREO), the compositions are extremely useful and the processing route is very straightforward.

Table III. Rare earth contents of major source minerals (percentage of TREO)

	Bastnaesite		**Ion adsorption clays**		**Monazite**			**Loparite**
	Bayan Obo, Inner Mongolia	Mountain Pass, USA	Xunwu Jiangxi, China	Longnan Jiangxi, China	Mount Weld, Australia	Guangdong, China	SKK,* South Africa	Lovozersky, Russia
La_2O_3	23.0	33.2	43.4	1.8	25.5	23.0	20.7	28.0
CeO_2	50.0	49.1	2.4	0.4	46.7	42.7	45.8	57.5
Pr_6O_{11}	6.2	4.3	9.0	0.7	5.3	4.1	5.1	3.8
Nd_2O_3	18.5	12.0	31.7	3.0	18.5	17.0	17.7	8.8
Sm_2O_3	1.1	0.8	6.3	5.0	2.4	3.1	2.8	0.1
Eu_2O_3	0.2	0.1	0.5	0.1	0.4	0.1	0.1	0.1
Tb_4O_7	0.1	trace	trace	1.3	0.1	0.7	0.2	0.1
Dy_2O_3	0.1	trace	trace	6.7	0.1	0.8	1.0	0.1
Y_2O_3	trace	0.1	8.0	65.0	0.3	2.4	4.2	trace

SKK*: Steenkampskraal, Western Cape
Sources: Roskill, IMCOA, Industry Research.

Table III illustrates a selection of the most important commercial sources of RE metals expressed in the typical fashion of the industry as a proportion of the RE oxide (REO) concerned as a percentage of the total rare earth mix from that source (REO/TREO). In this way the mineral resource can be assessed for its relative content of metals.

APPLICATIONS FOR RARE EARTHS

Permanent Magnets

This is an extremely important, but largely unobserved, part of modern technology.
Applications vary from small machines (milligram-sized magnets in wristwatch movements) to major installations (several tons per large direct drive wind turbine); through the essential drives in hard disc data storage with hundreds of millions of parts produced per year.

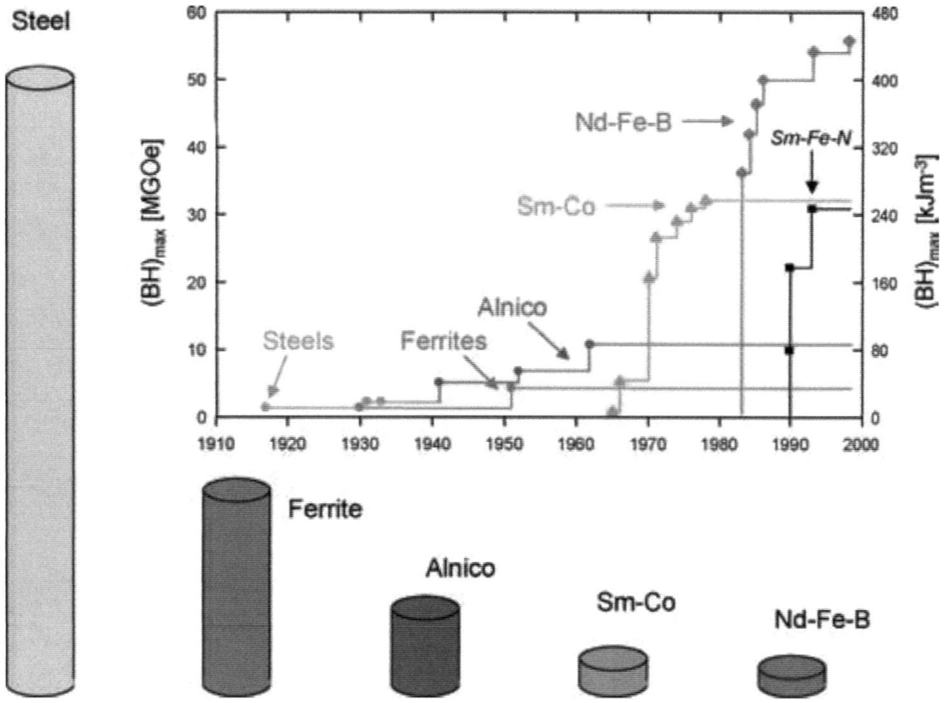

Figure 1. The development of magnetic materials over the last 100 years (Gutfleisch et al., 2010).

Figure 1 shows the $(BH)_{max}$ value for permanent magnets. $(BH)_{max}$ represents the maximum energy product, which is a combination of the magnetic field produced and the resistance to demagnetization. The coloured rods represent the size of magnet needed in the different materials to produce the same magnetic field.

NdFeB was discovered in 1983. Growth in application and use has been extremely fast and research has been intensive, firstly to find alternatives due to patent constraints and secondly to find a more powerful material. Despite all of the resources spent on the effort, there is no known material that will supersede NdFeB. The limit on energy product from any material in the periodic table is that for iron if it could be induced to resist demagnetization (Fe is actually a 'soft' magnetic material and is used for easy switching of the magnetic field in transformers), and this limit would be 920 kJ/m^3. $Nd_2Fe_{14}B$, which contains 70% Fe by weight and a contribution to the magnetic moment from the Nd, has a coercivity (resistance to demagnetization) that matches its residual flux density due to its magneto-crystalline anisotropy and can achieve more than 50% of the theoretical limit of permanent magnet energy product from any known materials.

The properties of NdFeB have been adjusted by additions of other metals. In particular, the performance of NdFeB at elevated temperatures is enhanced by additions of dysprosium or terbium (this is essential for automotive applications in the engine compartment). This has caused a high degree of concern for the supply of the relatively scarce Dy, and research continues to reduce its usage and find substitutes in commercial grades. An early pioneer and inventor of NdFeB (Masato Sagawa) has observed that if the average Dy requirement in magnets exceeds 3%, then the scarcity of Dy will suppress the development of the technology. Discovery of new resources containing Dy and research to reduce the level of Dy content are addressing this problem.

There are several types of NdFeB magnets, principally falling into sintered types or bonded types. The sintered magnets are anisotropic, fully dense, and have the highest magnetic properties. Bonded types, which, for example, can be compression moulded or injection moulded, can be isotropic or anisotropic depending on the manufacturing route. The bonded types of magnet are of lower energy product but may have advantages for industrial application imparted by the production method; an example might be near net shape moulding directly into a motor body.

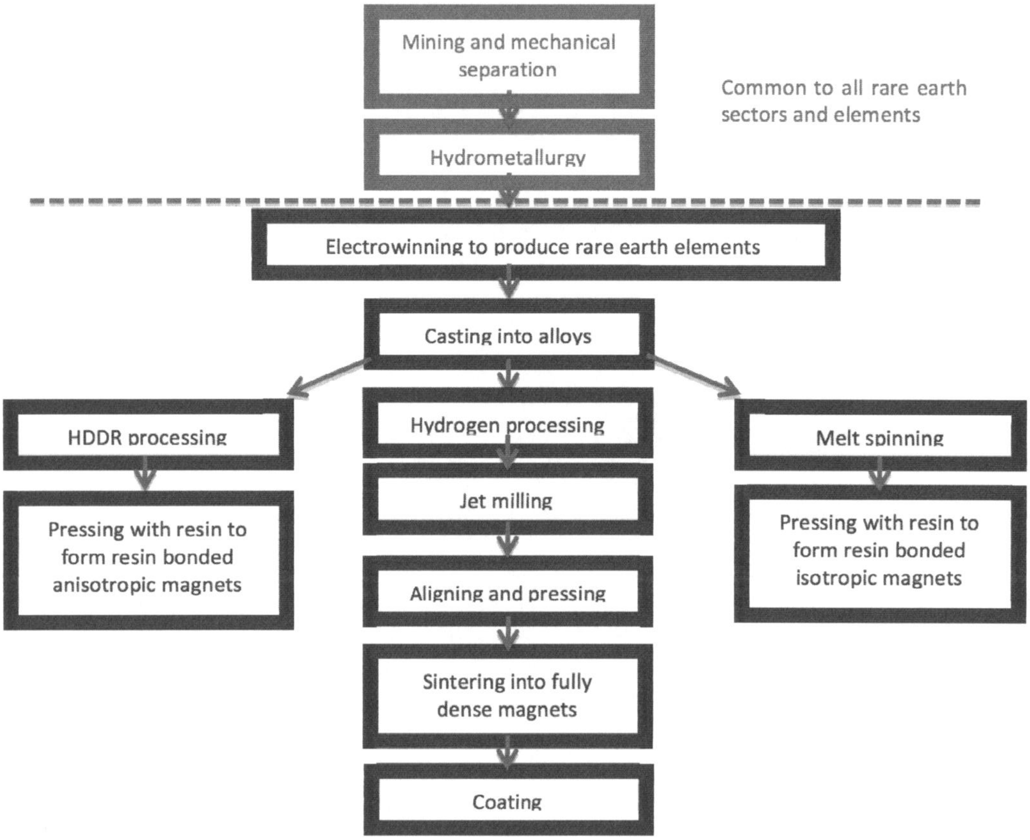

Figure 1. Processing routes for sintered and bonded magnets (Walton, 2015).

The dominant products are the sintered magnets, which are made by powder metallurgy methods. Figure 1 shows the steps required to take a RE chemical compound from the RE processing plant to a finished magnet. In essence, the magnet exploits the natural flux density of the intermetallic compound $Nd_2Fe_{14}B$ and combines this with the magneto-crystalline anisotropy of the tetragonal structure of the alloy. This necessitates reducing the size of particles to single crystals and sintering them together to form a bulk object. At all steps along the processing chain it is essential to minimize impurities, particularly avoiding contact with the atmosphere as the materials are highly reactive.

The RE metals used in permanent magnets are Nd, Pr (can be used as a direct substitute for Nd up to a certain level), Dy, Tb (which is even more effective than Dy in improving temperature resistance, but is significantly scarcer and has competing demands). Manufacturers of permanent magnets have developed alloy compositions that fulfil their product specifications and are adapted to their processing conditions. In general the total RE metal content of an NdFeB magnet is approximately 31wt%. This can be solely Nd, but commonly the Nd is diluted with Pr and there may be additions of Dy for greater temperature tolerance. In some applications, such as early versions of motor/generators for automotive applications, there could be as much as 11wt% substitution of Nd with Dy.

Sm and occasionally small additions of Pr, Gd, and Ce are used for samarium-cobalt-based magnets, which are particularly useful at higher temperatures and in corrosive environments.

Samarium-cobalt alloys for permanent magnets were discovered in 1966 by Karl Strnat and colleagues at the US Air Force Materials Laboratory, following an observation that yttrium-cobalt, possessed an extraordinary magnetic anisotropy. The high resistance to demagnetization (the 'coercivity' mechanism that differentiates permanent magnets from soft magnets) is due, in the case of SmCo, to the magneto-crystalline anisotropy of the hexagonal crystal structure of the $SmCo_5$ alloy. In 1972, a team at the University of Dayton, led by K.J. Strnat and Alden Ray, further developed SmCo by adding Fe to increase the saturation magnetization; in order to retain the coercivity they added copper and zirconium. The coercivity mechanism of the new alloys depends on a magnetic domain pinning mechanism, which is developed by heat treatment of the sintered magnet. Both of these types of SmCo are still in use as SmCo 1:5 and SmCo 2:17 types.

The positive aspects of the samarium-cobalt alloys are their high Curie temperature (the highest temperature limit of their ferromagnetic properties at around 650°C) combined with a relatively good corrosion resistance. Above 150°C, SmCo magnets exceed the energy products of NdFeB magnets. Their commercial application is hindered by their reliance on cobalt as the major element in composition (65wt% Co in $SmCo_5$ and 50wt% Co in 2:17) and ultimately the availability of Sm in the natural mix, which as illustrated in Table III is approximately one-tenth the availability of Nd plus Pr. Due to these constraints, large-scale commercialization of SmCo has been limited to specialist applications where stability and temperature resistance are at a premium, such as in sensors and aerospace.

Metals and Alloys

RE metals are difficult to produce owing to their extreme oxidation potential. The bulk of metals are produced by fused salt electrolysis. This is a process similar to aluminium production as it relies on a molten fluoride bath that dissolves REO and electrolysis results in oxygen formation at the carbon anode, which is consumed during the process. The liquid RE metal pools underneath a tungsten cathode and is collected periodically. In the case of DyFe master alloys, the temperature of the bath is kept within feasible levels by using an iron cathode which dissolves with the deposition of Dy and forms a relatively low-melting-point eutectic.

In addition to the permanent magnet alloys, other important materials are the hydrogen storage alloys based on lanthanum-nickel that are used in nickel-metal hydride (NiMH) batteries. There continues to be growth in the use of NiMH batteries in automotive applications despite competition from alternatives, due to their history of safe application.

Wider use of hydrogen as an energy 'vector' holds potential for lanthanum-nickel ($LaNi_5$-type) alloys due to the high volume of hydrogen that they can absorb at relatively low pressure. Each molecule of $LaNi_5$ can store 6.6 atoms of hydrogen. This has potential for stationary storage, but due to the low hydrogen content (maximum 1.3 wt%%) this is not a solution for transportation.

There are further alloys that have found limited application or are in various stages of development or early exploitation. As examples: an alloy of terbium, dysprosium, and iron of general composition $REFe_2$ known as 'Terfenol' has a giant magnetostrictive property that enables it to be used in sonar and sound generation; scandium–aluminium alloys are used in civilian applications but may offer materials developments for high strength and low density in aerospace applications. RE metals are also added to some magnesium alloys that promise weight saving solutions for aerospace.

RE metals and alloys may find application in solid-state refrigeration using the magneto-caloric effect. Initially this was demonstrated using gadolinium metal as spheres or as foil. More recently alloys of the type $La(Fe_{1-x}Si_x)_{13}$, possibly with hydrogenation, have shown promise.

Minor additions of REs are used in special steels and for modification of cast iron. In particular, lanthanum and yttrium are used in aerospace alloys for enhanced temperature resistance and may be used as oxides and in conjunction with zirconium for thermal barrier coatings.

Catalysts

There are two distinct types of catalyst applications for REs. They are used in 'fluid cracking catalysts' (FCC) in petroleum refineries and in catalyst support 'washcoats' in emission control in mobile and stationary applications. Generally this latter group is most commonly associated with autocatalysts, though there are many applications for abatement of harmful emissions; commonly driven by government regulation and tightening of standards.

Petroleum refinery operations require increasing yield of lighter hydrocarbons from heavy mineral oils. This is achieved with fluid cracking catalysts, whose purpose is to change the 'boiling range distribution' of feed and are thought to be used for approximately 7% of the output of refineries. The catalyst used in a refinery operates at 500°C to 530°C and is routinely regenerated at 650°C to 800°C; the pellets contain an alumina matrix, with a binder and clay for strength and RE-containing zeolite. The zeolite may comprise from 10% to 50% of the catalyst pellet. The REO content may constitute only 2% by weight of the catalyst. Yttrium is used in the manufacture of the zeolite and Ce and La replace Na in the alumina-silicate structure of the zeolite.

Emission control catalysts for removing pollutants from vehicle exhausts and other applications normally contain platinum group metals (PGMs), but they rely on catalyst supports that have a coating of RE/zirconia ceramics. These materials are under continuous development to keep abreast of changing regulations and for use with different fuels and for 'low-temperature light-off' to improve the efficiency in startup conditions. The predominant RE used in the application is Ce for engines run on gasoline, with lesser requirements of La. The compounds are produced by co-precipitation of metal compounds with other transition metals such as zirconium from solution and are proprietary formulations. The function of the RE component is oxygen storage and release, enabled by the nonstoichiometric stability of REO, which in turn enables improved efficiency of the PGM catalytic conversion of exhaust emissions.

Polishing

RE compounds have been used for polishing operations (for example on glass or semiconductor wafers) for many years and are commonly known as 'cerium rouge'. The process is chemical-mechanical polishing or planarization. The RE compounds are a combination of cerium, lanthanum, and praseodymium oxides; typically the ratio of Ce to La is 2:1, with approximately 5% Pr. Ironically it is the Pr addition that creates the red pigmentation; due to the demand for Pr in magnet alloys polishing powders have been created with only Ce and La. Reports say that the revised composition performs well but acceptance is slow due to the unfamiliar colour.

Glass

Rare earths have applications in glass as colourizers and decolourizers. Cerium oxide is used for decolourizing (to clarify glass by counteracting the effect of Fe impurities) and for producing a yellow colour at levels of 1% addition. Others RE useful in glass are particularly La for high refractory glass, with colours from Nd, Nd/Pr (didymium), Pr, Ho, and Er. Of these, Nd and Nd/Pr are important additions in UV absorption for welding glasses.

The adjustment of refractive index enabled by additions of RE to glass is exploited in fibre-optics where surface layers of the glass are altered to improve total internal reflection, resulting in longer transmission distances.

Phosphors

Rare earths have had wide application as colour phosphors for use in televisions (cathode ray tubes), and this the motivation for large-scale solvent extraction methods for separating REs. It was not until

1970 that the sales of colour televisions exceeded those of monochrome televisions. The use of RE phosphors has continued in computer displays and other visual displays, including liquid crystal and plasma technologies. The compounds used for colour phosphors are complex and diverse and are developed specifically for the type of display.

The key rare earths in phosphors are yttrium, europium, and terbium.

The general chemical formulae for key phosphors are:

Red	Y_2O_3:Eu
Green	$LaPO_4$:Ce,Tb
Blue #1	$(Ba,Mg,Al)_{10}O_{17}$:Eu
Blue #2	$(Sr,Ca,Ba)_5(PO_4)_3Cl$:Eu

In the green phosphor the terbium content is about 10%. In the blue phosphors the Eu content is about 5%. In a compact fluorescent lamp there is a 'tri-phosphor' for suitable colour distribution; the normal ratio is approximately 55% red; 35% green, and 10% blue. Applications in lighting are strongly impacted by low-energy-consuming LED technology; this is envisaged to reduce demand for the RE phosphors, though the extent and speed of substitution is not certain due to the continuing use of RE-containing phosphors to adjust the emission spectrum for LEDs, but quantities used per unit may be down to 1% of those used in compact fluorescent lamps.

Scandium iodide is used in high-intensity arc lamps for stadium lighting.

Though this application uses only an estimated 7% of rare earth production, it is a field that uses high-value and high-purity materials with a disproportionately high value of the phosphor segment of rare earth markets.

Ceramics

The important technological applications of rare earths in ceramics include medical, electronic, and mechanical applications. They are rarely the chief element in the composition but are a vital component in alteration of physical properties. Yttrium is the most widely used RE in ceramic applications, for example in yttria-stabilized zirconia and in yttrium-aluminium-garnet used as a carrier for Nd-doped lasers. Other notable applications are new materials for coating prosthetic implants, and yttria and other REs are used in thermal barrier coatings in jet engine turbine blades. Other REs (La, Ce, Pr, and Nd) are used to produce capacitors or to adjust temperature compensation and permeability characteristics of barium-titanate ceramic dielectric capacitors.

RE compounds may be used in bulk materials and as coloured coatings on ceramics (decorative tiles); Pr is used for yellows, Y for orange, and Nd as a light purple.

Other Applications

Rare earths have a number of other unusual properties and applications. Gadolinium has the highest neutron capture cross-section of any material and has application in nuclear power and research; similarly, dysprosium and erbium find application in nuclear science. Cerium compounds have uses in water sterilization and reduction of arsenic impurities in water supplies. Further to application in ceramics, Ce and Sc hold potential in solid oxide fuel cells. Thus far, REs have not been found to cause serious toxicological or environmental problems; indeed, they have been trialled as fertilizers in China with some record of success.

Given that the availability of these materials has been relatively short, it is expected that new developments will emerge, as happened with the first of the ceramic high-temperature superconductors made from yttrium-barium-copper-oxide.

Radioactivity and Thorium

Rare earth deposits are invariably linked to radioactivity and its management. The International Atomic Energy Agency (IAEO, 2011) reports the following main points.

Monazite deposits contain significant quantities of Th and its daughter products. In addition, monazite will contain a small proportion of uranium. The ThO_2 content can be from 5% to 12% of the total rare earth and thorium content. The Mount Weld deposit contains low levels of radioactivity (though generally classed as a monazite-type mineral) and a concentrate at 40% TREO contains approximately 0.17% ThO_2 and 0.006% U_3O_8.

Bayan Obo bastnaesite contains a proportion of monazite; the ratios are indicated as approximately 65–67% bastnaesite and 14–15% monazite. Overall ThO_2 contents have been reported in the region of 0.15–0.40%.

The Mountain Pass deposit has reported values of 0.02–1.0% ThO_2 and that a bastnaesite concentration of 60% TREO has a low activity level of 3–5 Bq/g. The operation is reported as being a 'Candidate for Exemption' based on the low exposure levels measured for workers at 0.3–0.4 mSv/a.

Loparite from the Kola Peninsula has been measured at 0.6% ThO_2 and 0.03% U_3O_8. Records from loparite operations indicate annual exposures for workers at 4 mSv/a from gamma exposure and 2 mSv/a from dust contamination. These values are well under the continuous allowable exposure to workers of 20 mSv/a; in perspective the annual exposure of the general public is 3 mSv/a from the natural environment.

The ion-adsorption clays have extremely low levels of radioactive impurities; measurements show trace levels of ThO_2 and U_3O_8 at 0.005%.

Underground mining of monazite vein deposits poses particular hazard management challenges due to the possible accumulation of radon in mine workings. Great emphasis must be placed on ventilation to remove daughter products of thorium decay.

Thorium as a by-product has had commercial applications in luminous gas mantles and as a minor addition in tungsten spot-welding tips (due to dispersion hardening of sintered tungsten and improved heat resistance). Thorium has potential for application in civil nuclear power generation with advantages over uranium due to the relatively short-lived waste legacy (estimated at 300 years) and the high efficiency of Th element consumption.

ACKNOWLEDGEMENTS

Dr A. Walton and Professor I.R. Harris, Magnetic Materials Group, University of Birmingham.

Mr M. Dawson. Former Production Manager, Thorium Ltd.

REFERENCES

Falconnet, P. (1985). The economics of the rare earths. *Journal of the Less-Common Metals*, 111 (1-2), 9-15.

Kennedy, D. (2014). *Strategische Rohstoffe – Risikvorsorge*. Springer Spektrum, Berlin. pp. 201-217.

Gutfleisch, O., Liu, J.P., Willard, M.A., Brueck, E., Chen, C.H., and Sankar, S.G. (2010). Magnetic materials for the 21st century: stronger, lighter and more efficient. *Advanced Materials*, 23 (7), 821-842.

Kooroshy, J., Tiess, G., Tukker, A., and Walton, A. (2015). Strengthening the European rare earths supply chain: Challenges and policy options. European Rare Earths Competency Network. http://ec.europa.eu/DocsRoom/documents/10882/attachments/1/translations/en/renditions/native

IAEO. (2011). Radiation and NORM residue management in the production of rare earths from thorium containing minerals. *Safety Report Series* no. 68. International Atomic Agency, Vienna.

David Kennedy

University of Birmingham, Honorary Research Fellow.

Currently Director, Cheshire Seals and Components Ltd.
Founded Less Common Metals Ltd, 1992 (UK). RE metals and alloys producer.
Former Director Great Western Minerals Ltd (Canada).
Former CEO Rareco, Western Cape, Owners of Steenkampskraal monazite mine.

Trevor Blench

Chairman, Steenkampskraal Thorium Ltd.

Currently Chairman, Steenkampskraal Thorium Ltd., South Africa, from 2010. and director of Thor Energy (Norway).
Formerly Chairman, Rare Earth Extraction Co. Ltd., South Africa, 2002-2010 and from 2015.
Prior career as a broker and trader in London, Geneva, Johannesburg and Mexico.

Solid-state reduction of an ilmenite concentrate with carbon

A. Bhalla, C.S. Kucukkaragoz, and R.H. Eric

School of Chemical and Metallurgical Engineering
University of the Witwatersrand

The solid-state reduction of ilmenite concentrate with graphite was studied under argon atmosphere at temperatures between 1050°C and 1350°C. The rate of reduction increases with increasing temperature and decreasing particle size, with the maximum extent of reduction found to be 80% at 1350°C for a particle size in the range of -106 + 90 μm. The activation energy for the reduction reaction of ilmenite with carbon was calculated to be 257 kJ/mole. Trivalent iron oxide was reduced first to divalent iron oxide and then to metallic iron. An annular ring of a partially reduced oxide phase containing metallic nuclei formed around the core of ilmenite and grew in size as the original ilmenite core shrank with increasing extent of reduction. Reduction of titanium dioxide also took place during this stage, which produced the Magnéli oxides in the form of Ti_nO_{2n-1}.. The metallic phase had very high Fe and low Ti contents throughout the reduction, whereas the partially reduced oxide phase adjacent to the metallic phase was high in Ti and depleted in Fe. The metallic phase globules formed during this stage were small and are considered suitable only for leaching and not for other concentration processes. The cores of the particles generally retained the composition of the original ilmenite. This work was aimed at the removal of iron and increasing the titanium oxide content in the ilmenite, hence producing a high-titanium charge for subsequent smelting .

INTRODUCTION

The titanium-bearing minerals, ilmenite ($FeTiO_3$) and rutile (TiO_2) are the primary natural sources used for the industrial production of titanium metal and titanium-based products. The common titanium-bearing minerals are ilmenite, rutile, anatase, brookite, leucoxene, arizonite, and perovskite. Although ilmenite is expressed as $FeTiO_3$, the mineral can accommodate Mn and Mg which are chemically bound in the form of $(Fe,Mn,Mg)TiO_3$. The concentrate can also contain Fe_2O_3 and Al_2O_3 in addition to the physically separable oxides such as SiO_2 and CaO. The crystalline structure of ilmenite is an ordered derivative of the corundum structure within a trigonal-rhomboheadral crystalline system. Fe^{2+} and Ti^{4+} ions occupy the octahedral sites, leaving the tetrahedral sites unoccupied. In each $FeTiO_3$ unit, there are three O^{2-} ions creating three octaheadral sites, two of which are occupied by Fe^{2+} and Ti^{4+} (Wilson *et al.*, 2005).

The system Ti-O is complex, involving various Ti_xO_y phases with y/x ratios varying from 2 to 1.0. The general chemical formula representing the Magneli oxides is interpreted in the form of Ti_nO_{2n-1} where *n* ranges from 4 to 10. The crystalline structures of Ti_2O_3, Ti_3O_5, and Ti_5O_9 are trigonal, monoclinic, and triclinic respectively. At higher temperatures, the Magnéli phases decompose to form a pseudo-brookite solid solution. (Pesl, 1997).

Titanium- and titanium-based products are produced from titanium ores by various pyrometallurgical, electrochemical, and hydrometallurgical processes. There are two main methods used in industry to upgrade ilmenite to prepare a feedstock that is later chemically treated to produce pure TiO_2. In the first method, a mixture of metallic iron and rutile is produced in the solid state. In the second method, a titania-rich slag and metallic iron are produced in the liquid state (Zietsman and Pistorius, 2004). Ilmenite reduction is an important stage in both methods, requiring a detailed knowledge of the mechanism of reduction of ilmenite with carbonaceous reducing agents under both inert and reducing gas atmospheres.

El-Guindy and Davenport (1970) found that the reduction of ilmenite was initiated by the reaction taking place at the contact areas between solid carbon and ilmenite particles, which controlled the reduction in the temperature range between 860 and 1020°C. Reduction was carried out by CO due to its high rate of regeneration through the reaction between graphite and CO_2 at temperatures above 1020°C. The diffusion of CO through product layer was found to be the rate-controlling step. In a study conducted by Wouterlood (1979), ilmenite was reduced with char instead of solid carbon so that the reduction conditions would closely simulate the plant conditions. The simulation conditions became complicated by the regeneration of CO through reaction between carbon and CO_2 as well as by the presence of hydrogen in the char. To overcome this difficulty, additional experiments were conducted using CO directly as the reducing gas. The results indicated that ferric iron was reduced to ferrous iron in the first, fast reduction stage, and the ferrous iron was reduced to metallic iron in the second, slower, reduction stage. Application of kinetic models proved that the unreacted core model (mixed control) was the most representative model. Grey *et al.* (1973) identified the stages of reduction taking place. Fe^{3+} was reduced to Fe^{2+} in the first stage, and to Fe (metal) in the second stage, followed by the reduction of TiO_2 to the lower oxide forms represented by Ti_nO_{2n-1} in the later stage. Merk and Pickles (1988) indicated that the metallic shell formed around the ilmenite grain due to the reduction of iron oxides by carbon monoxide in the temperature range 500° to 1100°C at later stages inhibited carbon monoxide diffusion and played an important role in the kinetics of reduction. Welham and Williams (1999) proposed that the reduction of ilmenite proceeded rapidly to form metallic iron and rutile. Rutile was then reduced to the lower oxides in the form of (Ti_nO_{n-1}), which ended in a more stable oxide, Ti_3O_5. The formation of TiC was observed at 1100°C, and was favoured only at higher temperatures. Wang *et al.* (2008) identified the reaction products as mostly reduced iron, iron carbide, rutile, and reduced rutile and its solid solution with Ti_3O_5. The reaction rate was chemical-reaction-controlled at lower temperatures and diffusion-controlled through the product layer at higher temperatures. Sai (2008) applied various mathematical models based on selected data obtained from the operational practices conducted in a rotary reactor, in which ilmenite was reduced with char. The results showed that the rate-controlling step is the solid-solid reaction during the initial stages, and the diffusion of CO through the product layers during the later stages of reduction.

Lobo *et al.* (2013) studied the gaseous pre-reduction of ilmenite using natural gas as the main reductant. It was shown that the iron oxides in ilmenite were reduced more effectively in the presence of hydrogen compared to CO. The shrinking core with multi-interface model was found to represent the reduction behaviour of the titanium oxides.

Preliminary results on the kinetics and mechanism of ilmenite reduction with carbon were reported previously, indicating the strong effect of temperature and particle size on the reduction rate (Kucukkaragoz and Eric, 2005, 2006).

The aim of the present work is to study the reduction sequence of ilmenite with carbon, particularly to define the reduction stage during which the majority of iron oxide is reduced to form the alloy phase before any significant reduction of titanium oxide in the solid state. To achieve the preferential reduction of iron oxides over titanium oxides becomes important, as also stated in the literature (Pistorius, 2002). The aim is to produce a product that has two physically separable phases, namely the metallic phase containing most of the iron and the partially reduced oxide phase containing most of the titanium present in the ore. The subsequent step then involves the development of a suitable process that can be used to separate the metallic phase from the titanium oxide phase in the solid state.

EXPERIMENTAL

Apparatus

The experimental system for thermogravimetric analysis (TGA) consisted of a vertical tube furnace, an electronic balance attached to a computer, a temperature controlling unit, a deoxidation furnace, and a gas regulator (manometer) unit for controlling the reaction gas atmosphere. The details of the experimental set-up can be found elsewhere (Soykan *et al.*, 1991). The furnace, which enclosed an alumina reaction tube, was heated using molybdenum wire as the heating element. The temperature of the hot zone (reaction zone) was controlled by a B-type thermocouple (Pt-6%Rh/Pt-30%Rh) within the range of 2°C, which was also checked with another B-type thermocouple positioned at the hot zone. The molybdenum heating element was protected from oxidation by purging anhydrous NH_3, which decomposed to nitrogen and hydrogen at reaction temperatures. An alumina pedestal attached to an electronic balance with ± 0.1 mg sensitivity was used to position the crucible in the reaction zone. A computer was used to record the mass change values measured with respect to time transmitted by the electronic balance. The system was closed and sealed with an expandable rubber bellow fitted between the work tube and the balance case before the work tube was purged with argon.

Experimental Procedure

A South African ilmenite concentrate from a local plant was used in the experiments. The composition of the concentrate, as provided by the plant, is given in Table I. The sample was crushed, ground, and screened into three particle size fractions: -125+109 µm, -109+90 µm, and -90+63 µm. Graphite, as the carbon source for the reduction reaction, was ground and screened to the same particle size as the concentrate. The samples were prepared by mixing homogeneously 4.5 g of ilmenite concentrate with 3.5 g of graphite, which was 50% in excess of the stoichiometric requirement to form Fe_3C and TiC according to expected reaction. After the furnace had been heated and the temperature of the reaction zone controlled at the required level, each sample was placed in an alumina crucible and positioned on top of the pedestal which was attached to the balance. The expandable rubber bellow was connected to form a sealed zone between the work tube and the balance before the reaction tube was purged with argon at a rate of 1 litre per minute for a period of 15 minutes. The sample was then lifted over a period of around 5 minutes and located in the reaction zone under sealed conditions. The start of each experimental run was recorded as the zero-time for reaction as soon as the sample reached the reaction zone and the recording of mass with respect to time was started by the computer. A series of experiments was conducted to study the influence of temperature and particle size on the reduction process at temperatures ranging from 1050°C to 1350°C.

Another group of experiments was conducted to identify the phases formed at different stages of reduction by reacting the samples of -106+90 µm particle size at 1350°C. Each experimental run was conducted until a predetermined mass change in the sample was observed. The experimental run was stopped at this stage in order to obtain the sample with the reduction percentage corresponding to the mass change recorded. The reduction plots were examined carefully to determine the regions of the curves exhibiting different reduction rates. The reduction stages were identified with the support of the analytical results obtained by scanning electron microscopy using energy-dispersive X-ray spectrometry (SEM-EDX) and X-ray diffractometry (XRD).

Table I. Chemical analysis of ilmenite ore.

Ilmenite	TiO_2	FeO	Fe_2O_3	Al_2O_3	MnO	MgO	SiO_2
Mass %	46.30	34.20	12.30	0.80	1.20	0.40	2.30
Mole %	48.35	39.68	6.43	0.65	1.41	0.83	2.65

Analytical Work

Each of the reacted samples was divided into two parts. One part was mounted in bakelite (suitable for SEM analysis), polished, and examined using a scanning electron microscope (SEM) and the other was subjected to XRD for phase identification. Point analysis for the distribution of elements throughout the phases was conducted by SEM-EDXs.

Stoichiometric Calculations

The samples were prepared so as to contain 50% excess C calculated on the basis of the stoichiometric mass of carbon required for completion of the reduction reactions of the following reducible oxides present in the ilmenite concentrate:

$$TiO_2 + 3C \rightarrow TiC + 2CO \quad [1]$$
$$3Fe_2O_3 + 11C \rightarrow 2Fe_3C + 9CO \quad [2]$$
$$3FeO + 4C \rightarrow Fe_3C + 3CO \quad [3]$$

The mass change recorded during an experiment was assumed to be due to the mass of CO evolved from the sample as a result of reactions [1], [2], and [3]. The percentage reduction was calculated by Equation [4]:

$$Reduction\ \% = \frac{Mass\ of\ CO}{Mass\ of\ oxygen * \frac{28}{16}} * 100 \quad [4]$$

where mass of oxygen indicates the total mass of removable oxygen present in the reducible oxides in the original ilmenite concentrate under the experimental conditions.

RESULTS AND DISCUSSION

The kinetic data obtained in the experimental work is presented in the form of reduction curves plotted as percentage reduction against reaction time. The kinetic curves indicating the influence of the change in temperature and the particle size are given in Figures 1 and 2.

The reduction rate and extent, as expected, increased with increasing temperature and decreased with increasing particle size. The differences in rates were noticeable at all temperatures and particle size ranges and the results indicate a strong influence of the change in temperature and particle size on the reduction rate during the course of reduction.

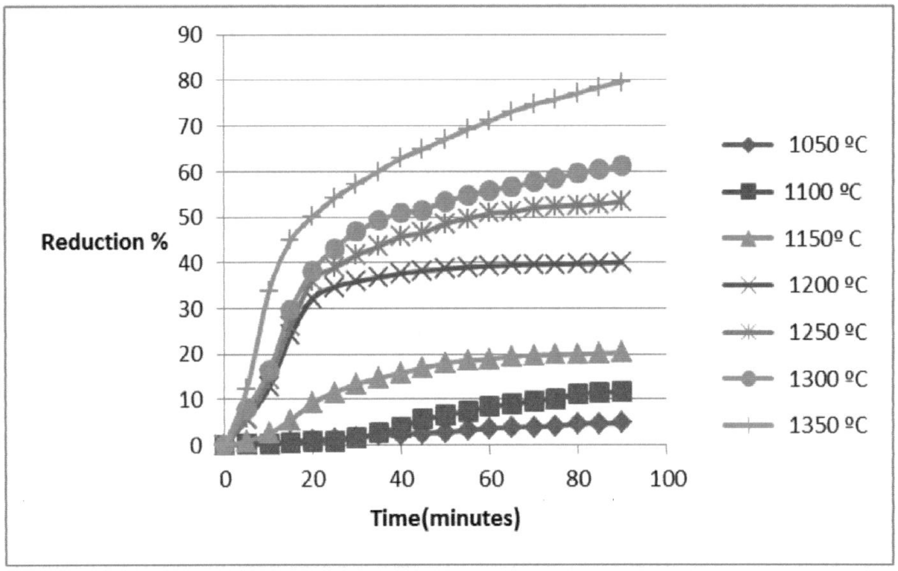

Figure 1. Effect of temperature on reduction rate at particle size (-106 + 90) μm.

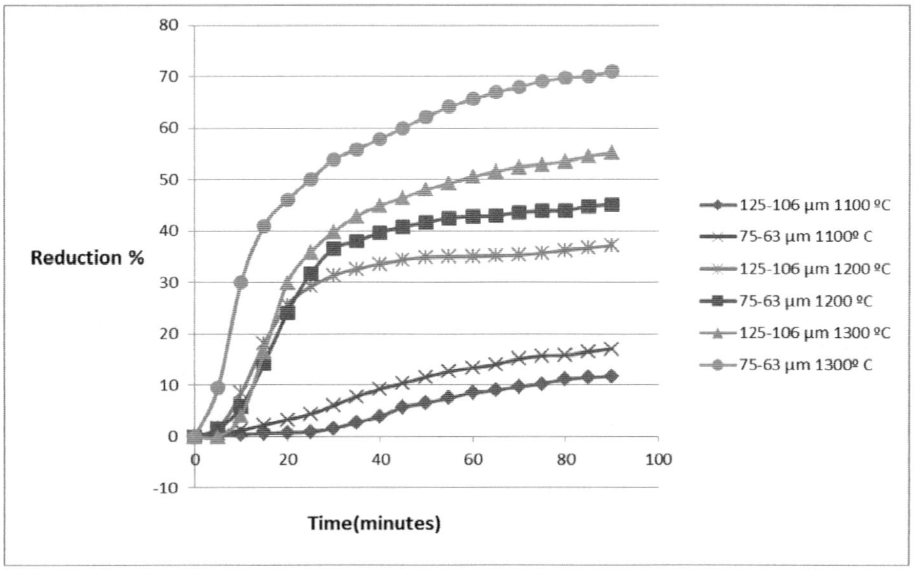

Figure 2. Effect of particle size on reduction rate at 1100 °C, 1200 °C, 1300°C.

Mechanism of Reduction

A series of experiments was by reducing -106+90 µm particle size samples at 1350°C and stopping the reaction at specific points to obtain varying degrees of reduction. The reduction curves obtained are presented in Figure 3.

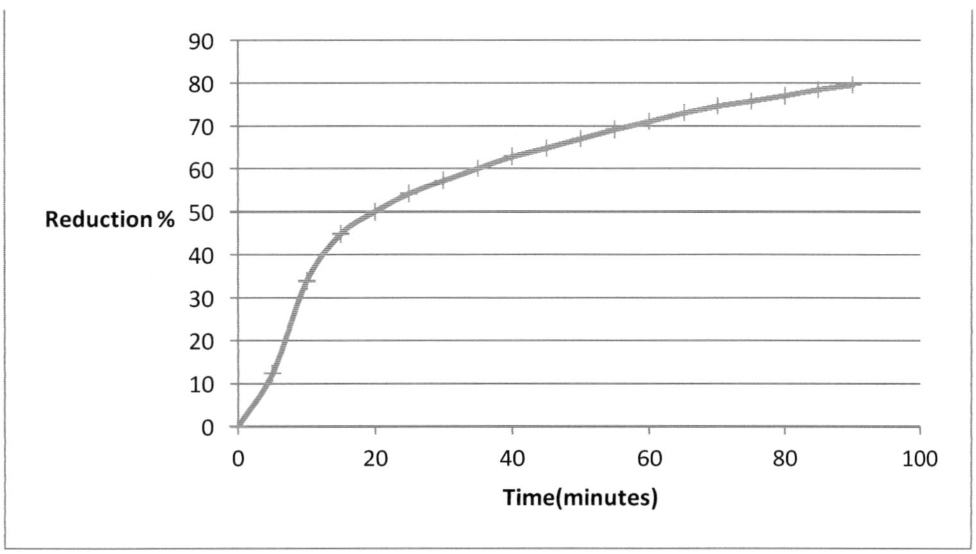

Figure 3. Reduction curve for -106+90 µm sample at 1350°C.

The phases formed in the reacted samples were examined by SEM and analysed by EDX to obtain the elemental point analysis. The SEM images and the elemental analytical results are given in Figures 4 to 11. The reduced samples were also analysed by XRD to identify the phases formed during reduction, as illustrated in Figure 12. The phases detected are given in Table III.

Formation of metallic iron was first seen at 11% reduction (Figure 4). The metallic phase and the partially reduced oxide phase containing the titanium oxide in the form Ti_nO_{2n-1} were detected between 11–40% reduction. The partially reduced oxide phase with a high titanium content and very low iron grew in size as the core of the particle shrank at the later stages of reduction.

According to the results, the metallic phase formed during initial reduction stages had a high iron content and low titanium content, whereas the oxide phase region adjacent to the metallic phase had a high titanium content and low iron content.

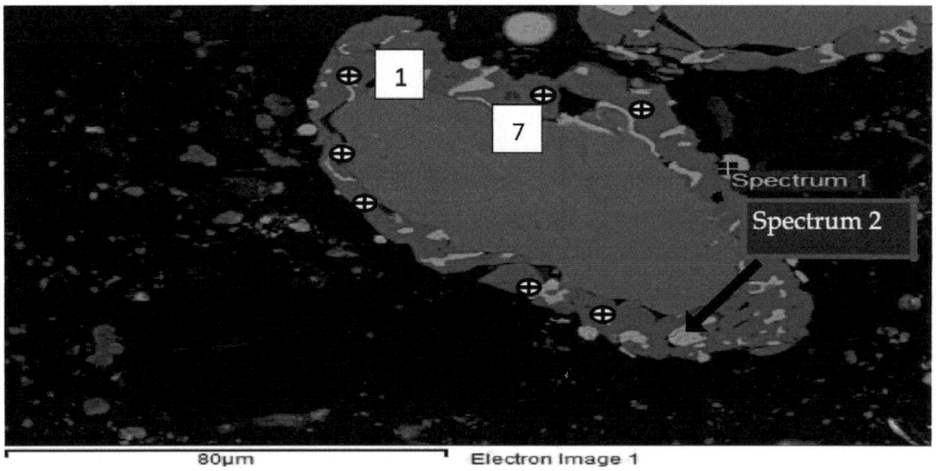

Figure 4. SEM backscattered image of particle at 11% reduction at 1350°C (-109+90 µm fraction).
⊕ Spot point (1-7) on reduced oxide phase (anticlockwise); spectrum 1 and 2 on metal phase

Table II. Chemical analysis (SEM-EDX) of metal phase at spectrum points (1 and 2).

Spectrum no.	Ti (mass %)	Fe (mass%)	C (mass%)
1	1.12	80.14	18.18
2	1.49	80.14	18.18

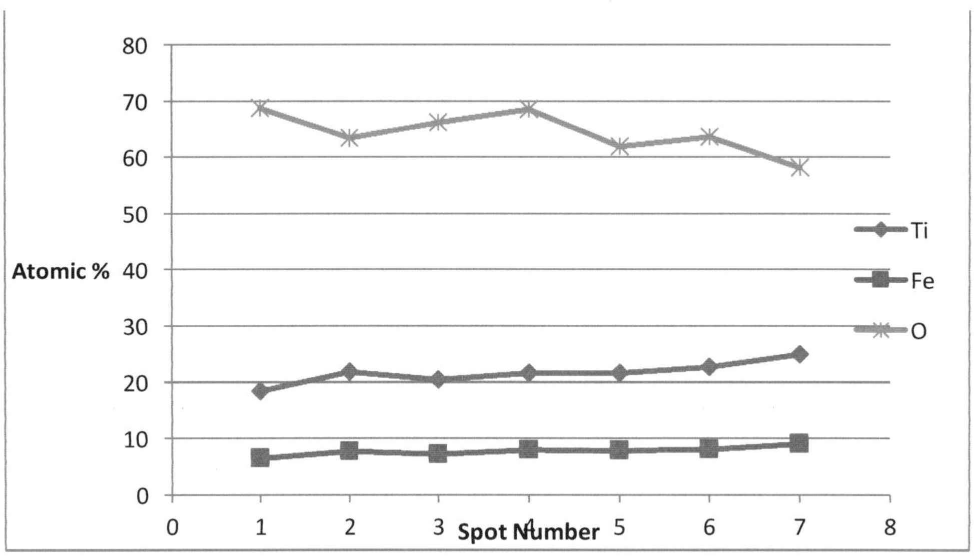

Figure 5. EDX elemental analysis of reduced oxide phase at spot points.

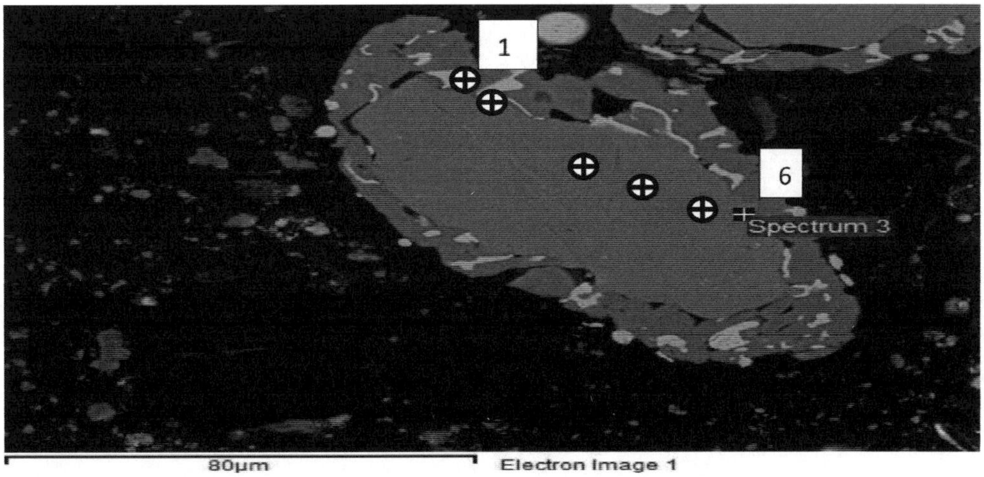

Figure 6a. SEM nackscattered image of particle at 1% reduction at 1350°C (-109+90 µm fraction). ⊕ *Spot number: 1–6 anticlockwise*

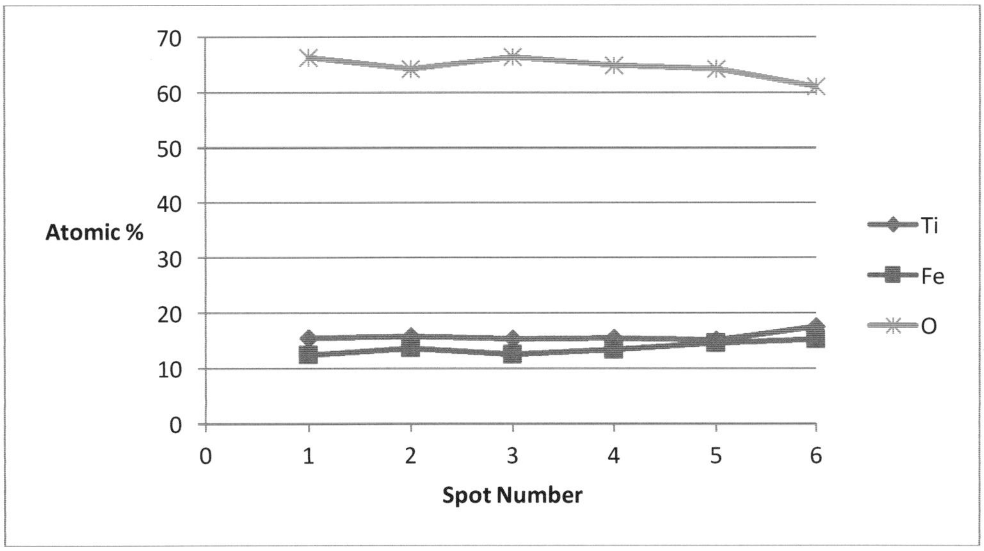

–

Figure 6b. ***SEM-****EDX elemental analysis of reduced oxide phase/ilmenite at spot points.*

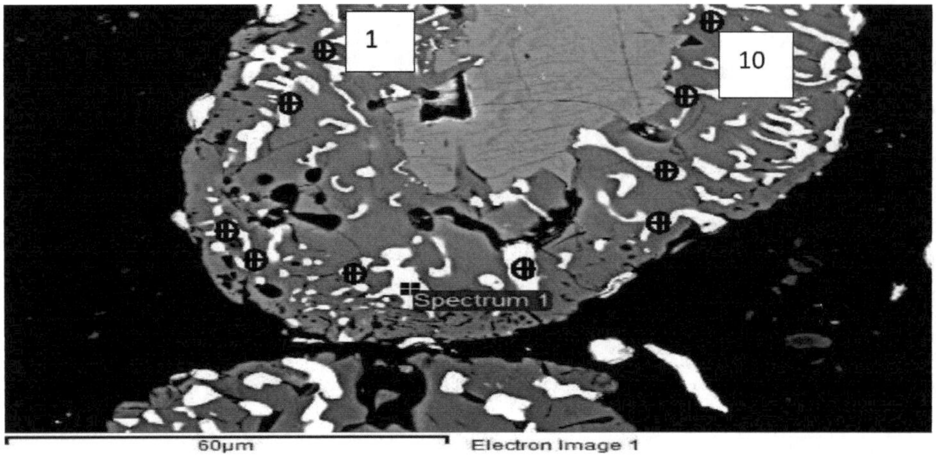

Figure 7a. SEM backscattered image of particle at 40% reduction at 1350°C (-109+90 µm fraction. ⊕ Spot number: 1–10 anticlockwise

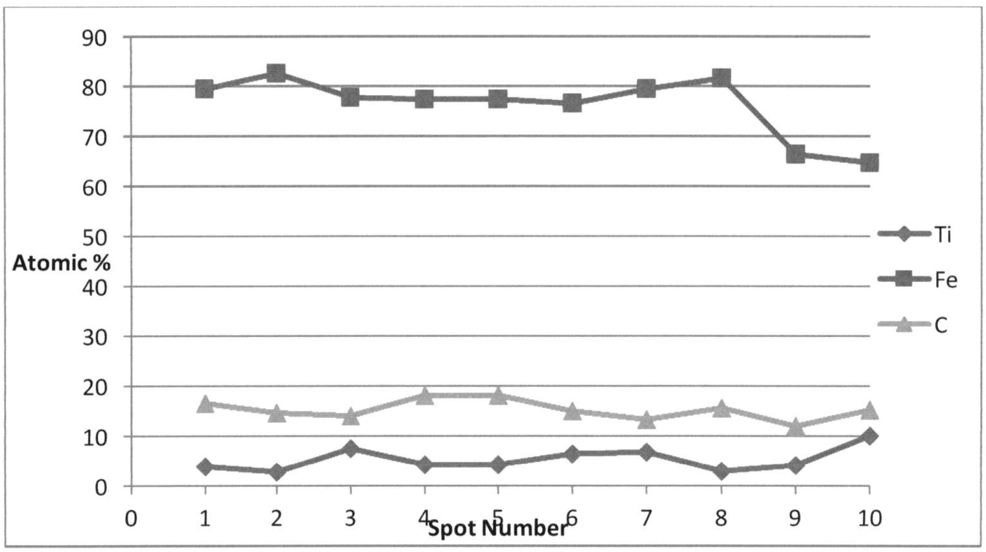

Figure 7b. SEM-EDX) elemental analysis of metal phase (spots 1–10) across the particle.

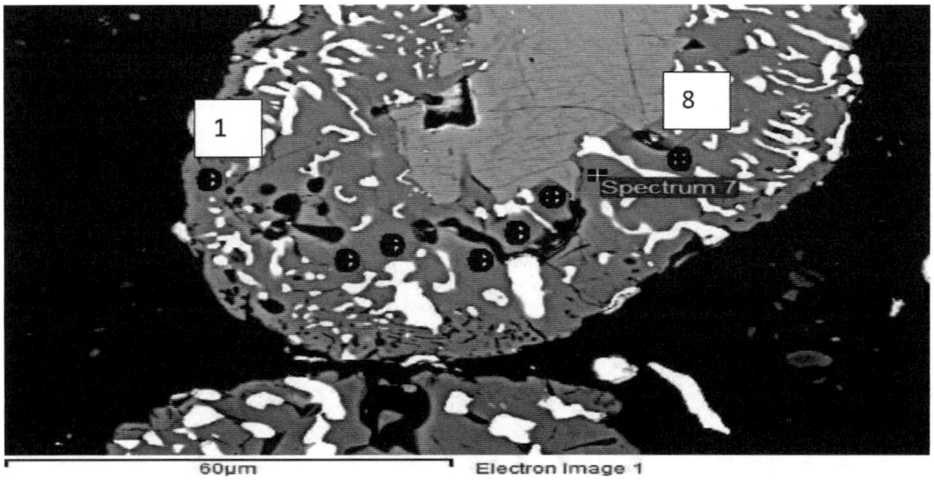

Figure 8a. SEM backscattered image of particle at 40% reduction at 1350°C (-109+90 µm fraction). ⊕ *Spot number :1–8 anticlockwise.*

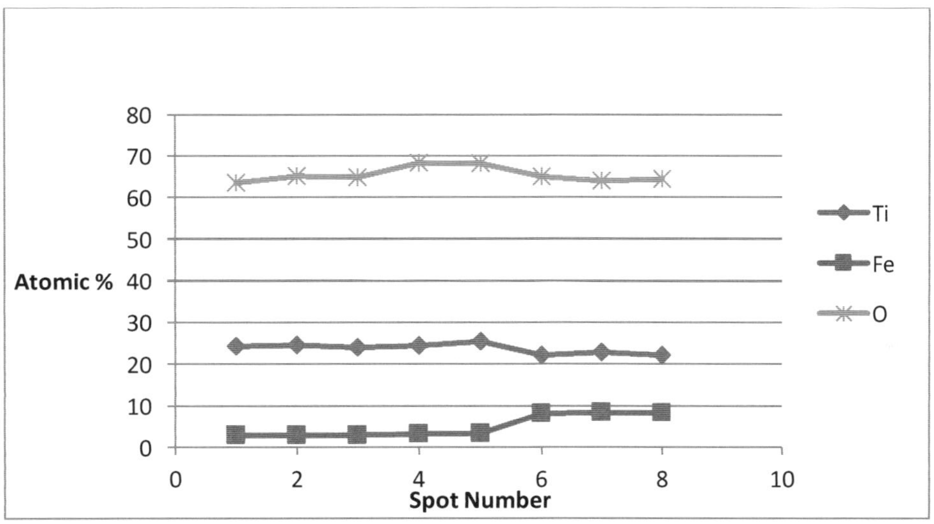

Figure 8b. SEM-EDX elemental analysis of reduced oxide phase at spot points across the particle.

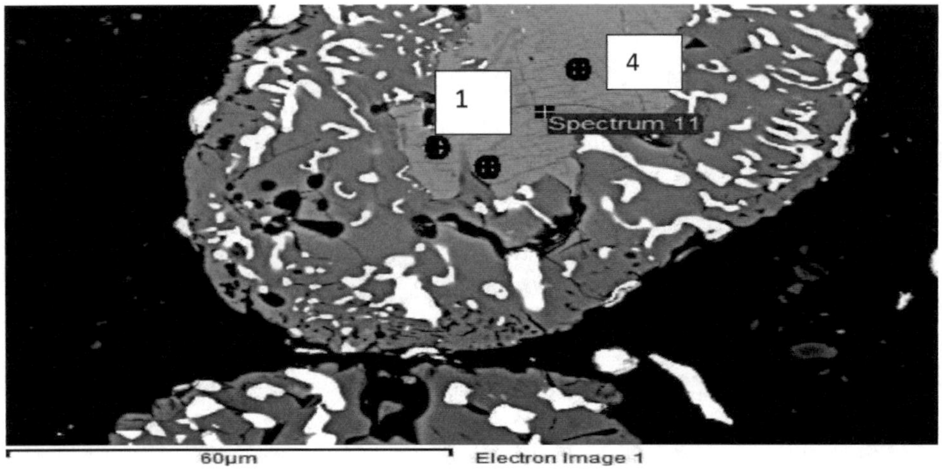

Figure 9a. SEM image of particle at 40% reduction at 1350°C (-109+90 µm fraction. ⊕ (spot number: 1–4 anticlockwise.

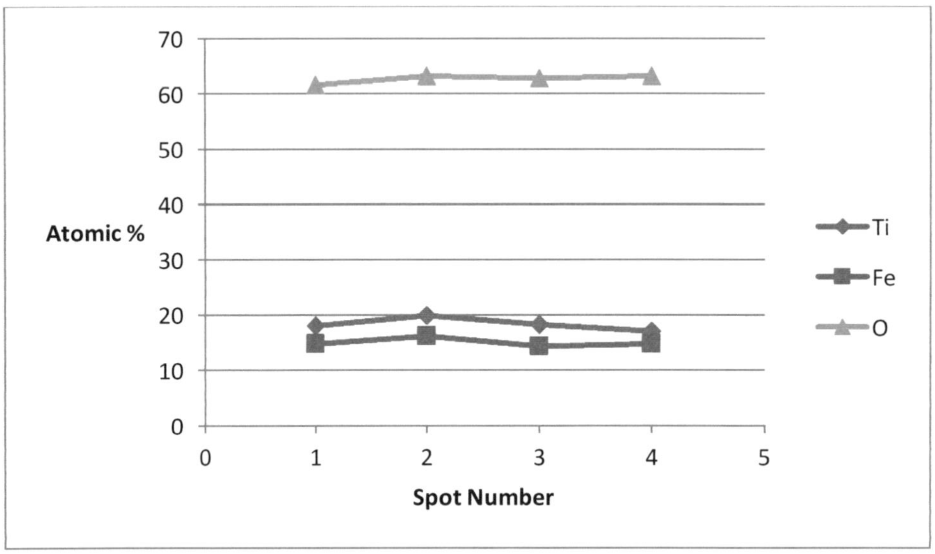

Figure 9b. SEM-EDX elemental analysis of the core ilmenite at spot points

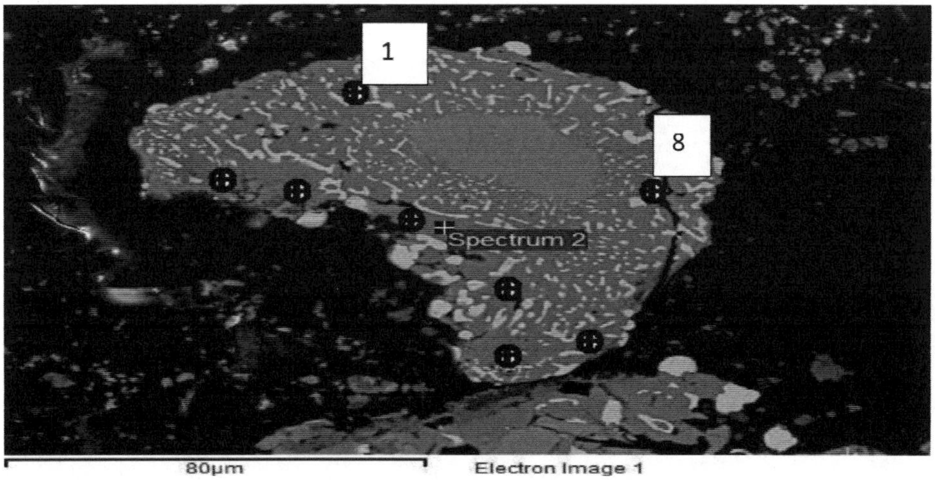

Figure 10a. SEM backscattered image of particle at 61% reduction at 1350°C (-109+90 µm fraction).
⊕ Spot number: 1–8 anticlockwise.

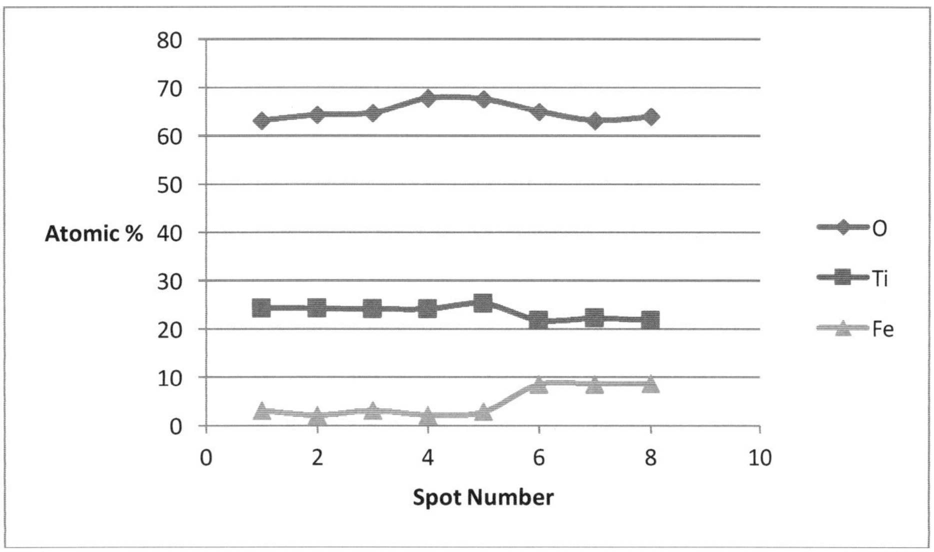

Figure 10b. SEM-EDX elemental analysis of the partially reduced oxide phase surrounding the core across the particle.

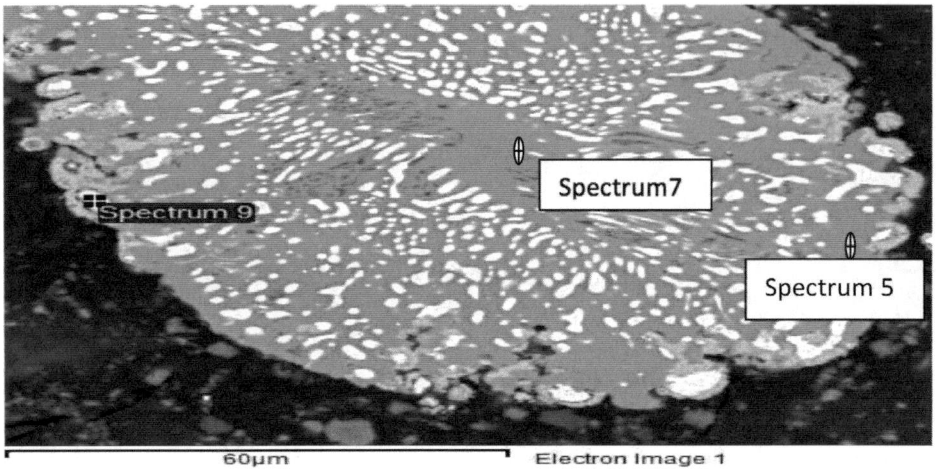

Metal phase (spectrum 9)

Element	Mass %	At. %
C	3.37	13.93
Ti	1.40	1.45
Fe	95.23	84.62
Totals	100	100

Reduced oxide phase (spectrum 5)

Element	Mass%	At. %
C	1.38	2.89
O	43.64	68.57
Ca	0.15	0.10
Ti	50.21	26.35
Mn	2.14	0.98
Fe	2.48	1.11
Totals	100	100

Centre core (spectrum 7)

Element	Mass%	At. %
O	41.89	68.49
Ti	55.05	30.06
Mn	1.90	0.91
Fe	1.16	0.55
Totals	100	100

Figure 11. SEM image and EDX analysis (at spot point) of particle at 80% reduction at 1350°C (-109+90 µm fraction).

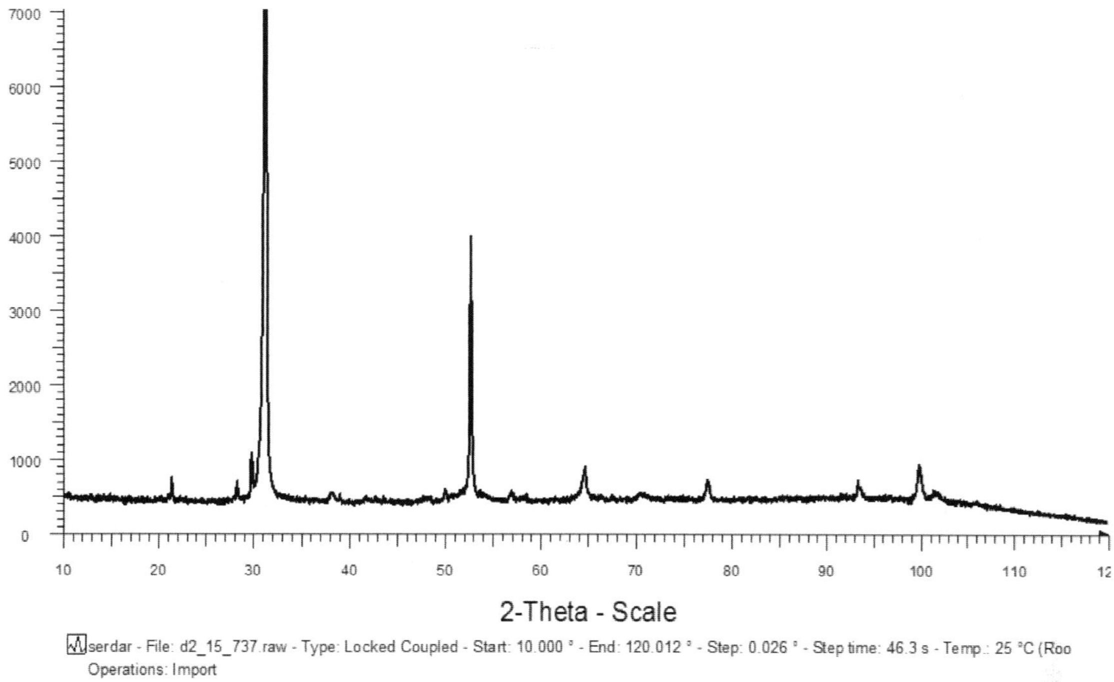

Figure 12. X-ray diffraction pattern of the sample at 50% reduction.

Table. III. Phases detected by X-ray diffraction.

T (°C)	Reduction (%)	Phases detected
1100	12	C, $FeTi_2O_5$, TiO_2 , Fe
1200	40	C, Ti_7O_{13} , Fe, Fe_3C,
1300	60	C, Fe_3C , Ti_6O_{11},Ti_3O_5,
1350	80	C, Fe_7C_3 , TiO

Stages of reduction

The reduction of ilmenite concentrate with carbon is reported take place in the following sequence (Pesl, 1997) :

1. $Fe_2O_3 \rightarrow Fe_3O_4 \rightarrow FeO \rightarrow Fe(metal)$ [5]
2. $TiO_2 \rightarrow$ Magnéli oxides (Ti_7O_{13} - Ti_6O_{11} - Ti_3O_5 - Ti_2O_3) $\rightarrow TiO \rightarrow Ti$ [6]

The phases detected by XRD in the reacted samples at varying reduction percentages are presented in Table III. According to the results, Fe was first detected at 11% reduction, Ti_7O_{13} at 40%, both Ti_6O_{11} and Ti_3O_5 at 60%, and TiO at 80%, which is in good agreement with the following theoretical approach based on stoichiometric calculations on the reduction of the ilmenite concentrate with given chemical composition.

Reduction of Fe_2O_3 to FeO: 0.0–4.0% reduction range
Reduction of FeO to Fe: 4.0–34% reduction range
Reduction of TiO_2 to TiO: 35–65% reduction range
Reduction of TiO to Ti: 65–100% reduction range

The highest reduction level obtained in the present work was 80%, which indicates the possibility of the partial reduction of TiO to TiO_{1-x}, where the value of (x) is the corresponding mass of oxygen removed.

First Stage of Reduction

The initial reduction reaction started at the contact points between the solid carbon and the ilmenite particles. This was followed by the reaction between CO and ilmenite. The regeneration of CO via the Boudouard reaction became effective at temperatures above 1050°C. Nucleation of a metallic phase took place at the most favourable sites on the surface of ilmenite particles during this stage, as can be observed in Figure 4. The high number of metallic nuclei covering the surface of the particle is an indication of the gaseous-type reaction. Mostly iron oxides were reduced in this stage as the metallic nuclei, which were rich in Fe and low in Ti, formed and grew in size as the reduction progressed. During the initial stages of reaction (until 35% reduction), a high iron concentration gradient was developed due to reduction, which caused the driving potential for iron diffusion across the particle. As the reduction proceeded, most of the iron was reduced to the metallic state, causing a change in the Fe concentration profile across the particle (Figure 10). The initial stages of partial reduction of titanium oxide took place simultaneously during this stage; which together with the metallic nuclei formed a shell of a partially reduced oxide phase covering the core of ilmenite particle, as can be seen in Figure 4. This stage took place between zero and 35% reduction and involved the part of the reduction during which the reduced oxide phase, together with the metallic nuclei, grew in size until the shell reached a certain thickness. The reduction rate was more likely influenced by the formation and growth of the metallic nuclei, which was completed within this stage as seen in Figures 6a and 7a.

Second Stage of Reduction

This stage took place at greater than 35% reduction, during which the shell of the partially reduced oxide phase, together with the previously formed metallic nuclei, grew in size as the core ilmenite phase shrank. Theoretically, the reduction mechanism involved steps such as mass transfer of the reactants to the reaction interface, reaction between the oxides and the CO prevailing at the interface and the mass transfer of products from the reaction interface. The reduction reactions took place at the reaction interface between the partially reduced oxide and ilmenite, driven by the carbon monoxide gas. Microscopical examination of the reduced particles revealed that each particle retained its original shape and size during the course of reduction. The removal of oxygen from the oxide phase therefore created a porous structure that allowed the passage of CO to and CO_2 from the reaction interface. The removal of oxygen also resulted in an oxygen concentration in the unreduced ilmenite higher than that in the partially reduced oxide, thus caused the transfer of oxygen to react with CO at the reaction interface.

SEM-EDX analysis indicated the possibility of mass transfer of the reaction elements due to their high concentration gradients through the product layers across the particle. During the later stages of reduction, the original ilmenite was reduced almost entirely to the partially reduced oxide phase containing most of the titanium oxides present in the original ilmenite, as can be seen in the microstructures and elemental analysis given in Figure 11.

Kinetics of Reduction

Examination of the representative reduction curve given in Figure 3 reveals that the reduction process is characteristically separable into two stages, with each stage representing a different reduction rate. The first stage of reduction is defined as the linear part of the curve between 0–35% reduction and the second stage of reduction as the part of the curve representing 35% reduction and above. These stages are proposed to represent the fast initial stage followed by the slower second stage.

The activation energy for the reduction in the first stage (0–35% reduction) can be calculated according to the following procedure.

Application of Arrhenius Equation

The temperature dependence of the rate of a chemical reaction can be expressed by Arrhenius equation:

$$\text{Rate of reaction (k)} = A \exp(-E_A/RT) \qquad [7]$$

where k is the reaction rate constant, A is frequency factor (constant/s), R is the gas constant

(J/mol.K), T is temperature (K), and E_A is activation energy(J/mol).

The calculation procedure is as follows (El-Guindy and Davenport, 1970; Kekkonen *et al.*, 1995). The following assumptions were made for calculation of the activation energy:

- The activation energy is valid for the reduction stage between zero and approximately 35% reduction, during which only iron oxide is reduced
- The set of kinetic data taken from the reduction curves obtained at 1050, 1100, and 1150°C were used in the activity energy calculations (Figure 1).

Taking the logarithm of each side of the Equation [7] yields:

$$\ln (k) = \ln (A) - (E_A/R)(1/T) \qquad [8]$$

which is a straight line with a slope equal to ($-E_A/R$). Rate constant (k) values were derived from the straight line portions of the kinetics curves given in Figures 1 for the first stage of the reduction. The slopes of the straight lines representing ln K against $1/T$ were used to calculate the average activation energy for the reduction reaction. In this work, the ln(k) and $1/T$ values were substituted into Equation [8] in order to calculate the activation energy (E_A) for the reduction reaction.

Using the experimental data given in Figure 1 for temperatures of 1050, 1100, and 1150°C, the activation energy was calculated as 257 kJ/mole for the initial stages of reduction (0–35% reduction). This value was found to be close to the activation energy value of 64 kcal/mole reported in the literature (El-Guindy and Davenport, 1970).

CONCLUSION

The reduction rate increased with increasing temperature and decreasing particle size. The metallic phase contained higher Fe and low Ti, whereas the oxide phase adjacent to the metallic phase contained high titanium oxide and low iron oxide.

The reduction of ilmenite proceeded through the stages during which iron oxides were first reduced to form metallic iron and later to form iron carbides, whereas titanium oxides were reduced to Ti_nO_{2n-1} during the initial stages with the possible formation of lower oxides at the later stages. The core of the particle (the original ilmenite grain) retained its composition with slight reduction in both iron and titanium oxides content during the initial stages of reduction, and was reduced partially in the later stages.

The metallic nuclei formed in the partially reduced oxide phase were found to be very small and are therefore not considered suitable for separation by mineral concentration processes. The kinetics of reduction were greatly influenced by the formation and growth of metallic nuclei during the initial stages of reduction and by the mass transfer (diffusion) process at the later stages. The overall activation energy for the initial stages of reduction of ilmenite with carbon was calculated as 257 kJ/mole in the temperature range 1050 to 1150 °C.

ACKNOWLEDGEMENTS

The authors wish to express their thanks to the School of Chemical and Metallurgical Engineering, University of the Witwatersrand for the laboratory facilities and the analytical services used. Thanks are also due to O. Netshifamadi, T.J. Molokwane, and S. Mokgadi (fourth-year project students) for carrying out some of the laboratory work, and to B. Nelwalani (SEM Technician) for the SEM-EDX analytical work.

REFERENCES

El-Guindy, M.I. and Davenport, W.G. (1970). Kinetics and mechanism of ilmenite reduction with graphite. *Metallurgical Transactions*, 1, 1729-1734.

Grey, I.E., Jones, D.G., and Reid, A.F. (1973). Reaction sequences in the reduction of ilmenite (1): Introduction. *Transactions of the Institution of Mining and Metallurgy C*, 82, C 151-152.

Kekkonen, M., Xiao, Y., and Holappa, L. (1995). Kinetic study on solid state reduction of chromite pellets. *Proceedings of INFACON 7*, Trondheim, Norway, June 1995. FFF, Trondheim.

Kucukkaragoz, C.S. and Eric, R.H. (2005). Phase changes in solid state reduction of a natural ilmenite. *Proceedings of Heavy Minerals 2005*. Society for Mining, Metallurgy and Exploration, Littleton, CO. pp. 229-233.

Kucukkaragoz, C.S. and Eric, R.H. (2006). Solid state reduction of a natural ilmenite. *Minerals Engineering*, 19 (3), 334-337.

Lobo S., Kolbeinsen, L., and Seim, S. (2013). Pre-reduction of ilmenite with natural gas-model development and use. *Proceedings of the 13th International Ferroalloys Congress, Efficient Technology in the Ferroalloy Industry*, Almaty, Kazakhstan, 9-13 June 2013. pp. 857-868.

Merk, R. and Pickles, C.A. (11988). Reduction of ilmenite by carbon monoxide. *Canadian Metallurgical Quarterly*, 27 (3), 179-185.

Pesl, J. (1997). Thermodynamics and phase equilibria in the Fe-Ti-O system at 1500°C and 1600°C and metal-slag-equilibria pertinent to ilmenite smelting. PhD thesis, University of the Witwatersrand, Johannesburg.

Pistorius, P.C. (2002). The relationship between FeO and TiO_2 in ilmenite smelter slags. *Scandinavian Journal of Metallurgy*, 31 (2), 120-125

Sai, P.S.T. (2008). Evaluation of mathematical models for the reduction of ilmenite with char in a rotary reactor. *Indian Institute of Chemical Engineers*, 50 (4), 312-322.

Soykan, O., Eric, R.H., and King, R.P. (1991). Kinetics of reduction of the Bushveld Complex chromite ore at 1416°C. *Metallurgical and Materials Transactions B*, 22B , 801-810.

Wang, Y.M., Yuan, Z.F. ,Guo, Z.C., Tan, Q.Q., Li, Z.Y., and Jiang, W.Z. (2008). Reaction mechanism of natural ilmenite with graphite. *Transactions of the Nonferrous Metals Society of China*, 18, 962-968.

Welham, N.J. and Williams, J.S. (1999). Carbothermic reduction of ilmenite ($FeTiO_3$) and rutile (TiO_2). *Metallurgical and Materials Transactions B*, 30B (December), 1075-1081.

Wilson, N.C., Muscat, J., Mkhonto, D., Ngoepe, P.E., and Harrison, N.M. (2005). Structure and properties of ilmenite from first principles. *Physical Review*, B 71, 075202. 14 February 2005.

Wouterlood, J.H. (1979). The reduction of ilmenite with carbon *Journal of Chemical Technology and Biotechnology*, 29, 603-618.

Zietsman, J.H., and Pistorius, P.C. (2004). Process mechanisms in ilmenite smelting. *Journal of the South African Institute of Mining and Metallurgy*, 104 (11), 653–660.

Amit Bhalla

Ph.D Scholar
University Of Witwatersrand, Johannesburg, South Africa.

Worked in Industry & lectured after completing B.TECH from I.I.T Kanpur, India, did Masters and PH.D (submitting)in Pyro metallurgy from Department of Chemical and Metallurgical Engineering , University of Witwatersrand , Under Supervision of Professor R. Hurman Eric.

In-depth characterization and preliminary beneficiation studies of heavy minerals from beach sands in Brazil

C. C. Gonçalves and P.F.A. Braga

Center for Mineral Technology, Rio de Janeiro, Brazil

Brazil's heavy minerals deposits are still barely explored, being restricted to the Guaju Mine (Cristal Group) located in northeastern Brazil. Thus, to start research into the characterization and beneficiation of heavy minerals from beach sand, two samples were collected in northeastern coastal Brazil. According to the heavy liquid separation results, the total heavy minerals content of the samples ranged from 6.45% to 10.14%. The mineralogical compositions were almost the same for both studied samples, consisting of magnetite, goethite, ilmenite, pseudorutile, staurolite, monazite, actinolite, albite, epidote, xenotime, zircon, kyanite, rutile, anatase, quartz, muscovite, other silicates, amphiboles, and feldspars, with the heavy minerals mostly in the -75+53 µm size fraction for both samples. Ilmenite is the principal Ti-bearing mineral and is highly altered, mainly in pseudorutile and/or anatase. Beneficiation tests using a flow sheet consisting of screening followed by gravity separation and two stages of magnetic separation demonstrated promising results, especially for improved separation in all stages and the ilmenite and zircon recovery. This process can be simplified by omitting one stage of magnetic separation and the inclusion of an electrostatic separation stage. The Brazilian heavy mineral deposits studied thus far have very distinct characteristics, especially regarding particle size, total heavy minerals, and slimes content due to the environmental dynamics and geological factors. This suggests that it may be difficult to identify a specific beneficiation process for Brazilian beach sands, and the necessity of more in-depth studies with other samples.

INTRODUCTION

The main characteristic of heavy minerals is that their specific gravity (SG) is higher than that of common and/or gangue minerals. They can originate from igneous, metamorphic, and sedimentary rocks and are concentrated in coastal and erosion areas. In Brazil, heavy minerals deposits are still barely explored, being restricted to the Guaju mine (Cristal Group) located in northeastern Brazil. The principal minerals beneficiated are ilmenite ($FeTiO_3$) and zircon ($ZrSiO_4$). Previously, Brazilian Nuclear Industries (BNI) produced ilmenite, zircon, rutile (TiO_2), and monazite [$(Ce,La,Nd,Th)PO_4$] concentrates aiming at their *in natura* commercialization (Schnellrath *et al.*, 2001; Worobiec *et al.*, 2007) mentioned the importance of beach sands as a source of rutile, ilmenite, and zircon, and their worldwide use.

Usually, heavy minerals from beach sand are concentrated firstly using gravity separation, as expressed in many studies of pilot plant tests and in industrial plants (Carcagiu *et al.*, 1990; Sampaio *et al.*, 2001; Schnellrath *et al.*, 2001; Laxmi *et al.*, 2013), magnetic separation, electrostatic separation, and also flotation (Routray and Rao, 2011; Routray *et al.*, 2013).

The Tenth International Heavy Minerals Conference
Sun City, 16–17 August 2016
Southern African Institute of Mining and Metallurgy

Thus, to start research into the characterization and beneficiation of heavy minerals from beach sands in Brazil, some viable new deposits were selected in order to determine a specific beneficiation process suited to the characteristics of Brazilian coastal deposits.

EXPERIMENTAL

We collected samples from two different localities in the Luis Correia district, Piauí, on the eastern Brazilian coast. They are identified as PL1 and PL2 in this work. Firstly, we prepared, homogenized, and quartered the samples. The composition of the samples was practically unknown; thus we conducted particle distribution analysis, chemical analysis, mineralogical analysis, heavy liquid separation, magnetic separation, and laboratory beneficiation tests in order to facilitate the identification of important minerals and the principal concentrates/products. We based the methodology on the standard method proposed by Reyneke and van der Westhuizen (2011), with modifications, and on other research (Carcangiu *et al.*, 1990; Laxmi *et al.*, 2013).

Characterization of Raw Material

We determined the elemental composition of the raw material using X-ray fluorescence spectrometry (XRF) (Panalytical Axios Max, Rh emission) and X-ray diffraction (XRD) (D4 Endevor Co emission). We then wet-screened the samples using 425, 300, 212, 150, 106, 75, 53, and 45 µm apertures and further examined each size fraction by XRF and XRD. We measured other physical characteristics such as bulk density and specific gravity. To complete the preliminary information, we tested other fractions of the two samples with bromoform dense liquid (SG 2.89) and methylene iodide (SG 3.32). We also analysed the sink fractions from methylene iodide by XRD and XRF.

Characterization of Heavy Minerals

We separated the methylene iodide sink fraction (-150 µm) using a Frantz mineral separator (model L-1 Isodynamics, forward slopes 15°), using magnetic field strengths of 0.150 T (in order to remove the magnetite), 0.450, 0.675, 0.950 and 1.300 T. We then analysed each fraction by XRD and XRF, and characterized them in detail using a scanning electron microscope (SEM Hitachi TM 3030 Plus Tabletop Microscope) and micro-Raman equipment (LabRAM HR Ar emission), when necessary.

Preliminary Beneficiation Studies

Simulating a real flow sheet with a gravity separation and magnetic separation (see Figure 1), we subjected each sample to gravity separation using a Knelson concentrator (KC) and two stages of magnetic separation (0.450 T and 1.300 T) using a BoxMag Rapid magnetic separator (MS). We chose the magnetic field strength to firstly recover the more susceptible magnetic products such as ilmenite.

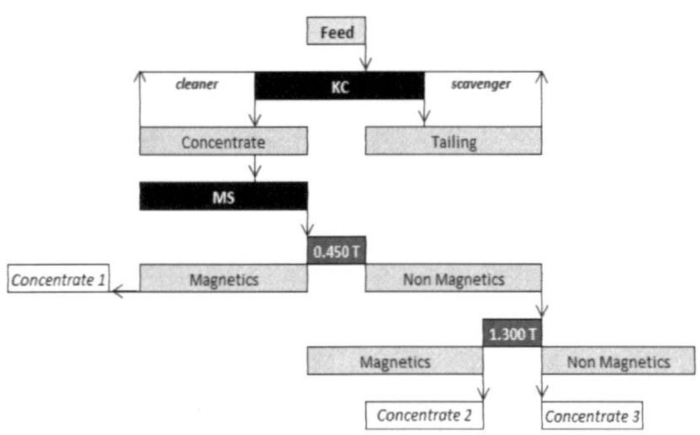

Figure 1. Beneficiation flow sheet.

Based on the results of the gravity separation by Knelson concentrator and the characterization studies, we decided to review the preliminary beneficiation flow sheet (Figure 1). This change meant the replacement of the Knelson concentrator by a screening stage, with a cut-off size of 150 µm, gravity separation using a shaking table (model Super Duty Diagonal Deck, Deister Concentrator

Company), before two stages of magnetic separation as shown in (Figure 2). We controlled all steps with semi-quantitative mineralogical analysis (Rietveld method).

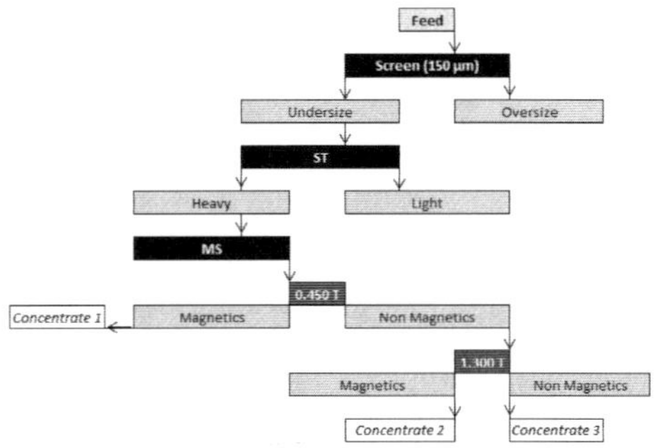

Figure 2. Second beneficiation flow sheet.

RESULTS AND DISCUSSION

Characterization Studies of Raw Material

The physical properties of the samples and the preliminary results for ore concentration are given in Table I. In the present study we considered as total magnetic heavy minerals (TMHM) all minerals which have a SG higher than that of methylene iodide (3.32) and which are recovered up to 1.300 T.

Table I. Physical properties of samples.

Property	PL1 values	PL2 values
Bulk density of raw material (g/cm^3)	1.46	1.62
Specific gravity of raw material (g/cm^3)	2.73	2.76
d_{80} passing size (μm)	190	400
Slimes (-45 μm) (%)	0.31	0.11
Total heavy minerals - THM (%)	6.45	10.14
Heavy minerals - HM (SG. > 3.32) (%)	4.03	8.82
Heavy minerals (2.89 < SG < 3.32) (%)	2.42	1.33
Total magnetic heavy minerals - TMHM (%)	65.6	64.4
Total nonmagnetic heavy minerals - TNHM (%)	34.4	35.6
Specific gravity of sink fraction of methylene iodide (g/cm^3)	4.17	4.36

Some properties such as bulk density, specific gravity, and percentage of magnetic heavy minerals (MHM) and nonmagnetic heavy minerals (NHM) are almost equal for both samples, and the more marked differences are the grain size distribution and total heavy minerals (THM). The difference in grain size distribution can be decisive for beneficiation efficiency by gravity separation as this method is affected by differences in the minerals' specific gravities, and also particle shape and size (Sampaio and Tavares, 2005; Lins, 2010). This finding can necessitate the application of enhanced separation forces such as centrifugal (Figueiredo and Chaves, 1998).

We considered as the principal oxides for process control SiO_2, Fe_2O_3, Al_2O_3, TiO_2, ZrO_2, and REE_{oxides} (CeO_2, La_2O_3, Nd_2O_3 and ThO_2) + P_2O_5. According to the elemental chemical analysis of the raw material, PL1 is composed of 88.5% SiO_2, 1.4% Fe_2O_3, 5.0% Al_2O_3, 1.5% TiO_2, 0.36% ZrO_2, and less than 0.1% REE oxides + P_2O_5. For PL2, the values are: 78.2% SiO_2, 2.2% Fe_2O_3, 2.3% Al_2O_3, 2.6% TiO_2, 1.1% of ZrO_2, and 0.2% of REE_{oxide} + P_2O_5. In addition, XRD analysis verified the presence of quartz, rutile, and actinolite in PL1; and quartz and staurolite in PL2. The elemental chemical analysis by size fraction, as show in Figure 3, indicates, for both samples, that SiO_2 is concentrated in the coarse fractions, and elements such as Ti, Zr, REE, and P are concentrated in the finer fractions. Minerals such as monazite, ilmenite, rutile, and zircon are concentrated preferentially in the -150 μm size fraction (Figure 3). It was possible to see differences between the samples in terms of particle size and mineralogical composition caused by the environmental dynamics and geological factors, even with the geographical proximity of the sampled areas (Figure 3 and Table I). The size fraction where the greater part of the minerals of economic interest is concentrated is -150+53 μm for both samples.

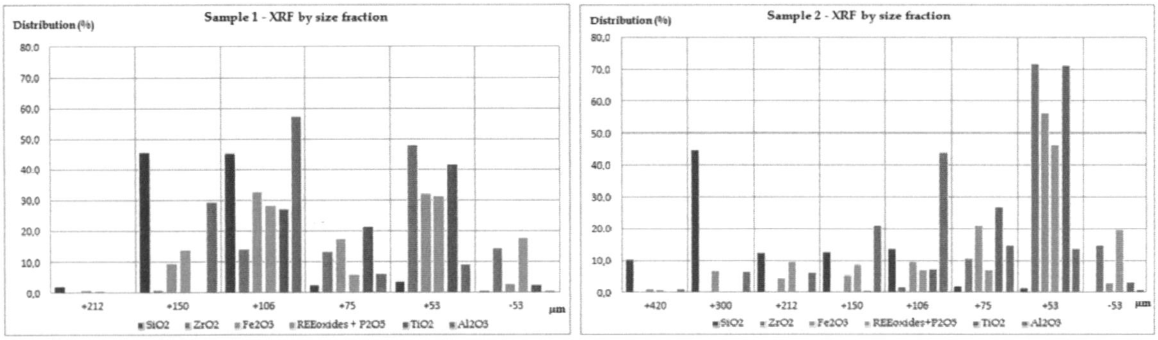

Figure 3. Elemental chemical composition of PL1 and PL2 by XRF.

The results of heavy liquid separation with methylene iodide are presented in Table II. The Ti-, Fe-, P- and Zr-bearing minerals reported to the sink fraction for both samples, as expected.

Table II. Elemental chemical composition of sink and float fractions by XRF.

Sample	Content (%)					
	SiO_2	Fe_2O_3	Al_2O_3	$REE_{oxide}+P_2O_5$	ZrO_2	TiO_2
PL1 sink	15.8	21.5	13.7	1.0	9.3	33.8
PL1float	81.0	3.6	8.4	<0.1	0.1	0.8
PL2 sink	12.4	23.6	7.5	2.9	13.9	35.3
PL2 float	76.6	3.0	6.7	0.1	0.3	0.7

Characterization of the Different Grades of Heavy Minerals Products obtained from Magnetic Separation

An average of 65%, by weight, of the sink fraction from methylene iodide (-420 μm) is composed of magnetic minerals (retained in 1.300 T magnetic field strength) as ilmenite, magnetite, goethite, epidote, monazite, actinolite, and staurolite. The nonmagnetic fraction contains zircon, rutile, kyanite, anatase, and quartz.

The greater part of the samples is recovered from 0.450 to 0.675 T (56.97% of PL1 and 55.90% of PL2 by weight). According to Pereira *et al.* (2005), in this interval, the heavy minerals of beach sands are magnetite, ilmenite, garnets, xenotime, epidote, some amphiboles, and some tourmalines. We confirmed part of this information by the Raman spectroscopy result, which demonstrated the presence of epidote in the magnetic fraction retained at 0.675 T.

As the minerals of interest are in the finer fractions (-150 μm), we carried out magnetic separation, followed by chemical analysis by XRF, on the -150+106 μm, -106+75 μm, and -75 μm size fractions for both samples. The -420+150 μm material is considered just in terms of distribution calculus as well as the material loss on ignition (LOI).

The elemental chemical results for each size fraction after magnetic separation are shown in Tables III and IV for PL1 and PL2, respectively. In accordance with the mass availability, we divided the products into magnetic and nonmagnetic fractions or products recovered at some specific magnetic field strength. The principal minerals found by XRD of PL1, by size fraction, are shown in Figure 5.

Table III. Magnetic separation results for PL1

		TiO_2		ZrO_2		Fe_2O_3		REE_{oxide} + P_2O_5		Al_2O_3		SiO_2	
PRODUCTS	**Weight** (%)	**Cont.** (%)	**Distr.** (%)	**Cont.** (%)	**Distr.** (%)	**Cont.** (%)	**Distr.** (%)	**Cont.** (%)	**Distr.** (%)	**Cont.** (%)	**Distr.** (%)	**Cont.** (%)	**Distr.** (%)
-420+150 µm	0.59												
Total	0.59		3.5		2.2		0.0		0.6		0.7		0.3
-150+106 µm													
Mag	5.50	37.2	6.1	0.0	0.0	27.3	6.9	0.1	0.7	14.7	5.9	12.3	4.2
Nonmag	4.35	11.5	1.5	4.4	2.0	0.5	0.1	0.0	0.0	42.5	13.5	38.1	10.3
Total	9.85		7.5		2.0		7.1		0.7		19.4		14.5
-106+75 µm													
0.450 T	10.15	52.3	15.7	0.1	0.1	40.4	19.0	0.0	0.0	1.9	1.4	2.2	1.4
0.675 T	22.38	46.6	30.9	0.0	0.0	33.3	34.5	0.3	6.7	4.8	7.9	8.0	11.1
0.95–1.3 T	8.34	11.7	2.9	0.2	0.2	15.0	5.8	0.7	5.5	28.4	17.3	29.2	15.1
Nonmag	17.95	16.9	9.0	15.4	29.6	0.6	0.5	0.4	7.0	33.7	44.2	30.0	33.4
Total	58.81		58.5		29.9		59.8		19.3		70.8		61.1
-75 µm													
0.450 T	9.36	49.6	13.7	0.2	0.2	43.4	18.8	0.0	0.0	1.9	1.3	2.1	1.2
0.675–1.3 T	9.37	45.4	12.6	0.4	0.4	32.4	14.1	2.2	21.1	4.5	3.1	5.1	3.0
Nonmag	12.02	11.7	4.2	50.7	65.2	0.5	0.3	4.8	58.4	5.4	4.7	26.7	19.9
Total	30.75		30.5		65.8		33.2		79.4		9.1		24.1

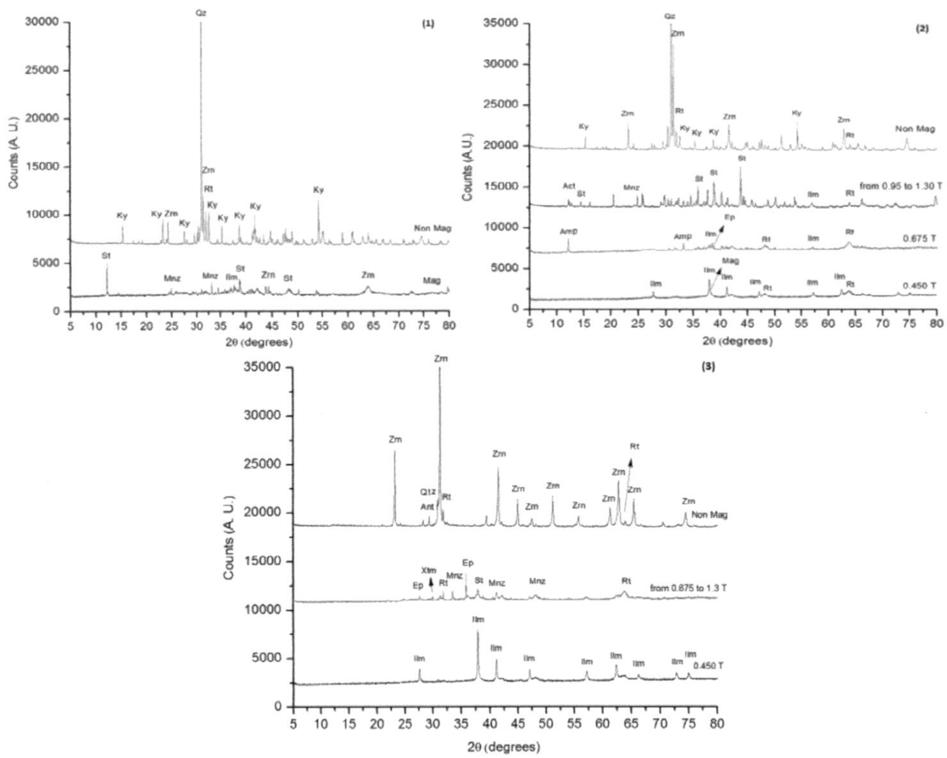

Figure 5. XRD spectra of magnetic products of PL1 by size fraction: (1) -150+106 μm, (2) -106+75 μm, and (3) -75 μm.

The -150+106 μm magnetic fraction contains mostly ilmenite, staurolite, monazite, and zircon; the +106 μm nonmagnetic fraction is composed of kyanite, zircon, and rutile. The fraction -106+75 μm (0.450 T) contains ilmenite, rutile, and magnetite; the same fraction at 0.675 T is composed of amphibole, ilmenite, epidote, and rutile; and at 0.950–1.300 T, of actinolite, staurolite, monazite, ilmenite, and rutile. The nonmagnetic fraction of -106+75 μm contains kyanite, zircon, quartz, and rutile. The -75 μm fraction (0.450 T) is composed of ilmenite; and at 0.675–1.300 T ,of epidote, xenotime, monazite, rutile, and staurolite; the nonmagnetic fraction of -75 μm is composed of quartz, zircon, rutile, and anatase. The details of the magnetic products are shown in Figure 6.

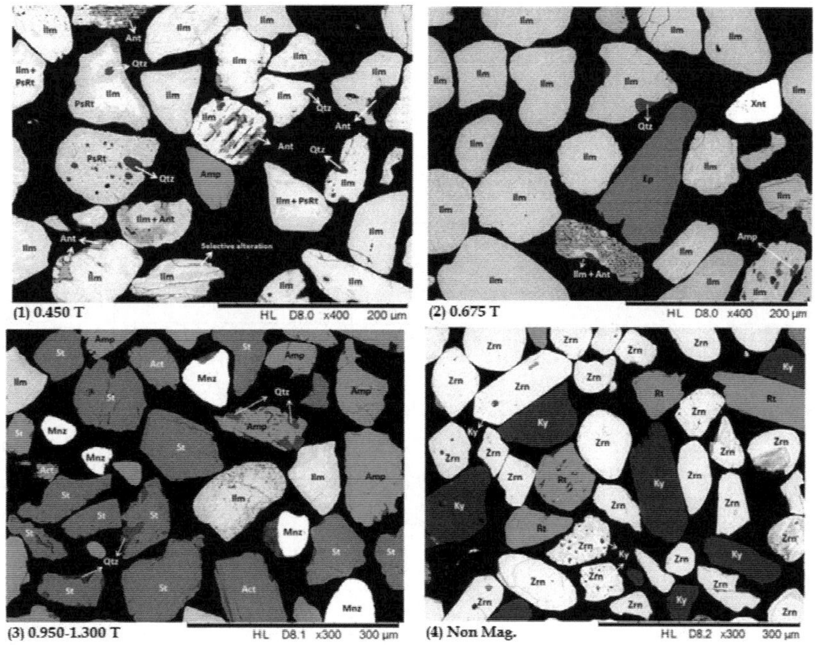

Figure 6. SEM images of magnetic products of PL1.

In Figure 6.1 we can see altered ilmenite, especially pseudorutile (PsRt) and anatase (Ant); the ilmenite grains contain quartz, silicates, and zircon inclusions and are mostly subrounded to subangular. Figure 6.2 shows many altered ilmenite grains, ilmenite with quartz and amphibole inclusions, xenotime (Xnt), and epidote (Ep); the grain shapes are variable. Minerals such as monazite (Mnz), staurolite (St), actinolite (Act), other amphiboles (Amp) with quartz inclusions, and highly altered ilmenite can be seen in the 0.950-1.300 T fraction (Figure 6.3). The nonmagnetic fraction (Figure 6.4) is composed of kyanite (Ky), prismatic and rounded zircon (Zrn) with kyanite inclusions, and rounded and prismatic rutile (Rt).

The elemental chemical composition for PL2 is shown in Table IV and the principal minerals detected by XRD, by size fraction, are shown in Figure 7. In the -150+106 µm fraction the magnetic fraction comprises ilmenite, staurolite, zircon, and actinolite; the -150+106 µm nonmagnetic fraction is composed mostly of kyanite, zircon, anatase, and rutile. The -106+75 µm (0.450 T) fraction contains ilmenite, rutile, and amphibole; the same fraction at 0.675 T is composed of amphibole, ilmenite, epidote, and rutile; and at 0.950–.300 T of actinolite, staurolite, monazite, ilmenite, and rutile. The nonmagnetic fraction of -106+75 µm is composed of kyanite, zircon, quartz and rutile. The -75 µm fraction at 0.450 T is composed of ilmenite and rutile; the same fraction at 0.675 T is composed of actinolite, ilmenite, and rutile; and at 0.950–1.300 T, of monazite, epidote, ilmenite, and rutile; the nonmagnetic fraction of -75 µm is composed of zircon and rutile.

Table IV. Magnetic separation results for PL2

PRODUCTS	Weight (%)	TiO_2		ZrO_2		Fe_2O_3		REE_{oxide} + P_2O_5		Al_2O_3		SiO_2	
		Cont. (%)	Distr. (%)	Cont. (%)	Distr. (%)	Cont. (%)	Distr. (%)	Cont. (%)	Distr. (%)	Cont. (%)	Distr. (%)	Cont. (%)	Distr. (%)
-420+150 µm	2.00												
Total	2.00		6.8		0.7		2.3		0.9		11.0		7.2
-150+106 µm													
Mag	2.39	41.9	2.8	0.1	0.02	29.6	3.0	0.1	0.1	9.9	3.1	9.8	1.9
Nonmag	1.79	11.6	0.6	4.7	0.6	0.6	0.05	0.0	0.0	39.0	9.3	41.4	6.0
Total	4.19		3.4		0.6		3.0		0.1		12.4		7.9
-106+75 µm													
0.450 ⅂	14.33	51.3	20.8	0.1	0.1	42.2	25.6	0.0	0.0	1.5	2.9	1.9	2.2
0.675 ⅂	20.34	47.6	27.4	0.5	0.6	34.8	30.0	0.1	0.7	4.6	12.4	6.3	10.3
0.95 – 1.3 ⅂	3.48	14.7	1.4	0.5	0.1	14.6	2.2	8.2	9.4	21.6	10.0	25.1	7.0
Nonmag	15.06	18.2	7.8	28.8	28.9	0.6	0.4	0.4	1.9	20.5	41.0	25.0	30.3
Total	53.22		57.4		29.8		58.1		12.0		66.3		49.9
-75 µm													
0.450 ⅂	13.72	49.7	19.3	0.2	0.2	44.5	25.8	0.0	0.0	1.2	2.2	1.6	1.8
0.675 ⅂	6.18	52.1	9.1	0.2	0.1	37.8	9.9	0.7	1.4	2.1	1.7	2.9	1.4
0.95 – 1.3 ⅂	3.91	5.4	0.6	1.0	0.3	3.5	0.6	59.3	76.4	3.2	1.7	4.3	1.4
Nonmag	16.79	7.2	3.4	61.1	68.4	0.4	0.2	1.7	9.2	2.1	4.7	22.5	30.4
Total	40.59		32.4		68.9		36.5		87.0		10.3		35.0

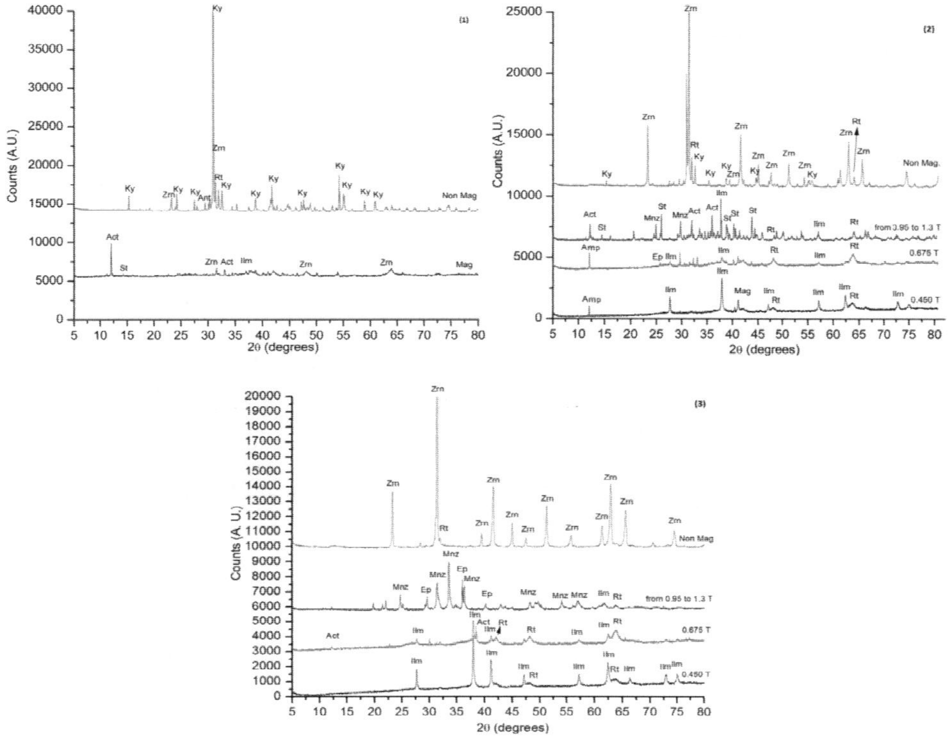

Figure 7. XRD traces of magnetic products of PL2 by size fraction: (1) -150+106 μm, (2) -106+75 μm, and (3) -75 μm.

The details of the magnetic products are shown in Figure 8. In Figure 8.1 we see altered ilmenite + pseudorutile, and anatase as ilmenite alteration; the ilmenite grains contain quartz, feldspar, and zircon inclusions; the grains are mostly subrounded. Figure 8.2 shows altered ilmenite grains with anatase, pseudorutile, ilmenite with iron oxide and quartz inclusions, epidote, and actinolite; the grain shapes are variable. Minerals such as monazite, staurolite, actinolite with quartz inclusions, pseudorutile, anatase with feldspar inclusions, and rutile with feldspar inclusion are found in the 0.950-1.300 T fraction (Figure 8.3). The nonmagnetic fraction (Figure 8.4) is composed of kyanite, prismatic and rounded zircon with feldspar inclusions, quartz, and rounded rutile; fractures are present in the zircon grains.

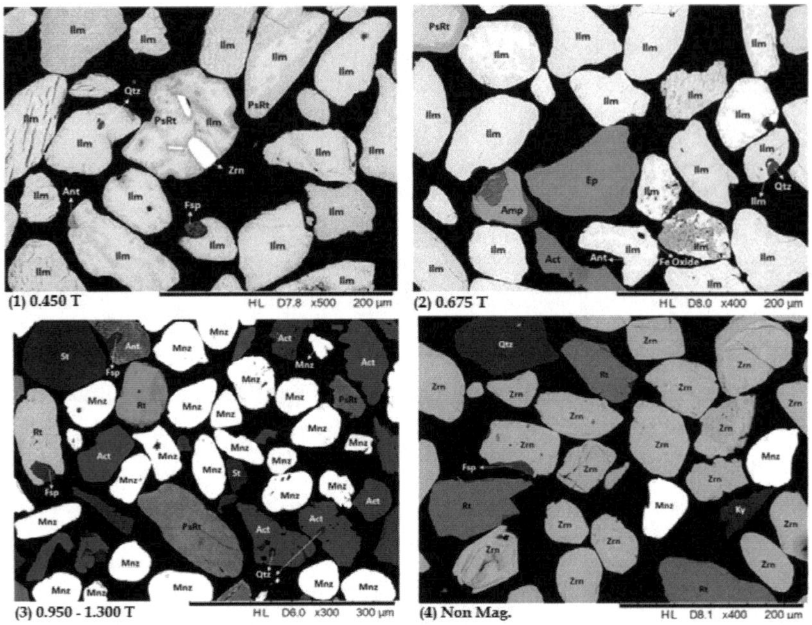

Figure 8. SEM images of magnetic products of PL2.

After magnetic separation by the Frantz mineral separator (Tables III and IV), it can be seen that for PL1 and PL2, the principal Ti-bearing mineral is ilmenite, which is concentrated in the -106+75 μm fraction. Figures 6.1 and 8.1 exhibit ilmenite grains with quartz and zircon inclusions; these grains are extremely altered, sometimes to pseudorutile and also anatase (grain boundaries), as reported by Nair *et al.* (2009). This alteration is also seen in Figures 6.2 and 8.2; an important fact associated with this alteration is the large amount of ilmenite retained in the 0.675 T fraction. Ilmenite usually occurs in pure form, but we noticed the presence of Mn in some grains.

The majority of the rutile, for both samples, is concentrated in the same fraction (-106+75 μm nonmagnetic), but its content is three times less than the ilmenite content. A more in-depth investigation is necessary to confirm the precise concentration of rutile or anatase in the nonmagnetic fractions. The existence of ilmenite-rutile grains, a result of the replacement of ilmenite by rutile, explains its presence in all XRD magnetic products (Figures 5 and 7).

The zircon is concentrated in the -75 μm size fraction. The zirconium content of zircon (approx. 53.0%) is not comparable to the theoretical zirconium content of approximately 43.0% (Webmineral, 2016), because of the Nb and Al contents, but the Hf/Zr ratios, according to the chemical analyses, are 0.04 for both samples, corroborating to the results reported by Reyneke and van der Westhuizen (2001). Zircon grains sometimes contain quartz (Figure 6.4) and feldspar (Figure 8.4) inclusions and their fractures (Figure 8.4) can influence the quality of the final product, especially with the application of physical processing methods such as screening and scrubbing.

The Al silicates of the nonmagnetic fractions, mostly kyanite, are in the +75 μm size fraction, facilitating their separation by size from the zircon. In other deposits this separation is more complicated (Reyneke and van der Westhuizen, 2001) and a flotation stage is needed (Routray *et al.*, 2013).

The concentration of monazite (-75 μm size fraction) is notable in the interval of 0.950–1.300 T (Pereira *et al.*, 2005). We noticed the presence of radioactive elements such as U only in sample PL2. The Th is associated with the monazite structure, together with elements such as Ce, La, and Nd.

Beneficiation Studies

The mass balance for the beneficiation studies using a Knelson concentrator is given in Table V for sample PL1. The recovery (63.8%) was considered insufficient compared with the results of heavy minerals concentration by shaking table (Tables VI and VII). Carrisso and Braga (1995) successfully beneficiated heavy minerals from beach sand using a Knelson concentrator, but we did not confirm their results. The low recovery for sample PL1 can be explained by the high grade of HM in this sample (6.45%) compared with 2.5% in the sample evaluated by Carrisso and Braga (1995).

Table V. Mass balance for the recovery of heavy minerals from sample PL1 using a Knelson concentrator.

Fraction	Weight (%)	Weight distribution (%)	Ilmenite (%)	Monazite (%)	Rutile (%)	Zircon (%)	% HM	Recovery of HM (%)
Concentrate	19.7	19.7	2.4	1.0	2.0	2.5	7.9	63.8
Tailing	80.3	80.3	0.0	0.4	0.1	0.6	1.1	36.2
Total	**100.0**	**100.0**	**0.5**	**0.5**	**0.5**	**1.0**	**2.4**	**100.0**

The detailed mass balances for the beneficiation studies with the screening stage and shaking table are in Tables VI and VII for samples PL1 and PL2, respectively. The recoveries of HM after the screening stage are almost the same for both samples, but a considerable difficulty was detected during this step for sample PL1 due to its grain size distribution (97.4% weight retained at +75 μm).

Table VI. Mass balance for the beneficiation of sample PL1 with a screening stage.

Fraction	Weight (%)	Weight distribution (%)	Ilmenite (%)	Monazite (%)	Rutile (%)	Zircon (%)	% HM	Recovery of HM (%)
(a) Screening stage (+150 μm)								
Oversize	45.2	45.2	0.0	0.3	0.4	0.1	0.8	23.9
Undersize	54.8	54.8	0.2	0.7	0.7	0.5	2.1	76.1
Total	**100.0**	**100.0**	**0.1**	**0.5**	**0.6**	**0.3**	**1.5**	**100.0**
(b) Shaking table								
Light	54.2	29.7	0.0	0.3	0.1	0.6	1.0	27.0
Heavy	45.8	25.1	0.9	0.7	0.4	1.2	3.2	73.0
Total	**100.0**	**54.8**	**0.4**	**0.5**	**0.2**	**0.9**	**2.0**	**100.0**
(c) Magnetic separation								
Concentrate 1	3.7	0.9	6.6	1.3	2.9	3.4	14.2	12.9
Concentrate 2	14.6	3.7	5.7	1.2	3.4	2.5	12.8	46.5
Concentrate 3	81.7	20.5	0.1	0.3	0.8	0.8	2.0	40.6
Total	**100.0**	**25.1**	**1.2**	**0.5**	**1.3**	**1.1**	**4.0**	**100.0**

The most significant difference in the samples recoveries is seen in the shaking table step, probably due to the differences in feed grade; the key to a better separation for PL1 could be the improvement in the feed mass flow rates and the deck slope (Manser *et al.*, 1991). The second beneficiation process (Figure 2) demonstrated satisfactory results for PL2, but we expected a higher recovery and ilmenite concentration in the first step of magnetic separation, which was not attained due to ilmenite alteration. In order to simplify the beneficiation process, this stage can therefore be eliminated, and the ilmenite recovered from the monazite by electrostatic separation. The significant content of zircon in the magnetic fractions (Tables VI and VII) is due to the presence of zircon inclusions in the ilmenite/pseudorutile grains (Figure 8.1).

Table VII. *Mass balance for the beneficiation of sample PL2 with a screening stage.*

Fraction	Weight (%)	Weight distribution (%)	Ilmenite (%)	Monazite (%)	Rutile (%)	Zircon (%)	% HM	Recovery of HM (%)
(a) Screening stage (+150 μm)								
Oversize	78.30	78.30	0.2	1.4	0.1	0.1	1.8	23.7
Undersize	21.70	21.70	7.4	2.9	3.6	7.0	20.9	76.3
Total	**100.00**	**100.00**	**1.8**	**1.7**	**0.9**	**1.6**	**5.9**	**100.0**
(b) Shaking table step								
Light	40.23	8.73	0.1	1.2	0.2	0.5	2.0	4.4
Heavy	59.77	12.97	13.2	2.3	1.4	12.2	29.1	95.6
Total	**100.00**	**21.70**	**7.9**	**1.9**	**0.9**	**7.5**	**18.2**	**100.0**
(c) Magnetic separation								
Concentrate 1	9.68	1.26	23.5	1.7	2.2	14.7	42.1	11.7
Concentrate 2	54.12	7.02	20.1	2.8	6.6	14.2	43.7	67.7
Concentrate 3	36.20	4.70	4.5	2.3	1.0	12.1	19.9	20.6
Total	**100.00**	**12.97**	**14.8**	**2.5**	**4.1**	**13.5**	**34.9**	**100.0**

CONCLUSIONS

The heavy mineral deposits studied have very distinct characteristics, especially as regards particle size, THM, and slime contents, due to environmental dynamics and geological factors. This suggests that it may be difficult to establish a specific beneficiation process for Brazilian beach sands and the necessity of detailed studies with diverse samples.

The mineralogical compositions were almost the same for the two samples, comprising magnetite, goethite, ilmenite, pseudorutile, staurolite, monazite, actinolite, albite, epidote, xenotime, zircon, kyanite, rutile, anatase, quartz, muscovite, other silicates, amphiboles, and feldspars. The greater part of the magnetic minerals was recovered at a magnetic field strength of 0.450–0.675 T (56.97% of PL1 and 55.90% of PL2 by weight), and the fraction with the greatest heavy mineral content was -75 +53 μm for both samples.

Ilmenite is the principal Ti-bearing mineral and it is highly altered, mainly to pseudorutile and/or anatase. The beneficiation process can be simplified by omitting one stage of magnetic separation and the inclusion of an electrostatic separation step (Worobiec *et al.*, 2007).

ACKNOWLEDGMENTS

The authors acknowledge the assistance of the Center for Mineral Technology, which permitted the project development, carried out the chemical and mineralogical analyses and the beneficiation tests. We thank the National Council for Scientific and Technological Development (CNPq) for the research grant.

REFERENCES

Carcangiu, G., Palomba, M., Peretti, R., and Zucca, A. (1991). The heavy minerals of some coastal sand deposits in Southwest Sardinia. Application of electron microanalysis and beneficiation tests. *Ore Geology Reviews*, 6 (91), 365-380.

Carrisso, R.C.C. and Braga, P.F.A. (1996). Estudo de variáveis no desempenho do concentrador centrífugo KNELSON. *Proceedings of XVI Encontro Nacional de Tratamento de Minérios e Hidrometalurgia,* Rio de Janeiro. pp. 88-99.

Figueiredo, P.C. and Chaves, A.P. (1998). Recuperação de partículas minerais finas e ultrafias no âmbito da concentração gravítica. *Proceedings of XVII Encontro Nacional de Tratamento de Minérios e Metalurgia Extrativa,* São Paulo. pp. 483-492.

Laxmi, T., Srikant, S.S., Rao, D.S., and Rao, R.B. (2013). Beneficiation studies on recovery and in-depth characterization of ilmenite from red sediments of badlands topography of Ganjam District, Odisha, India. *International Journal of Mining Science and Technology,* 23 (13), 725-731.

Lins, F.A.F. (2010). Concentração gravítica. *Tratamento de Minérios.* Luz, A.B., Sampaio, J.A., and França, S.C.A. (eds.). CETEM/MCTI, Center for Mineral Technology, Rio de Janeiro. pp. 301-326.

Manser, R.J., Barley, R.W., and Wills, B.A. (1991). The shaking table concentrator - the influence of operating conditions and table parameters on mineral separation - the development of a mathematical model for normal operating conditions. *Minerals Engineering,* 4 (3/4), 369-381.

Nair, A.G., Babu, D.S.S., Damodaran, K.T., Shankar, R., and Prabhu, C.N. (2009). Weathering of ilmenite from Chavara deposit and its comparison with Manavalakurichi placer ilmenite, southwestern India. *Journal of Asian Earth Sciences,* 34 (9), 115-122.

Pereira, R.M., Ávila, C.A., and Lima, P.R.A.S. (2005). *Minerais em Grãos: Técnicas de Coleta, Preparação e Identificação,* 1st edn. Oficina de Textos, São Paulo.

Reyneke, L. and van der Westhuizen, W.G. (2001). Characterization of a heavy mineral-bearing sample from India and the relevance of intrinsic mineralogical features to mineral beneficiation. *Minerals Engineering,* 12 (01), 1589-1600.

Routray, S., Laxmi, T., and Rao, R.B. (2013). Alternate approaches to recover zinc mineral sand from beach alluvial placer deposits and bandlands topography for industrial applications. *International Journal of Material and Mechanical Engineering,* 4 (13), 80-90.

Routray, S. and Rao, R. B. (2011). Beneficiation and characterization of detrital zircons from beach sand and red sediments in India. *Journal of Minerals and Materials Characterization and Engineering,* 15 (11), 1409-1428.

Sampaio, J.A., Luz, A.B, Alcantera, R.M., and Araújo, L.S.L. (2001). Minerais pesados Millennium. *Usinas de Beneficiamento de Minérios no Brasil.* Sampaio, J.A., Luz, A.B., and Lins, F.F. (eds.). CETEM/MCTI, Center for Mineral Technology, Rio de Janeiro. pp. 233-239.

Sampaio, C.H. and Tavares, L.M.M. (2005). *Beneficiamento Gravimétrico,* 1st edn. UFRGS, Porto Alegre.

Schnellrath, J., Monte, M.B.M., Veras, A., Jr, H.R., and Figueiredo, C.M.V. (2001). Minerais pesados INB. *Usinas de Beneficiamento de Minérios no Brasil.* Sampaio, J.A., Luz, A.B., and Lins, F.F. (eds.). CETEM/MCTI, Center for Mineral Technology, Rio de Janeiro. pp. 189-197.

Webmineral. (2016). Zircon mineral data. http://webmineral.com/data/Zircon.shtml#.V1AjrTUrK70 [accessed 2 Jun. 2016].

Worobiec, A., Stefaniak, E.A., Potgieter-Vermaak, S., Sawlowicz, Z., Spolnik, Z., and Grieken, R.V. (2007). Characterization of concentrates of heavy minerals sands by micro-Raman spectrometry and CC-SEM/EDX with HCA. *Applied Geochemistry,* 22 (07), 2078-2085.

Caroline Da Costa Gonçalves

Temporary research (Bachelor Degree)
Center for Mineral Technology (CETEM)

Graduated in Mining Engineering from Unifal, Brazil (2015). Scientific Initiation Student at Unifal (2012) with the project "Study of the flotation methods of nepheline syenite from Poços de Caldas region for potassium concentration". Intern at Votorantim Metais Aluminium (2012/2013), working in the development and optimization of bauxite processing coordination, Brazil. Scholarship student at Université du Québec à Chicoutimi, Canada (2013/2014) in Bachelor of Geological Engineering. Intern at the Study Center for Mineral Resources - CERM (2014), Canada. Currently Mining Engineer Researcher at CETEM (2015/now), working with mineralogical and chemical characterization and mineral processing of heavy minerals.

Paulo Fernando Almeida Braga

Senior Technologist
Center for Mineral Technology (CETEM)

Degree in Chemical Engineering from UFRRJ, Brazil (1979). Specialization in Minerals and Metallurgy Processing by CIMM Processing, Santiago, Chile (1997). Master's degree (1999) and PhD (2013) in Mineral Engineering from USP, Brazil. Process engineer at Remetálica, working in technical and economical projects for lithium and gold recovery (1986/1987). Project and industrial manager at Lithium Brazilian Company, responsible for the implementation and operation of carbonate and lithium hydroxide plant (1988/1992). Currently Senior Technologist at CETEM (1993/now), working in mineral processing - emphasizing methods of ore concentration and enrichment (gravity separation and flotation), and management of multidisciplinary projects

Making test work dollars count

G. Diener, S. Gilman, E. Williams, and G. Williams

TZ Minerals International, Perth, Western Australia

The traditional project development route to plant design is iterative and involves multiple stages of test work. However, there is a perception that test work programmes deliver little value relative to the cost. The new economic reality therefore requires a smarter approach to test work. It is critical to understand the factors that must be considered in optimizing the development of a process flow sheet from test work information while at the same time reducing investment risk and minimizing costs. This requires targeted test work and project development. Inexperienced project developers spend disproportionate amounts on drilling programmes where significant savings could be realized by completing an appropriate amount of drilling and supplementing the resource development work with process and market thinking. By following an iterative approach and checking project viability against the market at each stage of the development, test work dollars can be spent in a way that provides maximum benefit while enhancing the outcomes of work in all areas of the development of the project.

INTRODUCTION

In a changing global market and in a world where project funding is in short supply there is increasing pressure on companies that develop mineral sands projects to take the shortest, lowest cost route from exploration to operation.

Under increased time and cost pressures all areas of project development are impacted. In an attempt to bring forward the delivery date, the resource definition, engineering, environmental management, infrastructure provision, and process design project streams are run in parallel. However, this can introduce significant risk to the end value of the overall project. With the traditional iterative approach that results in an increasing level of detail at each stage, information was constantly fed back into the decision-making process. In the new order, with less iteration and a fast track to decision-making, it is essential that an integrated approach to project development is followed.

The specific focus of this paper is the integration of test work and flow sheet development with other disciplines to ensure that test work dollars spent are appropriate and deliver maximum benefit.

PROJECT DEVELOPMENT BY PHASES

Project implementation is the culmination of a series of studies, ideally with gateway decision points after each phase to confirm the economic viability of the project. These studies aim to develop the project from exploration to operation in the most cost-effective way possible.

The common project phases for resource investment and project development are shown in Figure 1. Test work is a necessary part of project development from the initial exploration phase through to the feasibility study, with information gathered for input into detailed design.

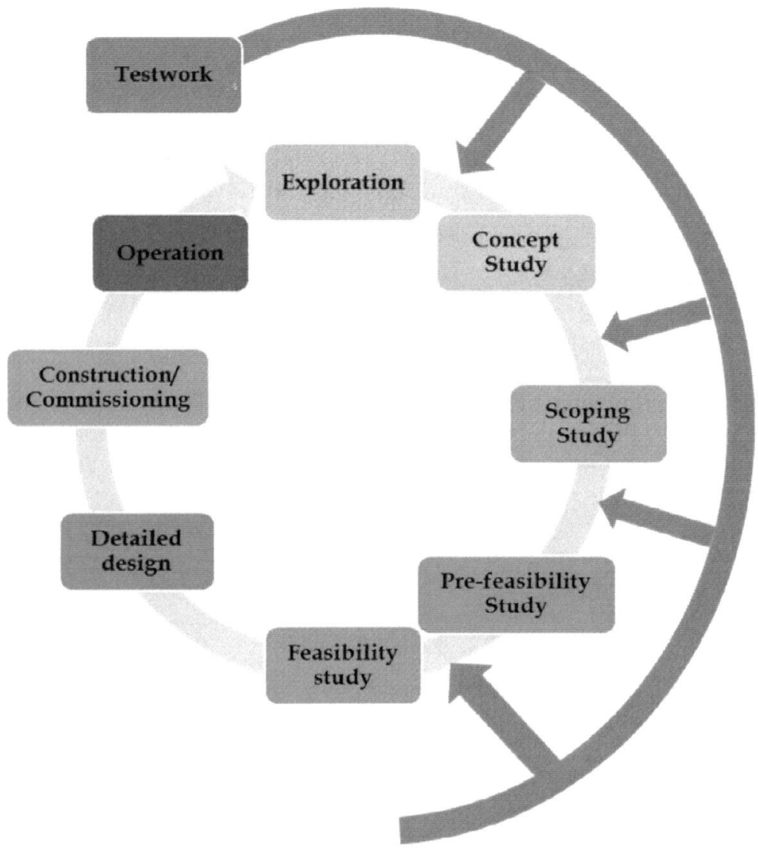

Figure 1. Phases of project development.

Traditionally, projects have followed an extended pathway passing through each of the project stages, with multi-disciplinary work conducted. Each advance in the project required an increase in the level of detail, which resulted in an exponential increase in the costs of each stage.

PHASES OF METALLURGICAL TEST WORK

Metallurgical test work needs to carry a lot of emphasis early in project development as the initial aim is to support the resource definition. Additionally, the majority of the test work has to be conducted before engineering design can commence. Consequently, test work makes up a high proportion of the project development costs in the early stages, with the result that these programmes are often targeted for cost reduction as project owners try to stretch available resources. This can result in critical metallurgical work being delayed or omitted to meet short-term goals.

As with other project development aspects, the process flow sheet design and supporting metallurgical test work programme are undertaken in multiple stages with the typical progression shown in Figure 2. The focus for data collection at each of these stages aims to answer specific questions.

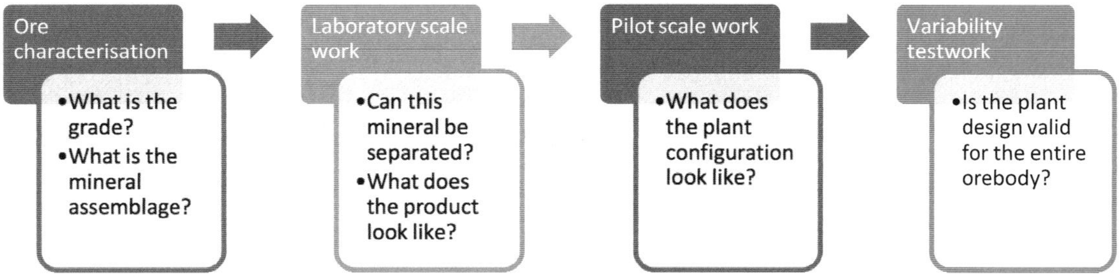

Figure 2. Typical progression of metallurgical test work programmes for process design.

By understanding the outcomes of each of the test work stages and the links with other disciplines, significant value can be realized over the project lifetime by ensuring that test work programmes are streamlined and targeted to deliver the 'must know', rather than the ' interesting to know'.
Some of the important aspects and the most common mistakes made in the integration of test work into project development are discussed below.

Ore Characterization
Metallurgical assessment should start during the exploration phase with small samples used for ore characterization to determine orebody grade and assemblage. This is required to support resource assessment and geology.

Mineralogy is a key aspect of mineral sands evaluation, a fact that is often missed by those not familiar with the industry. A detailed understanding of mineralogy and product quality issues should be given greater attention early on the project development than excessive definition drilling. Drilling programmes should be reviewed at each stage to ensure that information collected is relevant, contributes to the overall knowledge base, and is useful for development of inputs to other disciplines at that stage of the project.

An analytical method specific to the deposit should be developed to determine the mineral assemblage, including both economic and gangue material. In addition to quantifying the contribution of key economic minerals to revenue streams, it is important to identify the presence of gangue minerals that are known to interfere with separation processes early in the project. This allows early identification of minerals that would require unusual process technology to produce marketable products or to assess whether product characteristics do not easily fit into existing market niches and require special markets.

A recent study presented to TZMI clearly illustrates the importance of understanding the requirements for mineralogical studies early in the exploration cycle, as well as highlighting the unique knowledge base that is necessary to develop mineral sands orebodies.

Case Study 1 – Whole Sample Analyses
Company 1 completed an extensive drilling programme on a mineral sands deposit. Samples were collected and reported in line with the guidelines in Australian reporting standards (JORC, 2012). The work was directed and signed off by a Competent Person, *i.e.* a geologist with extensive experience in iron ore. The drill sample assay process involved analysing the bulk sample by XRF to determine the metal content. The end result was an assessment that met all of the criteria required for assessing a commodity such as iron ore, but failed to provide much of the data relevant to understanding a mineral sands deposit.

Table I shows average values from the data-set provided to TZMI for comment on project viability.

Table I. Whole ore analyses results on drill samples in mass %.

SiO_2	Al_2O_3	Fe_2O_3	K_2O	Na_2O	CaO	TiO_2	ZrO_2
77.9%	9.2%	3.30%	2.50%	2.00%	1.50%	1.04%	0.08%

On inspection, some obvious problems were identified with the data-set:

- The whole ore sample chemical analyses did not provide the mineralogical data needed to assess a mineral sands deposit
- The high grades of SiO_2 and Al_2O_3 reported diluted the contribution of economic minerals
- The ZrO_2 content of more than half of the samples analysed was reported as below the detection limit due to dilution with light minerals
- Due to the dilution, the grades of TiO_2 were low and therefore less precise
- No mineralogical data was collected. As a result, no assessment could be made of the relative proportions of economic minerals such as ilmenite and rutile or the potential contributors to the high Al_2O_3 grades reported.

This is one of several examples where initial expenditure on drilling and sampling was wasted by conducting incorrect and incomplete sample analysis. The total set of samples had to be reprocessed to provide information on slimes content, heavy mineral content, and mineral assemblage, resulting in rework and significant additional expense.

The other extreme encountered is the excessive use of costly mineralogical techniques that provide little of the vital information regarding separability and final product quality.

Case Study 2 - Use of Expensive Mineralogical Techniques

A typical example was presented through a study completed by *Company 2*. Hundreds of samples from an extensive drilling programme on a mineral sands deposit were submitted for analysis by QEMSCAN. The results provided detailed information on modal mineralogy, but at a high cost and without providing information regarding the potential for separation of saleable products. Due to poor documentation on mineral classification used in the data analyses, this extensive data-set could not be integrated with information collected through other sampling programmes and therefore did not comply with requirements for reporting of resources.

In decision-making it is important to assess 'What we need to know' *versus* 'How we can use this bright new gadget'. Analytical methods have to be selected to provide data relevant to the particular stage of development.

Laboratory Scale Test Work

Once the ore characterization methodology has been developed, larger samples are required to conduct laboratory-scale test work. This work is typically started during the concept phase and continued through the various feasibility study phases, and is essential to provide information to transform orebody information from Resources to Reserves.

The primary goal of the laboratory-scale work is to assess the ability to separate and isolate the economic minerals. The test work programme should also include an assessment of physical separation processes to support the development of a conceptual flow sheet.

Indicative flow sheet variables are defined by laboratory-scale work to guide later, more detailed work. In order to optimize the laboratory work it is critical to assess the information collected in the light of market requirements. The importance of this can be demonstrated through the following case study.

Case Study 3 – Including Market Assessments in Early Project Development

Company 3 was developing a deposit with a complex mineral assemblage. Due to significant alteration in the titanium minerals, the laboratory isolated a number of titanium products, ranging from ilmenite (approx. 50% TiO_2) to rutile (approx. 94% TiO_2), with 'HiTi products' identified for intermediate TiO_2 ranges. Extensive further test work was proposed to improve the TiO_2 recovery by isolating these intermediate products into separate product streams.

This included processing samples to produce a complex product suite with narrow specification ranges. While this can be done in the laboratory it is not feasible to do this on a plant scale. In addition, it is unlikely that sufficient volumes of these products would have been produced to place commercially. This approach inevitably leads to an iterative test work process due to poor understanding of market requirements, resulting in increased costs and a redefining of the project economics.

During laboratory work it is important to understand what the information can and should be used for and also what the limitations are. This is especially important when transitioning from laboratory- to pilot-scale work to define the scope of work that has to be completed during pilot-scale testing.

Pilot-scale Test Work

The pre-feasibility phase is supported by pilot-scale test work. This aspect of flow sheet development is often neglected in mineral sands project development due to the assumption that mineral sands flow sheets comprise known technology and that it is therefore possible to predict process outcomes based on laboratory-scale tests.

Figure 3 shows a comparison of processing a plant feed sample using a laboratory-scale wet table and testing the same material using two production-scale spirals from different suppliers, as typically used in the rougher duty.

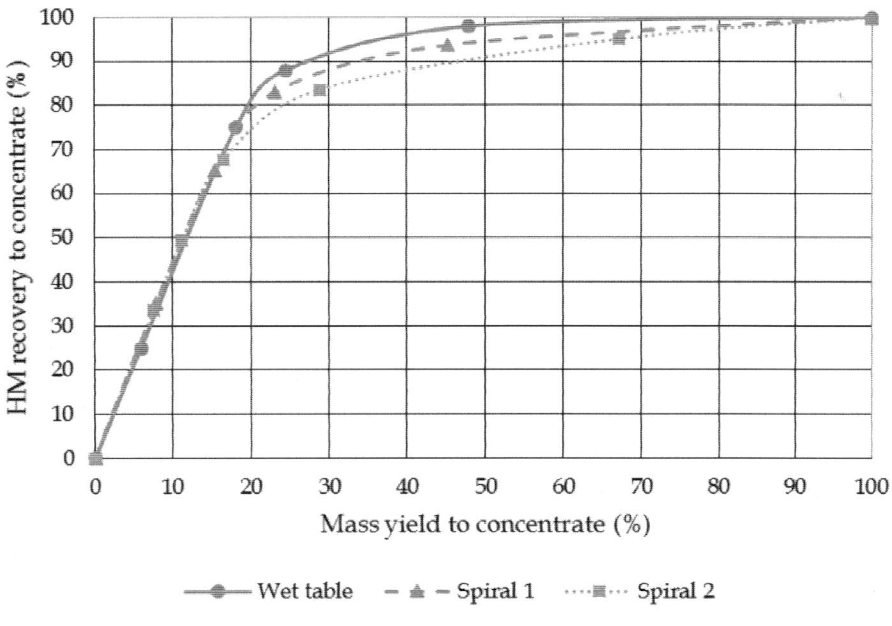

Figure 3. Heavy mineral recovery to concentrate vs yield.

At a mass yield of 20% to the concentrate, the wet table separation curve predicted a recovery of 85% of the heavy minerals to concentrate. This decreased to 80% when using spiral 1 and to 75% when using spiral 2. If no pilot-scale test-work had been carried out, it would not be possible to accurately predict the recovery in this duty, or for that matter the composition or grade of the product recovered for further processing. As shown in this example, the assumptions that have to be made will depend on the equipment selected. This in turn will have a significant impact on the design of downstream processes.

A major stumbling block in executing pilot-scale work, especially for mineral separation plants, is the large volume of sample that is required to provide sufficient material to complete detailed process development work at the tail end of the flow sheet. This means that the metallurgical test programme is often a large component of the cost associated with scoping studies and pre-feasibility studies.
Well executed pilot-scale work provides critical information regarding the metallurgical response on plant-scale equipment such as spirals for gravity separation and pilot-scale magnets for magnetic separation. The data that is collected is used to construct the mass balance, which is a key input into the engineering design and overall project evaluation. Other information is also collected, *e.g.* particle size distributions for the design of equipment such as cyclones and specific gravity of process streams.
A theme often repeated in this paper is the importance of understanding the integration between the test work, resource development, and especially the market. In the following case study, extensive work was done at the pilot scale with no consideration for downstream markets.

Case Study 4 - Pilot-Scale Testing without Market Input
Company 4 is developing a hard rock deposit. The project development was supervised by professionals with extensive experience in hard rock processing and metallurgy. The proposed flow sheet incorporated multiple stages of crushing and grinding, followed by flotation and magnetic separation. A high-grade product was produced that met the chemical product specifications. However, the product size distribution achieved in the test programme resulted in more than 50% of the material being too fine to meet current market acceptance criteria.

This approach failed in the following ways:

- Multiple large pilot-scale test work campaigns were executed to develop the flow sheet, including grinding and lock cycle flotation testing
- No engineering data was collected during the initial campaigns, resulting in additional work required to firm up engineering design data
- Parts of the project were developed to feasibility stage, including detailed engineering and process design, without consideration for the market requirements
- The product produced did not meet market specifications and as a result only half of the product was suitable for market placement.
-

Having an understanding of the market early in the test programme will prevent spending or process development on products that cannot be sold.

Pilot-scale test work has a very important role to play in the development of projects to feasibility stage. For this reason, it is important that the campaigns are planned with a clear goal to ensure that repeat work is not needed and that plant modelling and engineering can be completed to the expected standard.

Variability Programmes
From the initial assessment of a deposit it is often clear that the material is hosted in different geological zones. Material sourced from these zones can differ in metallurgical response due to mineral alteration, surface coatings, or changes in mineral assemblage. In order to finalize the flow sheet it is important to confirm whether the selected process flow sheet would be suitable for material sourced from all zones within the deposit.

One of the most common misconceptions in mineral sands projects is the belief that the deposit is homogeneous. This erroneous belief is often used to justify bypassing the evaluation of the metallurgical response of material from different zones within the deposit and has led to significant delays in the start-ups and commissioning of a range of plants and processes.

CONSIDERING THE INTEGRATED APPROACH

Integration of all aspects of project development is the only approach that will allow for an efficient pathway in project development, leading to a reduction in the cost of development and final implementation. In order to minimize the number of iterations while still optimizing the project outcomes, all study disciplines should be aligned within the constraints of what the orebody can deliver and what the market can receive.

By understanding how the various disciplines influence the final outcome, the test work and flow sheet development can be streamlined and directed towards providing maximum benefit. An effective way to reduce costs, yet still meet the overall information requirement, is by thinking first about what data is needed. This should cover all disciplines, including reporting, process design, financial evaluation, and market requirements at each stage of project development.

Understanding the factors that have to be considered will guide an optimized process design that reduces investment risk for the lowest possible costs. This means taking a smart approach to generate the required knowledge without investing time and effort on test work that doesn't reduce project risk. This means focusing data collection on identifying the 'fatal flaw' versus the 'nice to know' and means *always* asking 'how does this data add value to the project?'

There are numerous examples where the integrated approach and considering multiple aspects in the decision-making fundamentally changed the financial evaluation and consideration of a project.

Some of these examples include:

- Redefining a large-scale mineral sands prospect with multiple processing stages to deliver an intermediate product acceptable to the market, resulted in a significant reduction in capital cost, infrastructure requirements, and a simplified flow sheet
- Reducing the cost of project development by early elimination of products that could not be placed in the market
- Incorporation of mineralogical investigations early in the assessment, allowing for streamlined test work to target rejection of gangue minerals.

A deep understanding of mineral sands processing technology along with TiO_2 feedstock and zircon market requirements assists the rapid development of a process flow sheet to generate products that will be readily accepted by the market. Understanding the implications of flow sheet options also allows development of process flow sheets that produce relevant products for current and future markets, and which can be commercially implemented.

CONCLUSIONS

The development of an effective streamlined metallurgical test work programme can be achieved only by using a multidisciplinary approach and ensuring that the project team has the relevant experience to understand and interpret the data to combine geology, processing, engineering, and - very importantly - an understanding of the market.

Experience in knowing where to focus efforts early on in a project's life helps project developers to minimize development costs. For example, developing a targeted drilling and analytical programme that brings the Resource to the right point at each stage of the project, supplemented by the appropriate level of process development work, helps to steer the project in the right direction. Inexperienced project developers spend disproportionate amounts on drilling programmes where significant savings could be realized by completing an appropriate amount of drilling and supplementing the resource development work with process and market thinking.

Executing relevant metallurgical test work at the right time in the project cycle can add significant value, while inappropriate test work spend adds cost it can also detract from the project goal.
By following an iterative approach and checking project viability against the market at each stage of the development, test work dollars can be spent in a way that provides maximum benefit while enhancing the outcomes of work in all areas of the development of the project.

REFERENCES

Dunlop, J.S. (2013). Mineral project management - a perspective from four decades in the industry. *Proceedings of Metplant 2013, Metallurgical Plant Design and Operating Strategies.*Australasian Institute of Mining and Metallurgy, Perth, WA.

Gilman, S.K. and Hugo, V.E. (2003). Minerals sands mineralogy and metallurgy - can they speak the same language. *Proceedings of the Southern African Heavy Minerals Conference,* October 2003. Southern African Institute of Mining and Metallurgy, Johannesburg.

JORC. 2012. Australasian Joint Ore Reserves Committee. Australasian Code for Reporting of Exploration Results, Mineral Resources and Ore Reserves. The Joint Ore Reserves Committee of the Australasian Institute of Mining and Metallurgy, Australian Institute of Geoscientists, and Minerals Council of Australia. http://www.jorc.org/docs/JORC_code_2012.pdf

McCarthy P. 2003. Managing technical risk for mine feasibility studies. *Proceedings of the Mining Risk Management Conference,* September 2003. Australasian Institute of Mining and Metallurgy, Perth, WA.

Elana Williams

Senior Consultant
TZ Minerals International

Elana joined TZMI in 2011 where she consults on testwork program management and metallurgical process development. She has developed mineral processing flowsheets for both Wet Concentrators and Mineral Separation Plants for projects in Africa, India and Australia. This work included aspects of process design from concept studies through to implementation at existing operations.

Prior to joining TZMI, Elana worked as a Senior Project Manager at Bureau Veritas Mineral processing where she managed large scale testwork programs and data collection for clients covering a range of commodities including minerals sands, iron ore and uranium.

Before relocating to Australia, Elana spent 12 years working at Richards Bay Minerals in South Africa in roles covering technical, operations management and business improvement.

Gavin Williams

Managing Consultant
TZ Minerals International

Gavin joined TZMI in 2008 and has participated in various consulting assignments including the execution of due diligences, plant process reviews, project feasibility studies, technical and financial evaluations including cost modelling and industry cost analysis. He has been involved in the development of mineral sands projects in Africa, Australia and Madagascar which involved concept development in the form of scoping studies through to management and execution of feasibility studies and supporting testwork programs.

Prior to joining TZMI, Gavin spent 12 years at Richards Bay minerals where he held both technical and operational management roles focusing on mineral sands processing with emphasis on ilmenite smelting.

Review of titanium feedstock selection for fluidized bed chlorinators

A. Kale and K. Bisaka

Pyrometallurgy Division, Mintek

Titanium tetrachloride ($TiCl_4$) is the basic raw material for the production of titanium dioxide (TiO_2) pigment and titanium (Ti) metal. Worldwide, around 3.5 Mt of TiO_2 pigment is produced annually by the chloride process, consuming in the region of 4.5 Mt of titania feedstock. Titania feedstock is chlorinated in the presence of a carbonaceous reductant at temperatures of approximately 1000°C for the production of $TiCl_4$.

Mintek has been associated with research work on titania feedstock since the early 1970s. The DC arc smelting process for production of titania slag from ilmenite was jointly developed and patented by Mintek and Anglo American. This technology is successfully used on a commercial scale at the Tronox-operated plants in South Africa. Over the years, Mintek has been working on various aspects of the chlorination of titania feedstock. Feedstock such as titania slag, synthetic rutile, upgraded slag, ilmenite, and rutile have all been evaluated in Mintek's laboratory-scale chlorination facility. Chlorination kinetics of various species present in the feedstock as well as the hydrodynamic properties of various titania feedstock materials, for example the effect of density and particle size on the quantity of blowovers from the chlorinator, have been investigated.

This paper provides a review of the characteristics of the titania feedstock used in industrial fluidized bed chlorinators, the effects of the chlorination kinetics of various species on the chlorinator bed composition, and the impact of the metal chlorides produced on the downstream operations in the production of $TiCl_4$. The information from the review may be useful for selecting titania feedstock for the production of $TiCl_4$ using fluid bed chlorination technology.

INTRODUCTION

Titanium dioxide is the most widely used white pigment in the coating industry. It has high brightness and a very high refractive index. It efficiently scatters visible light, thereby imparting whiteness, brightness, and opacity when incorporated into a coating. TiO_2 is also an effective opacifier and is widely used in applications such as paints, plastics, paper, and ink. TiO_2 pigment is produced by two commercial processes - the chloride and the sulphate process. The chloride process has become the more dominant process as it produces a superior quality pigment with significantly fewer waste products.

Titania feedstock consisting mainly of ilmenite, rutile, and titania slag is the basic source for the production of TiO_2 pigment. This paper reviews the effect of selecting various titania feedstocks used in the chlorination process for producing $TiCl_4$, which is an intermediate product in the production of TiO_2 pigment and Ti metal. In the chlorination process, titania feedstock is chlorinated in the presence of a carbonaceous reductant at temperatures of 1000°C for the production of titanium tetrachloride ($TiCl_4$). The effect of various impurities present in the titania feedstock on the operation of the fluid bed chlorinator and the downstream operations in the production of $TiCl_4$ is also discussed.

The Tenth International Heavy Minerals Conference
Sun City, 16–17 August 2016
Southern African Institute of Mining and Metallurgy

Titania ores occur in both primary magmatic and secondary placer deposits. Shoreline placer or beach sand deposits are by far the largest titanium deposits. Titanium is generally bonded to other elements in nature and occurs primarily as an oxide. Ilmenite is the most abundant titanium mineral. The other common titanium minerals from which TiO_2 is extracted include rutile, anatase, altered ilmenite, leucoxene, and titanomagnetite. The TiO_2 content depends on the source and extent of weathering of the mineral. The other major constituents include iron oxides (FeO, Fe_2O_3), MgO, MnO, CaO and silica. (Tittle *et al.*, 1973). Rutile ore extracted from natural deposits is too impure to be used directly as a pigment and is thus further processed. Ilmenite, the main titania ore, can be used directly in the chlorination process for manufacturing of TiO_2 pigment or upgraded to produce either titania slag or synthetic rutile, which may then be used as a titania feedstock in the chlorination process.

Mintek is involved in the investigation of methods for the production and purification of $TiCl_4$ from various titania feedstocks - this is one of the components of the titanium beneficiation programme driven by the South African Department of Science and Technology. As a result of this work, Mintek has established a knowledge base in the chlorination of titania feedstock. Feedstocks such as titania slag, synthetic rutile, upgraded slag, ilmenite, and rutile have been evaluated in Mintek's laboratory-scale chlorination facility. Chlorination kinetics and hydrodynamic properties of these titania feedstock have been evaluated. Mintek has published literature on the study of chlorination kinetics and behaviour of various species present in the titania feedstock. A laboratory-scale chlorination set-up and a fluid bed chlorination pilot plant (Kale *et al.*, 2010) at Mintek are used extensively to conduct the research activities.

PRODUCTION OF TITANIUM TETRACHLORIDE BY THE CHLORIDE PROCESS

Process

The production of $TiCl_4$ used for TiO_2 pigment or Ti metal involves chlorination of titania feedstock in a bubbling fluid bed chlorinator using petroleum coke as a reductant. Bubbling fluid bed chlorination technology is widely used commercially for the chlorination of titania feedstock due to its high rate of heat and mass transfer.

A schematic of the chloride process is shown in Figure 1. The mixture of titania feedstock and petroleum coke is fed into the chlorinator bed, where chlorine gas and oxygen are used to fluidize the chlorinator bed and maintain the desired operating conditions. The process operates at high temperatures, typically between 900 and 1100°C. The fluid bed chlorinator is lined with silica/alumina-based refractory in order to withstand the harsh operating conditions.

Various elemental species present in the titania feedstock are chlorinated at different stages, depending on the reactor conditions and their individual thermodynamic properties and chlorination kinetics. The metal chlorides formed in the chlorinator either leave the chlorinator as gaseous chlorides or accumulate in the chlorinator, depending on their individual boiling points and the operating conditions of the chlorinator. The lower boiling point metal chlorides leave the chlorinator and are condensed in the first condenser.

The condensed metal chlorides are treated before being discarded as waste. The treatment involves neutralization with lime. The $TiCl_4$ from the chlorinator is condensed in a separate stream in the main condenser and collected as liquid. The liquid $TiCl_4$ contains impurities that are removed by a complex proprietary process involving chemical treatment and distillation. The purified $TiCl_4$ may then be further oxidized to produce TiO_2 pigment.

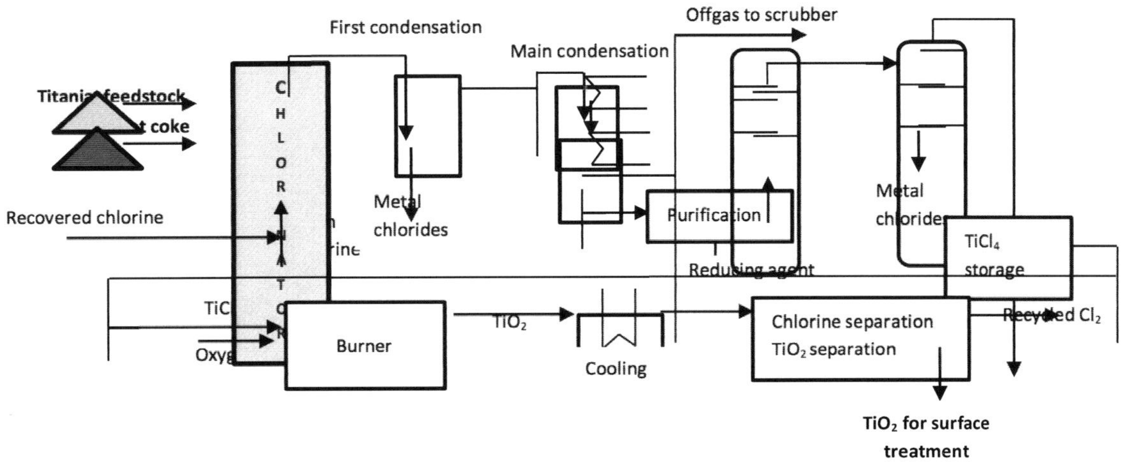

Figure 1. Schematic flow diagram of the chloride process.

Titania Feedstocks and their Upgrading

Titania feedstock is commercially available in the form of ilmenite, titania slag, rutile, synthetic rutile, and upgraded slag. Ilmenite and natural rutile are natural resources, while synthetic rutile and titania slag are obtained by the processing of ilmenite and upgraded slag (UGS) is obtained by further processing of the titania slag. Typical compositions of titania feedstock are tabulated in Table I.

Minerals separation plants utilize screening, magnetic separation, and electrostatic and gravity separation circuits to separate valuable minerals from the non-valuable minerals and produce different grades of ilmenite, rutile, and zircon products. The TiO_2 content of commercially available ilmenite can vary from 40% to 60%.

Iron oxide is the major impurity present in ilmenite, and can range from 30 to 50% by mass. Ilmenite also contains various impurities such as SiO_2, Cr_2O_3, V_2O_5, Al_2O_3, ZrO_2, MgO, MnO, CaO, and the radioactive elements U and Th. The composition of the impurities varies widely, depending on the source and geographical location of the ilmenite deposits. Some of these impurities occur as discrete minerals (for example ZrO_2 and SiO_2, which are found as $ZrSiO_4$). Particles of this mineral find their way into concentrates of rutile and ilmenite. Many impurities occur in solid solution in the principal phase (rutile, ilmenite, or M_3O_5[ss] of slag). Examples include iron and niobium in rutile, MnO and MgO in ilmenite, and MnO and Al_2O_3 in pseudobrookite (M_3O_5) solid solution [ss] (Bessinger *et al.*, 1997).

Table I. Typical chemical composition (wt%) of various titania feedstocks that Mintek has evaluated.

Oxide	Feedstock			
	Ilmenite	Rutile	Ti Slag	UGS
TiO_2	60.1	95.0	86	94.8
Fe_2O_3	26.3	0.6	-	2.4
FeO	10.4	-	8.9	-
SiO_2	0.7	1.4	1.7	2.1
V_2O_5	0.2	0.3	0.4	0.5
MnO	0.4	0.01	1.2	0.1
MgO	0.4	0.05	1.0	1.1
CaO	0.04	0.05	0.3	0.2
ZrO_2	0.2	0.08	0.1	<0.01
Al_2O_3	0.6	0.3	0.8	0.5
Cr_2O_3	0.1	0.35	0.01	0.08
Nb_2O_5	<0.01	0.4	<0.01	<0.04
U+Th (ppm)	<150	<50	<50	<50

Natural rutile is mined from various deposits and is mostly associated with ilmenite mining. The TiO_2 content in natural rutile can range from 90 to 97%. The iron content can vary between 0.5 to 2 %. SiO_2 and ZrO_2 are the major impurities associated with rutile.

Synthetic rutile is produced by roasting the ilmenite, followed by treatment of the roasted ilmenite with hydrochloric acid or sulphuric acid to remove iron and other impurities, thereby increasing the TiO_2 content. Typically, synthetic rutile contains in the region of 90 to 93% TiO_2. Other impurities present in synthetic rutile depend on the geographical source of the ilmenite deposit.

Titania slag is produced by the smelting of ilmenite. The iron is removed from ilmenite as pig iron, thereby increasing the TiO_2 content of the remaining slag. Typically the TiO_2 content of the slag can be in the range of 80 to 90%. The iron content in the slag can vary from 5 to 15%. Upgraded slag is produced by further treating the titania slag to increase the TiO_2 content to 90–95% by roasting and leaching out the impurities with hydrochloric acid.

IMPACT OF VARIOUS PARAMETERS ON THE CHLORINATION REACTION

Process Economics

Titania feedstock is a major cost in the production of TiO_2 pigment, contributing around 40–45% of the total variable cost for a chloride route pigment plant (Adams, 2003). Selection of an appropriate titania feedstock is therefore a critical factor affecting the profitability of the TiO_2 pigment business. There is a significant difference in the price of various titania feedstocks - typical titania feedstock containing about 85% TiO_2 is about 2.5 times the price of ilmenite, for example. Natural rutile, synthetic rutile, and UGS are almost five times the price of ilmenite. Cost considerations therefore play a major role in the selection of an appropriate feedstock for the chloride process.

The economics of feedstock selection is further affected by selection of technology used for certain types of feedstock. A plant using ilmenite as a feedstock needs to be designed to cater for the higher quantity of iron chlorides produced in the chlorinator. Due to the lower TiO_2 content (typically about 60%) in ilmenite, a larger quantity of feedstock is required to produce a ton of $TiCl_4$; hence, the raw material handling facility and wrap-around process plant will be significantly larger than that of a chlorination plant using titania slag or rutile as feedstock. As can be seen in Table II, approximately 850 kg of ilmenite is required to produce a ton of $TiCl_4$, as compared to 530 kg of rutile. A plant using ilmenite as a feedstock will therefore be required to handle 60% more raw material than a plant handling rutile as feedstock. Higher material handling requirements will lead to increased wear and tear of equipment, and higher capital and maintenance costs. The overall footprint of a production facility using ilmenite will be larger compared to a plant using rutile as feedstock to produce the same quantity of $TiCl_4$. This will result in a higher capital investment for a $TiCl_4$-producing facility using ilmenite as a feedstock. The quantity of chlorine required will be significantly greater due to the chlorine consumed for chlorinating the iron impurities in ilmenite. The quantity of iron chloride waste generated will also be significantly higher, hence using the comparatively cheaper ilmenite feedstock will result in higher waste treatment and waste disposal costs. As a result, many chloride process plants use a blend of feedstock to increase the TiO_2 content to 88–92% to optimize their operations.

Table II. Titania feedstock required to produce one ton of $TiCl_4$.

Ilmenite	850 kg
Rutile	530 kg
Ti Slag	600 kg
UGS	530 kg

Operational Challenges

The various impurities present in the titania feedstock play an important role in the production of $TiCl_4$ in a bubbling fluid bed chlorinator (van Dyk, 1999). These impurities also affect the quality of TiO_2 pigment (in most cases adversely). In order to produce a high-quality TiO_2 pigment, the impurities in $TiCl_4$ must be removed to as low as <50 ppm.

The bubbling fluid bed chlorinator is typically operated at temperatures around 900°C to 1100°C. At these temperatures, most of the volatile metal chlorides (M_xCl_y, having boiling points below the chlorinator operational temperature) formed as a result of the reaction between chlorine and the titania feedstock will leave the chlorinator in gaseous form. The gas stream from the chlorinator is cooled to a temperature that permits selective separation of various undesired metal chlorides by condensation. The melting and boiling points of various metal chlorides are therefore critical when designing the condenser. Table III shows the boiling and melting points of various metal chlorides formed during the chlorination of titania feedstock.

The physical properties of titania feedstock can also affect the operation of bubbling fluid bed chlorinators. Particle size, density, and sphericity of the particles are important parameters for the design of an industrial chlorinator. Different titania feedstocks have appreciably different particle size distributions. As can be seen from Table IV, titania slag has particle size distribution from 850 μm to below 75 μm, while rutile and ilmenite has a particle size from 300 μm to below 75 μm.

Studies on the behaviour and hydrodynamic properties of the particles in a fluid bed chlorinator have shown that finer particles have a tendency to elutriate from the chlorinator. The particle size of the titiania feedstock has an effect on the terminal velocity of the particles. According to calculated terminal velocity values, particles finer than 75 μm will be highly susceptible to entrainment and elutriation from the chlorinator (Moodley *et al.*, 2012). It is therefore important to limit the quantity of fine particles in the titania feedstock. The elutriation of particles from the chlorinator makes these particles unavailable for chlorination, and is a direct loss of TiO_2 value. The elutriated material from the chlorinator also adds to the cost of solid waste disposal.

Table III. Boiling and melting points of various metal chlorides (Wikipedia).

Metal oxides	Metal chlorides	Melting point °C	Boiling point °C
Al_2O_3	$AlCl_3$	180 (sublimes)	180
CaO	$CaCl_2$	772	1935
Cr_2O_3	$CrCl_3$	1152	1300
Fe_2O_3	$FeCl_3$	306	316
MgO	$MgCl_2$	714	1412
MnO	$MnCl_2$	654	1225
Nb_2O_5	$NbCl_5$	204.7	248.2
SiO_2	$SiCl_4$	-68.4	57.6
TiO_2	$TiCl_4$	-24	136.4
V_2O_5	$VOCl_3$	-76.5	126.7
ZrO_2	$ZrCl_4$	331 (sublimes)	331

Table IV. Particle size distributions for various titania feedstocks.

Particle size (μm)	**Ilmenite (wt%)**	**Rutile (wt%)**	**Ti slag (wt%)**
+850	-	-	1.2
-850+600	-	-	18.7
-600+425	-	-	23.2
-425+300	-	-	19.5
-300+212	1.7	2.0	15.8
-212+150	29.3	19.4	8.7
-150+106	54.9	49.0	7.7
-106+75	10.7	28.0	3.5
-75	3.4	1.5	1.6

Chlorinator Bed De-fluidization

Metal chlorides with boiling points higher than the operating temperature of the chlorinator will tend to remain and accumulate in the chlorinator bed. If the melting point of these metal chlorides is below the chlorinator operating temperature, then these metal chlorides will affect the fluidizing properties of the chlorinator bed. From Table III, it can be seen that $CaCl_2$, $MgCl_2$, $MnCl_2$, and $CrCl_3$ will remain in the chlorinator bed as liquids. These metal chlorides may cause particle agglomeration and clumping in the chlorinator bed, resulting in the de-fluidization of the bubbling fluid bed and associated reduction in heat and mass transfer. This will ultimately lead to gas channelling, and unreacted chlorine and oxygen gas will leave the chlorinator. This may cause additional operational problems in the downstream process. The reduced heat transfer will lead to a drop in temperature of the chlorinator bed, slowing the rates of the chlorination reactions.

The accumulation of these metal chlorides in the chlorinator bed must be controlled by periodic draining of the bed. The disposal of the solid metal chlorides containing material from the chlorinator bed is also a concern due to their toxicity issues. Hence, while selecting the titania feedstock it is important to make sure that the total content of CaO and MgO is below 1% (Cong *et al.*, 2006).

Chlorinator Refractory Lining

The high operating temperatures (between 900°C and 1100°C), together with the use of chlorine, makes it necessary to use a refractory brick lining inside the chlorinator for protecting the walls of the vessel. The refractory bricks used are typically silica/alumina-based. Over time, the refractory brick lining becomes worn and eventually needs replacement. Replacement is an expensive exercise, and involves manually removing the old refractory bricks and replacing them with new ones. The new refractory bricks need curing time, and as a result the entire operation leads to appreciable downtime, which can be more than a week.

Typically, the life of the refractory brick lining in the chlorinator can vary between 9 months and 24 months, depending on the operational parameters of the chlorinator and the type of impurities present in the feedstock. Damage to the refractory lining occurs mainly as a result of mechanical erosion, thermal shock, and chemical corrosion. In typical chlorinators, the titania feedstock and petroleum coke mixture is fed horizontally into the vessel over the chlorinator bed using a transporting gas at high pressure. The abrasive nature of the titania feedstock then causes mechanical erosion of the refractory lining. If ilmenite is used as a feedstock, the quantity of material fed into the chlorinator is significantly higher than when using rutile feedstock. This may lead to more severe mechanical erosion of the refractory brick lining. Some manufacturers have opted for a vertical feeding system in order to address the issue of erosion caused by the horizontal feeding of feedstock into the chlorinator.

Ferrous and ferric chloride formed by the chlorination of iron in the feedstock are a major source of chemical corrosion of the refractory bricks. Iron chlorinates vigorously compared to the TiO_2 in the feedstock (den Hoed *et al.*, 2003). The higher iron content of the titania feedstock may therefore lead to increased chemical corrosion of the refractory bricks. Although silica-based refractory is well known for its corrosion-resistance properties, it undergoes phase changes in the chlorinator, especially in the presence of iron at high temperature. This may lead to additional corrosion problems in the refractory lining.

The chlorinator bed needs to be drained periodically due to the accumulation of impurities in the bed. The draining disturbs the normal chlorinator operation, and leads to changes in chlorinator conditions such as temperature. This temperature variation imparts thermal shock to the refractory brick lining, adversely affecting its performance and increasing the likelihood of a lining failure.

Impact of Minor Impurities

Various minor impurities such as vanadium, niobium, alumina, and silica have significant impact on the chlorinator and downstream operations.

The silica and zirconia in the titania feedstock accumulate in the chlorinator bed due to the unfavourable chlorination kinetics of these species compared to iron oxide and titania (den Hoed *et al.*, 2003). Over time, the accumulation of silica and zirconia leads to changes in the bed composition. A critical point is reached when the percentage of these two components in the chlorinator bed leads to chlorine leaving the chlorinator without reacting, a phenomenon commonly referred to as 'chlorine slip' in the TiO_2 pigment industry. To mitigate this behaviour, once the silica content in the bed reaches a critical level of above 45%, the chlorinator bed is drained and replaced with fresh titania feedstock. Draining involves operational downtime, resulting in production loss. Frequent bed draining also affects the temperature inside the chlorinator and imparts thermal shock to the refractory lining. The higher the SiO_2 and ZrO_2 content in the titania feedstock, the more frequently the bed will need to be drained.

Small amounts of $ZrCl_4$ formed in the chlorinator leave the chlorinator bed, as $ZrCl_4$ sublimes at 301°C. The $ZrCl_4$ reacts with the organic polymers used for purification of crude $TiCl_4$ and forms a coating on the $TiCl_4$ distillation trays. This reduces the heat and mass transfer and hence the performance of the distillation column.

Alumina in titania feedstock reacts with chlorine and leaves the vessel in gaseous form along with $TiCl_4$ and other metal chlorides. $AlCl_3$ is highly corrosive and may cause damage to equipment used in the downstream operation. To eliminate the presence of $AlCl_3$ in the downstream operation, $TiCl_4$ producers use a variety of proprietary techniques in order to remove $AlCl_3$ from the gaseous stream leaving the chlorinator. Some $TiCl_4$ is lost in the process of removing $AlCl_3$, and hence higher quantities of alumina in the feedstock will result in increased losses of $TiCl_4$ associated with the removal of $AlCl_3$.

Niobium in the titania feedstock reacts with chlorine, producing niobium pentachloride ($NbCl_5$), which must be condensed and removed from the operation before it reaches the water-cooled condensers. If it is not removed, $NbCl_5$ forms a scale over the surfaces in the heat exchangers, affecting the efficiency of the units. This scaling needs periodic cleaning, the frequency of which increases with higher niobium content in the feedstock.

Vanadium in the feedstock is chlorinated to form vanadium oxychloride ($VOCl_3$), which has a boiling point of 127°C, close to the boiling point of $TiCl_4$ (134.4°C). The $VOCl_3$ therefore condenses with $TiCl_4$ in the product stream. Vanadium causes severe colour and brightness problems in the TiO_2 pigment. As can be seen in Table V, it is recommended that the V content in the purified $TiCl_4$ product should be less than 10 ppm. Vanadium compounds are removed from $TiCl_4$ by proprietary chemical treatments and careful distillation control. Titania feedstock with lower vanadium content is therefore preferred in order to reduce the cost of purification of crude $TiCl_4$ and improve the quality of TiO_2 pigment.

Table V. Typical composition of commercial $TiCl_4$ (http://KMML.com)

$TiCl_4$ (%)	V (ppm)	Fe (ppm)	Al (ppm)	Si (ppm)
>99.5	<10	<5	<20	<50

Radioactive elements such as U and Th are present in different quantities in various titania feedstocks. The high operating temperature of above 1000°C in the chlorinator will result in some U and Th leaving the chlorinator and precipitating in the condenser along with other undesirable metal chlorides (http://v.pl.3.eu-norm.org/index.pdf). These metal chlorides are mixed with water and neutralized (typically with lime). Neutralization precipitates all the radionuclides and the level of neutralization will determine the losses to the effluent stream. In the case of complete neutralization stream the U and Th will end up in the solid waste stream. Higher levels of U and Th in the feedstock will therefore lead to higher values of radionuclides in the solid waste, and hence it is desirable to control the U and Th levels in the feedstock as per the specific solid waste disposal requirements for the TiO_2 production site.

CONCLUSION

The selection of titania feedstock for the production of TiO_2 using chloride technology is a complex process. Major criteria affected by the selection are:

- Process plant design
- Economics of the process due to differences in price of feedstocks and various impurities present
- Loss in production due to process and equipment downtime
- Waste treatment and solid waste disposal cost
- Quality of the TiO_2 pigment product.

Process design is very much dependent on the type of feedstock used for production of TiO_2 pigment. A plant using ilmenite (60% TiO_2) as feedstock will have a larger footprint compared to a plant using rutile (95% TiO_2) for the same production capacity.

The difference in the price of feedstock is a major criterion in raw material selection. The low price of titania feedstock often comes with the associated cost of technological challenges and operational issues. If the technological challenges and operational issues are effectively managed, the low price of feedstock may constitute an advantage for producing the TiO_2 pigment at low cost.

The effect of physical properties such as particle size distribution of the titania feedstock should be considered when evaluating the fluid bed chlorinator operation. Finer particles below 75 μm have a tendency to elutriate from the chlorinator, leading to loss of titanium values and additional cost of treating the solid waste.

Impurities like iron present in the titania feedstock lead to operational challenges resulting from damage caused to the refractory brick lining in the chlorinator. Any decrease in the lifetime of the refractory lining decreases the availability of the chlorinator for production of $TiCl_4$, while also incurring additional operating costs.

Minor impurities such as SiO_2 and ZrO_2 that accumulate in the chlorinator bed lead to more frequent draining of the chlorinator bed. This reduces the availability of chlorinator and causes thermal shock to the refractory lining. CaO and MgO in the titania feedstock lead to the formation of high-boiling-point chlorides that remain in the liquid form in the chlorinator bed at the operating temperature of the chlorinator. If the content of CaO + MgO is higher than 1%, the metal chlorides formed will cause defluidization of the chlorinator bed and adversely affect the chlorinator operation.

The scaling caused by metal chlorides such as $NbCl_5$ and $ZrCl_4$ on the surfaces of heat exchangers and distillation columns may result in lower production efficiencies. The presence of these impurities, even in small quantities, therefore plays a major role in the operational problems in the production of TiO_2 pigment and reduces the production capacity of the plant. The direct loss of $TiCl_4$ associated with removal of $AlCl_3$ from the chlorinator product stream is proportional to the amount of Al_2O_3 in the titania feedstock, hence lower Al_2O_3 content in the feedstock is desirable.

Higher quantity of impurities in titania feedstock will in general result in higher waste treatment and solid waste disposal costs. Hazardous impurities such as radionuclides end up in the solid waste - this can be a critical factor in deciding the suitability of titania feedstock. Hence, it is important to consider the U+Th content in the titania feedstock.

Impurities such as vanadium should be removed to concentrations below 10 ppm in the final product, as they can impart colour to the TiO_2 pigment. The higher the content of V_2O_5 in the titania feedstock, the greater the quantity of treatment chemicals required for the removal of V from the TiO_2 pigment.

It is anticipated that a better understanding of the impact of various impurities titania feedstocks will help operators and businesses to make informed decisions in the selection of which feedstock to choose, and in optimizing the design and operation of their process plants for $TiCl_4$ production.

ACKNOWLEDGEMENTS

The authors wish to thank Mintek for support and the opportunity to publish.

REFERENCES

Adams, R. (2003). Focus on pigments. *Monthly Report*, May 2003. ISSN 0969-6210.

Bessinger,D., du Plooy, H., Pistorius, P.C., and Visser, C. (1997). Characteristics of some titania slags. *Proceedings of Heavy Minerals 1997.* Southern African Institute of Mining and Metallurgy, Johannesburg. pp. 151-156.

Den Hoed, P. and Nell, J. (2003). The behaviour of individual species in the carbochlorination of titaniferous oxides. *Proceedings of Heavy Minerals 2003.* Southern African Institute of Mining and Metallurgy, Johannesburg.

McNulty, G.S. (2010). Production of titanium dioxide. http://v.pl.3.eu-norm.org/index.pdf [accessed 22 April 2016].

Jones, R.T. (2014). DC arc furnaces - past , present, and future. *Celebrating the Megascale – Proceedings of the Extraction and Processing Division, Symposium on Pyrometallurgy*. The Minerals, Metals & Materials Society, Warrendale, PA.

Kale, A. and Bisaka, K. (2010). Fluid bed chloriation pilot plant at Mintek. *Proceedings of the Light Metals Conference.* Southern African Institute of Mining and Metallurgy, Johannesburg and the Advanced Metals Initiative.

Kerala Minerals & Metals Ltd. (Not dated). http://kmml.com/uploaded_files/productManual/manual_24.pdf [accessed 23 April 2016]

Pistorius, P.C. and Ferdus le Roux, J.T. (2007). Thermal and chemical structure changes during initial chlorination of titania slag. *Canadian Metallurgical Quarterly*, 41 (3), 289-298.

Moodely, S., Kale, A., Bessinger, D., Kucukkaragoz, C., and Eric, R.H. (2012). Fluidization behaviour of various titania feedstocks. *Proceedings of IFSA 2011, Industrial Fluidization South Africa*. Luckos, A. and den Hoed, P. (eds). Southern African Institute of Mining and Metallurgy, Johannesburg. pp. 297-305..

Tittle, K. and Floey, E. (1973). Investigation of the segregation of iron from ilmenite. Transactions of the Institution of Mining and Metallurgy Section C, Mineral Processing and Extractive Metallurgy, 83, 135-139.

Van Dyk, J.P. (1999). Process development for the production of beneficiated titania slag. PhD thesis, University of Pretoria, South Africa.

Cong, X., Zhangfu, Y., and Wang, X. (2006). Preparation of $TiCl_4$ with titanium slag containing magnesia and calcia in a combined fluid bed. *Chinese Journal of Chemical Engineering*, 14 (3), 281-188.

Aditya Kale

Chief Engineer
Mintek

MS - Materials Science and Engineering; B.Tech - Chemical Engineering. Twenty years of versatile industrial experience in production, process development, technical sales and research activities. Worked for organizations like Heubach Colours, India; Cristal, Saudi Arabia and Mintek, South Africa.

Hands on experience in production of pthalocynine pigments, aluminium hydroxide, titanium dioxide pigment and technical due diligence for technologies. Evaluated various grades of titania feedstock for industrial chlorinators. Associated with technical sales and market development for activated carbon regeneration technology.

Involved in various process development activities from lab scale to commercial scale production with specialization in high temperature fluid bed chlorination technology. Exposure to industrial fluid bed chlorinator operations, evaluation of various ilmenite beneficiation processes, and techno commercial activities. Working with Mintek, since 2007 and coordinating the minerals beneficiation cluster at Pyrometallurgy division.

Southern African Institute of Mining & Metallurgy
P.O. Box 61127
Marshalltown 2107
South Africa

ISBN 978-1-5108-3199-5

CURRAN ASSOCIATES INC.
proceedings.com